Topology

TOPOLOGY

Murray Eisenberg

UNIVERSITY OF MASSACHUSETTS, AMHERST

HOLT, RINEHART AND WINSTON, INC.

NEW YORK   CHICAGO   SAN FRANCISCO   ATLANTA   DALLAS
MONTREAL   TORONTO   LONDON   SYDNEY

Eisenberg, Murray, 1939–
    Topology.

    Bibliography: p. 405
    1. Topology.
QA611.E53   514   73-19879
ISBN 0-03-091366-7

4   5   6   7   8   038   9   8   7   6   5   4   3   2   1

To the memory of my father

Preface

This book introduces the most basic concepts, facts, and techniques of general topology at a level appropriate to a student's first exposure to the subject. It is suitable as a text for a variety of undergraduate courses of differing lengths and emphases, and for classes having varying backgrounds; some possible course outlines are suggested below. It may even be used for beginning graduate students who have not previously studied topology as an introduction to one of the standard advanced texts.

The only mathematical prerequisite for reading this book is calculus. No knowledge of the topology of Euclidean spaces or metric spaces is assumed. Some slight prior experience with "epsilontics" is desirable, but not indispensable; the greater the prior experience, the more material can be covered. Neither Zorn's lemma nor ordinal numbers are used.

One of our aims has been to assist the student's mathematical maturation. Hence careful attention has been paid to motivating new notions. For example, eleven pages of examples of metrics precede the actual definition of a metric, and a proof of the compactness of the unit interval (together with the corollary that continuous functions on it are bounded) precede the definition of compactness. There are examples galore of everything. Special pains have been taken to explain the significance of theorems and to write enough proofs in enough detail to provide models for the student's own proof making. In using a preliminary version of the manuscript the author has found that students can read much of the text themselves with only minimal guidance by the instructor, so that classroom time can be devoted mainly to the exercises and discussion of more difficult points.

Chapter 0 concisely presents the necessary preliminaries on sets, maps, countability, order-completeness of the real numbers, and equivalence relations. The time spent on this material will, of course, depend on the student's background, but it is suggested that all students at least read the chapter rapidly for review and to fix terminology. It is a good idea to review the section on equivalence relations in conjunction with the study of quotient spaces in Chapter 3.

Chapter 1 introduces open sets, closed sets, neighborhoods, continuous maps, and convergent sequences in metric spaces. The terminology used here— d-closed set, (d,d')-continuous maps, and so on—calls attention to the particular metrics involved. At the same time the entire thrust of the chapter is to justify the later definition of a topological space by demonstrating that notions of continuity and convergence remain unchanged when the metrics are replaced by equivalent ones. The section on completeness includes the Baire category theorem and its application to the existence of nowhere differentiable, continuous functions. Owing to its greater technical difficulty as well as its treatment of uniform, as distinct from topological, ideas, this section may be postponed or even omitted; except for occasional mention, completeness does not appear again until Section 2 of Chapter 4, where it is used to characterize compact metric spaces and to prove the Tychonoff theorem for a sequence of compact metric spaces.

The study of topology proper is begun in Chapter 2, where topologies, neighborhoods, Hausdorff spaces, bases and local bases, and countability properties are discussed. Of the whole hierarchy of separation properties, which in its entirety is liable to confuse the beginner, only the property T_2 is dealt with at length in the body of the text, the others being relegated to the exercises.

Continuity is the theme of Chapter 3. Here product and quotient spaces are constructed and their mapping properties are emphasized. Here, too, the theory of convergence of nets is introduced at a level kept elementary by treating subnets only briefly and by avoiding universal nets entirely. This theory deserves to be included in a first topology course: It places sequential convergence in proper perspective, facilitates a later study of filter convergence, and reveals the diverse kinds of limits the student has previously encountered as instances of a single unifying concept. Nevertheless, the treatment of nets can be omitted without substantial loss: the only places nets are used again are the net characterization of compactness (4.27), which can itself be omitted, and the sequence characterization of compactness of a metric space (4.36), which requires only the equivalence of sequential clustering with subsequential convergence (3.69).

The elementary facts about compactness are developed in Chapter 4. Because we avoid Zorn's lemma, the Tychonoff theorem is proved only for the product of finitely many spaces and separately for the product of a sequence of metrizable spaces. Sequential compactness and other variants of compactness are not studied in their own right, only as equivalents of compactness in the metrizable case. Also considered for metric spaces is the relationship of compactness to uniform continuity and to completeness. The discussion of locally compact spaces includes the one-point compactification.

The first three sections of Chapter 5 present the standard facts about connected sets, components, locally connected spaces, and path-connected spaces. This material, which is technically if not conceptually simpler than that on compactness, can be read before Chapter 4. The final two sections, on homotopy, culminate in proofs of the Brouwer fixed-point theorem in dimension 2 and the fundamental theorem of algebra. The only thing about compactness needed in these two sections is the existence of a Lebesgue number (4.41) for an open cover of the unit interval or of the unit square. In order to keep the algebraic machinery to a minimum and make things geometrically more transparent, our treatment of homotopy avoids explicit mention of the fundamental group (except in the exercises).

The exercises, found at the end of each section, number 583 in all. Ranging in difficulty from the routine to the challenging, they are meant both to test comprehension of the ideas presented in the text and to provide applications, additional examples, and extensions of these ideas. Many call not just for proofs, but for answers to such questions as "Is it true that . . . ?" or "What can be said about . . . ?" or "Is there an analog . . . ?" Included in the exercises are a number of topics this author did not deem so essential to a first course in topology to have included them in the text proper, but which are interesting and important in their own right. These topics are as diverse as completion of a metric space, T_0 and T_1-spaces, Cartesian sum topologies, manifolds with boundary, topological groups, the closed-graph theorem, cut points, and the fundamental group.

There are surely more exercises than an instructor would want to assign to any one class. We have therefore appended a Guide to the Exercises in which we cite each exercise needed for exercises in subsequent sections.

All definitions, theorems, and examples within a single chapter are numbered consecutively, so that 3.15 refers to the fifteenth item in Chapter 3. The exercises in each chapter are separately numbered consecutively; a reference to the fifth exercise of Chapter 3 would be "Exercise 5" if made within that chapter, but "Exercise 3.5" if made in another chapter.

The Bibliography includes only those books and articles referred to in the text or suggested for further reading. References to bibliographic entries are made by numbers enclosed in brackets. Appended to the Bibliography is a list of suggested readings on special topics about which individual students might report to the class.

Suggested course outlines The list below is not meant to be exhaustive, but only suggestive of possible courses that can be based on this text. The portion of the text covered by any given class will, of course, depend on the students' preparation; it will also depend heavily on the number and difficulty of problems assigned from among the many we have provided.

A minimal course (1 quarter or 1 semester)

Chapter 0   Sections 1 through 5
Chapter 1   Sections 1 through 4
Chapter 2   omit 2.41, 2.42, and 2.56(5) and (6)
Chapter 3   Section 1 except 3.11(2); Section 2 except 3.22(3) and 3.23;
             Section 3 through 3.40, except 3.35(5);
             Section 6 of Chapter 0; Section 4 except 3.49(7) through (9);
             Section 5 through 3.54
Chapter 4   Section 1 except 4.27
Chapter 5   Section 1 except 5.25 and 5.26;
             Section 2 through 5.33 or 5.35(4);
             Section 3 through 5.51—optional

A second course in topology (1 quarter or 1 semester)

Chapter 1   Section 5
Chapter 2   2.41 and 2.42
Chapter 3   Section 3 from 3.41; Sections 4 and/or 5
Chapter 4   4.27 (if Section 5 of Chapter 3 is included);
             Sections 2 and 3, or 4.41 through 4.44 and Section 3 through 4.56
Chapter 5   Section 2 from 5.34; Sections 3 through 5
Additional readings or individual projects (see the Bibliography)

A complete course (2 semesters or 3 quarters)

Chapters 0 through 5
Additional readings or individual projects

A standard course—emphasis on geometry
(1 semester or 2 quarters)

Chapter 0   Sections 1 through 5
Chapter 1   Sections 1 through 4
Chapter 2   omit 2.41, 2.42, and 2.56(5) and (6)
Chapter 3   Section 1 except 3.11(2); Section 2 except 3.23;
             Sections 3 through 3.40; Section 6 of Chapter 0;
             Section 4; Section 5 through 3.53
Chapter 4   Section 1; Theorem 4.41
Chapter 5   omit 5.35(5); include Exercises 5.97, 5.98, 5.107

A standard course—emphasis on analysis
(1 semester or 2 quarters)

Chapter 0   Sections 1 through 5
Chapter 1   include Exercises 1.85 and 1.86
Chapter 2
Chapter 3   Sections 1 through 3; Section 6 of Chapter 0;
             Section 4 except 3.49(7) through (9); Section 5
Chapter 4
Chapter 5   Section 1; Section 2 through 5.33;
             Section 3 through 5.51—optional

A brief course in set theory (3 to 5 weeks)
Chapter 0

A short course on metric spaces (8 weeks)
Chapter 0   Sections 1 through 5
Chapter 1   include Exercises 1.13, 1.14, 1.68, 1.85 through 1.89

A course in special topics (variable time)
Chapter 1   1.71 through 1.73; 1.69, 1.70, and Exercises 1.85 and 1.86
Chapter 3   3.23 (include Example 1.9 and Exercise 1.23)
Chapter 4   4.28 and 4.29; 4.41
Chapter 5   Examples 5.35(5) and/or 5.52;
             Sections 3 through 5

Acknowledgments My students at the University of Massachusetts spotted numerous errors and detected defects in exposition in a preliminary version of this text. Professor Victor Klee and several anonymous reviewers corrected some infelicities of style and suggested many improvements. Miss Margo Vidrine and Mrs. Rita Warner speedily and accurately typed the manuscript. To all these people I express my gratitude for making this book possible.

MURRAY EISENBERG
Amherst, Massachusetts

Contents

Topology

CHAPTER

Sets and Maps

In this preliminary chapter we collect some essential facts about sets and maps used throughout the text. Much of this material will doubtless not be new to the reader and is therefore covered rapidly, with few examples or proofs, to remind him of what he already knows and to fix the particular terminology and notation adopted here. Those topics that are likely to be less familiar—countability, order-completeness of the real numbers, and equivalence relations—are covered in somewhat greater detail. For a fuller treatment of these preliminaries at an elementary level see Fairchild and Ionescu Tulcea [10] or Foulis [12]; for an advanced, axiomatic treatment see Eisenberg [9].

1. SETS

Two logical connectives will be used frequently:

$\Rightarrow$ means *implies* or *if . . . then,*

$\Leftrightarrow$ means *if and only if.*

A *set* is a collection of mathematical objects. If x is one of the objects comprising a set X, we write

$$x \in X$$

and say that x is an *element, member,* or *point of* X and that x *belongs to* X; in the contrary case we write

$$x \notin X.$$

Two sets X and Y are *equal* to one another, in symbols

$$X = Y,$$

precisely when they have the same elements, that is, when

$$x \in X \iff x \in Y.$$

When it is not the case that $X = Y$, we write

$$X \neq Y.$$

Similar use of a slash mark to negate a statement will be made in the future without further explanation.

Two notational devices are used to specify particular sets. The first simply lists or indicates the elements of the set between braces. For example,

$$\{-1, 1\}$$

is the set having the two elements -1 and 1, and

$$\{2, 4, 6, \ldots\}$$

is the set of all even positive integers (since the latter set is infinite, its elements cannot all be listed explicitly, but their identity is supposed to be implicit in the few actually listed in conjunction with the context).

The second device uses the notation

$$\{x \mid P\}$$

to specify the set consisting of those objects x having a given property P. For example, if R denotes the set of all real numbers, then

$$\{x \mid x \in \mathsf{R}, x^2 = 1\} = \{-1, 1\};$$

this set, which consists of those elements of the set R satisfying a certain condition, may also be specified by the modified notation

$$\{x \in \mathsf{R} \mid x^2 = 1\}.$$

Because the vertical bar $\mid$ is also used as part of other notations (for example, for absolute value), a colon is sometimes used in place of the vertical bar in $\{x \mid P\}$ to avoid confusion. Thus

$$\{x : x \in \mathsf{R}, |x| = 1\} = \{-1, 1\} = \{x \in \mathsf{R} : |x| = 1\}.$$

0.1 Special sets. If x is an object, then the *singleton*

$$\{x\}$$

is the set having the lone member x; the set $\{x\}$ is just as different from the object x as a caged lion is from a loose lion. The *empty set* is the set $\emptyset$ having no members at all. Thus

$$\{x \mid x \neq x\} = \varnothing = \{x \in \mathsf{R} \mid x < x\}.$$

Any other set is *nonempty*.

Some sets of numbers for which we reserve special notation are:

N = the set of all natural numbers = $\{0, 1, 2, \ldots\}$
Z = the set of all integers = $\{\ldots, -2, -1, 0, 1, 2, \ldots\}$
Q = the set of all rational numbers
R = the set of all real numbers
C = the set of all complex numbers
I = $\{x \in \mathsf{R} \mid 0 \leq x \leq 1\}$

0.2 Subsets. We say that a set X is *contained in* a set Y, call X a *subset of Y*, and write

$$X \subset Y$$

to mean each element of X is an element of Y, that is,

$$x \in X \implies x \in Y.$$

We also write

$$Y \supset X$$

to mean the same thing and then say that Y *contains* X. For example,

$$\mathsf{N} \subset \mathsf{Z} \subset \mathsf{Q} \subset \mathsf{R} \subset \mathsf{C}, \qquad \mathsf{R} \supset \mathsf{I},$$

but

$$\mathsf{I} \not\subset \mathsf{Q}.$$

If x is an object, then

$$\{x\} \subset X \iff x \in X.$$

The empty set is a subset of every set X:

$$\varnothing \subset X.$$

(*Proof:* Since $\varnothing$ has no elements at all, it does not have any element that fails to be an element of X.)

Evidently

$$X = Y \iff X \subset Y \text{ and } Y \subset X.$$

Thus the inclusion $X \subset Y$ does not preclude the possibility that $X = Y$. When $X \subset Y$ but $X \neq Y$, we call X a *proper* subset of Y.

The language of subsets may be used to state the *principle of mathematical induction:*

Let $E \subset \mathsf{N}$. Suppose $0 \in E$ and suppose $n + 1 \in E$ whenever $n \in E$. Then $E = \mathsf{N}$.

This principle, which we accept as a fundamental property of the natural numbers, is the basis for "proof by induction".

To illustrate proof by induction, let us prove

(*) $$2^n > n \qquad\qquad (n \in \mathbf{N}).$$

First, $2^0 = 1 > 0$. Next, suppose $n \in \mathbf{N}$ and

$$2^n > n.$$

If $n > 0$, then

$$2^{n+1} = 2 \cdot 2^n$$

$$> 2 \cdot n \qquad \text{(by the assumption } 2^n > n)$$

$$= n + n \ge n + 1;$$

if $n = 0$, then $2^{n+1} = 2 > 1 = n + 1$. Thus $2^{n+1} > n + 1$ whenever $2^n > n$. This proves (*), for if we let

$$E = \{n \in \mathbf{N} \mid 2^n > n\},$$

then $E \subset \mathbf{N}$, and we have shown that $0 \in E$ and that $n + 1 \in E$ whenever $n \in E$; hence from the principle of mathematical induction we can conclude that $E = \mathbf{N}$.

The *power set* of a given set X is the collection

$$\mathcal{P}(X) = \{A \mid A \subset X\}$$

consisting of all the subsets of X. For example,

$$\mathcal{P}(\{0, 1\}) = \{\varnothing, \{0\}, \{1\}, \{0, 1\}\}.$$

In general,

$$\varnothing \in \mathcal{P}(X), \qquad X \in \mathcal{P}(X)$$

for any set X.

0.3 Union and intersection of two sets. Let A and B be sets. The *union of A and B* is the set

$$A \cup B = \{x \mid x \in A \text{ or } x \in B\}$$

of all those objects that belong to at least one of the sets A and B. The *intersection of A and B* is the set

$$A \cap B = \{x \mid x \in A, x \in B\}$$

of all those objects that belong to both of the sets A and B. The set A is *disjoint from B* when

$$A \cap B = \varnothing,$$

that is, when A and B have no elements in common; A *intersects* B in the contrary case.

Some handy formulas concerning union and intersection are:

$$A \cap B \subset A \subset A \cup B$$

$$A \cup A = A = A \cap A$$

$$A \cup \varnothing = A \qquad\qquad A \cap \varnothing = \varnothing$$

$$A \cup B = B \cup A \qquad\qquad A \cap B = B \cap A$$

$$A \cup (B \cup C) = (A \cup B) \cup C \qquad A \cap (B \cap C) = (A \cap B) \cap C$$

$$A \cup (B \cap C) = (A \cup B) \cap (A \cup C) \qquad A \cap (B \cup C) = (A \cap B) \cup (A \cap C)$$

$$A \cup B = B \iff A \subset B \iff A \cap B = A$$

0.4 Complements. For sets A and X, the *complement of A in X* is the set

$$X \backslash A = \{x \in X \mid x \notin A\}$$

of those elements of X that do not belong to A. For any sets A and X:

$$X \backslash \varnothing = X \qquad\qquad X \backslash X = \varnothing$$

$$A \cup (X \backslash A) = X \qquad A \cap (X \backslash A) = \varnothing$$

$$X \backslash (X \backslash A) = A$$

If A and B are subsets of X, then

$$A \subset B \iff X \backslash B \subset X \backslash A.$$

For any sets X, A, and B,

$$X \backslash (A \cup B) = (X \backslash A) \cap (X \backslash B),$$

$$X \backslash (A \cap B) = (X \backslash A) \cup (X \backslash B).$$

These two *De Morgan's laws* will be generalized in Section 3.

0.5 Ordered pairs and products. The *ordered pair* (x, y) formed from objects x and y is a new object in which x is the *first coordinate* and y is the *second coordinate*. Equality of ordered pairs is governed by the rule

$$(x, y) = (a, b) \iff x = a \text{ and } y = b.$$

[It is interesting, but unnecessary for our needs, to know that (x, y) may be defined as $\{\{x\}, \{x, y\}\}$; then the preceding rule may be deduced from this definition.] Observe that (x, y) is not the same thing as $\{x, y\}$: although

$$\{x, y\} = \{y, x\}$$

for any x and y,

$$x \neq y \implies (x, y) \neq (y, x).$$

The *product of* two sets X and Y is the set

$$X \times Y = \{(x, y) \mid x \in X, y \in Y\}$$

of all ordered pairs whose first coordinate belongs to X and whose second coordinate belongs to Y. Clearly

$$X \times Y \neq \varnothing \iff X \neq \varnothing \text{ and } Y \neq \varnothing.$$

A subset R of $X \times Y$ is called a *relation in X to Y*; for $x \in X$ and $y \in Y$ we write

$$xRy$$

to mean

$$(x, y) \in R$$

and interpret this statement to say that R "relates y to x". For example, the *usual ordering of* R is the relation

$$R = \{(x, y) \in \mathsf{R} \times \mathsf{R} \mid x \leq y\}$$

which satisfies

$$xRy \iff x \leq y \qquad\qquad (x, y \in \mathsf{R}).$$

(Here the parenthetical expression on the right means "for every $x \in \mathsf{R}$ and for every $y \in \mathsf{R}$" and qualifies the statement to its left.)

EXERCISES

1. **(a)** Do there exist two sets each of which is a proper subset of the other?
   (b) If X is a proper subset of Y and Y is a proper subset of Z, must X be a proper subset of Z?

2. Given a real number $\epsilon > 0$, find a real number $\delta > 0$ such that
   $$\{x\colon |x - 1| < \delta\} \subset \{x\colon |(3x - 1) - 2| < \epsilon\}.$$

3. Let
   $$A = \{\varnothing\}, \qquad B = \{\varnothing, A\}, \qquad C = \{\varnothing, A, B\}.$$

   (a) Compute the union and intersection of each pair of these three sets.
   (b) Compute $\mathcal{P}(A)$, $\mathcal{P}(B)$, and $\mathcal{P}(C)$.
   (c) Considering A, B, and C together with all the sets you computed in (a) and (b), determine which are elements of others, which are subsets of others, and which are equal to others.

4. Let
   $$A = \{x \in \mathsf{R} \mid x^2 < 2\}, \qquad B = \{x \in \mathsf{R} \mid x^2 = 2\}.$$

   (a) Compute and draw pictures of the sets $A \cup B$, $A \cap B$, $\mathsf{R} \setminus A$, and $\mathsf{R} \setminus B$.
   (b) Compute $B \cap \mathsf{Q}$ and $A \cap \mathsf{Z}$.

5. Establish the *absorption laws*
   $$(A \cup B) \cap B = B, \qquad (A \cap B) \cup B = B.$$

6. Exhibit subsets A and B of R for which:

  (**a**) $\mathsf{R} \setminus (A \cup B) \neq (\mathsf{R} \setminus A) \cup (\mathsf{R} \setminus B)$.

  (**b**) $\mathsf{R} \setminus (A \cap B) \neq (\mathsf{R} \setminus A) \cap (\mathsf{R} \setminus B)$.

7. Let A and B be subsets of a set X. Prove:

  (**a**) $A = B \iff X \setminus A = X \setminus B$.

  (**b**) $A \subset B \iff A \cap (X \setminus B) = \varnothing$.

8. (**a**) Express the intersection $(A \times B) \cap (C \times D)$ of products as the product of two sets.

  (**b**) Show by example that a union $(A \times B) \cup (C \times D)$ of products is not necessarily a product of two sets.

9. For which sets X and Y does $X \times Y = Y \times X$?

10. Determine all relations in $\{0, 1\}$ to $\{0, 1\}$.

11. Let $X = \{x \mid x \notin x\}$. Is $X \in X$? If not, is $X \notin X$?

2. MAPS

A *map* (or *function*)

$$f: X \to Y$$

from (or *on*) X *to* (or *into*) Y consists of sets X and Y together with a rule f which assigns to each $x \in X$ a unique element $f(x) \in Y$ called the *value of f at x*. The set X is the *domain*, the set Y is the *codomain*, and the rule f is the *graph* of the map. [It is unnecessary for our purposes to know that such a "rule" is actually a relation $f \subset X \times Y$ such that for each $x \in X$ there is exactly one $y \in Y$ with $(x, y) \in f$, and then $y = f(x)$.]

Let $f: X \to Y$ be a map. If $x \in X$ and $y = f(x)$, we write

$$x \mapsto y$$

and say that f *sends* or *maps* x to y. When $f(x)$ is specified by a single formula involving x for arbitrary $x \in X$, we write

$$f: X \to Y$$

$$x \mapsto f(x).$$

For example,

$$f: \mathsf{R} \to \mathsf{R}$$

$$x \mapsto x^2$$

is the map from R to R such that

$$f(x) = x^2 \qquad\qquad (x \in \mathsf{R}).$$

Sometimes several formulas specifying $f(x)$ are needed for various parts of the domain to which x might belong. For example, if $A \subset X$, then the *characteristic function of A in X* is the map

$$\chi_A: X \to \{0, 1\}$$

such that

$$\chi_A(x) = \begin{cases} 1 & \text{if } x \in A, \\ 0 & \text{if } x \in X \setminus A. \end{cases}$$

Likewise, the absolute value function

$$\mathsf{R} \to \mathsf{R}$$

$$x \mapsto |x|$$

is the map such that

$$|x| = \begin{cases} x & \text{if } x \geq 0, \\ -x & \text{if } x < 0. \end{cases}$$

Two maps

$$f: X \to Y, \qquad g: A \to B$$

are equal precisely when

$$X = A, \qquad Y = B,$$

and

$$f(x) = g(x) \qquad\qquad (x \in X).$$

Thus the two maps

$$f: \mathsf{R} \to \mathsf{R} \qquad g: \mathsf{R} \to \{y \in \mathsf{R} \mid y \geq 0\},$$

$$x \mapsto x^2 \qquad\qquad x \mapsto x^2$$

which have the same domain and are defined by the same rule, are different because their codomains are different.

The *range of* a map $f: X \to Y$ is the set

$$\{ f(x) \mid x \in X \}$$

of all its values. The range is a subset of the codomain and, as the preceding examples show, can be a proper subset. A map is *real-valued* when its range is contained in R—that is, when each of its values is a real number. A map $f: X \to Y$ is *constant* when its range is a singleton $\{c\}$—that is, when there is some $c \in Y$ with

$$f(x) = c \qquad\qquad (x \in X).$$

0.6 Restriction and extension. The *restriction of* a map $f: X \to Y$ *to a* set $E \subset X$ is the map

$$f|_E: E \to Y$$

$$x \mapsto f(x)$$

obtained by applying the rule f only to elements of E; an alternative notation is $f \mid E$. (Occasionally we will also restrict the codomain of a map.)

For an example of restriction consider a set X and the *identity map*

$$i: X \to X$$

$$x \mapsto x$$

of X, which sends each element of X to itself. If $E \subset X$, then the restriction

$$i\mid_E: E \to X$$

$$x \mapsto x$$

is called the *inclusion map of E into X.*

A map

$$g: Z \to Y$$

is called an *extension of* a map

$$f: X \to Y$$

to Z and is said to *extend f to Z* when

$$X \subset Z, \qquad g \mid_X = f,$$

that is, when

$$X \subset Z, \qquad g(x) = f(x) \qquad (x \in X).$$

For example, the two maps

$$g_1: \mathsf{R} \to \mathsf{R} \qquad g_2: \mathsf{R} \to \mathsf{R}$$

$$x \mapsto x^2 \qquad\qquad x \mapsto x^3$$

both extend to R the inclusion map

$$f: \{0, 1\} \to \mathsf{R}$$

of $\{0, 1\}$ into R.

0.7 Composition. If

$$f: X \to Y, \qquad g: Y \to Z$$

are maps, their *composite* is the map

$$g \circ f: X \to Z$$

$$x \mapsto g(f(x))$$

which assigns to each $x \in X$ the value of g at the element $f(x)$ of Y. (Notice
that we form $g \circ f$ only when the codomain of f equals the domain of g.)

Thus the composite of

$$f \colon \mathsf{R} \to \{y \in \mathsf{R} \mid y \geq 0\}, \qquad g \colon \{y \in \mathsf{R} \mid y \geq 0\} \to \mathsf{R}$$
$$x \mapsto e^x \qquad\qquad\qquad\qquad\qquad y \mapsto 2y$$

is

$$g \circ f \colon \mathsf{R} \to \mathsf{R}$$
$$x \mapsto 2e^x.$$

Their composite in the opposite order is

$$f \circ g \colon \{y \in \mathsf{R} \mid y \geq 0\} \to \{y \in \mathsf{R} \mid y \geq 0\}$$
$$y \mapsto e^{2y}.$$

In the preceding example $f \circ g \neq g \circ f$, so there is no commutative law for composition of maps. We do have, however, the associative law

$$h \circ (g \circ f) = (h \circ g) \circ f$$

for maps $f \colon X \to Y$, $g \colon Y \to Z$, $h \colon Z \to W$, because

$$h \circ (g \circ f)(x) = h(g \circ f(x)) = h(g(f(x))) = h \circ g(f(x)) = (h \circ g) \circ f(x)$$

for all $x \in X$.

Three maps $f \colon X \to Y$, $g \colon Y \to Z$, $h \colon X \to Z$ may be displayed together in a single triangular diagram showing "where the maps are from and where they are to":

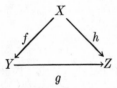

This diagram is said to *commute* (or to be *commutative*) when

$$h = g \circ f,$$

in other words, when the result $g(f(x))$ of applying to an arbitrary $x \in X$ first the map f and then the map g on the "path" from X to Z passing through Y is the same as the result $h(x)$ of applying to x the map h on the direct path from X to Z. Similarly, four maps $f \colon X \to Y$, $g \colon Z \to W$, $h \colon X \to Z$, $k \colon Y \to W$ may be displayed on a square diagram

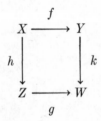

which is said to *commute* when

$$g \circ h = k \circ f.$$

A more complicated diagram of maps is said to *commute* when each of its constituent triangles and squares commutes.

As an example, consider the product $X \times Y$ of two sets. The *first* and *second projections* of $X \times Y$ are the respective maps

$$p: X \times Y \to X, \qquad q: X \times Y \to Y.$$
$$(x, y) \longmapsto x \qquad\qquad (x, y) \longmapsto y$$

Suppose $f: Z \to X$ and $g: Z \to Y$ are maps having the same domain Z. Then there is a unique map $h: Z \to X \times Y$ such that

$$p \circ h = f, \qquad q \circ h = g,$$

in other words, such that the following diagram commutes:

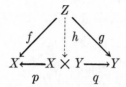

(Here the vertical arrow is dashed to indicate that the existence of h is in question.) This map h is defined by

$$h(z) = (f(z), g(z)) \qquad\qquad (z \in Z).$$

0.8 Images and inverse images. Let $f: X \to Y$ be a map. For $A \subset X$, the *image of A under f* is the set

$$f(A) = \{ f(x) \mid x \in A \}$$

of all values of f at elements of A. In particular, the image $f(X)$ of the domain of f is just the range of f. For $D \subset Y$, the *inverse image of D under f* is the set

$$f^{-1}(D) = \{ x \in X \mid f(x) \in D \}$$

of all elements of X that f maps to elements of D. In particular, $f^{-1}(\{y\})$ is the set of all $x \in X$ at which f takes the value y and is denoted more simply by $f^{-1}(y)$; note that $f^{-1}(y) = \emptyset$ unless $y \in f(X)$.

The inclusions

$$A \subset f^{-1}(f(A)) \qquad\qquad (A \subset X)$$
$$f(f^{-1}(D)) \subset D \qquad\qquad (D \subset Y)$$

both hold but need not be equalities (consider a constant map). However,

$$D \subset f(X) \quad \Rightarrow \quad f(f^{-1}(D)) = D.$$

With respect to union and intersection, images and inverse images behave according to the formulas:

$$f(A \cup B) = f(A) \cup f(B), \qquad f(A \cap B) \subset f(A) \cap f(B)$$

$$f^{-1}(D \cup E) = f^{-1}(D) \cup f^{-1}(E), \qquad f^{-1}(D \cap E) = f^{-1}(D) \cap f^{-1}(E)$$

Finally,

$$f^{-1}(Y \setminus D) = X \setminus f^{-1}(D)$$

for any $D \subset Y$.

0.9 Injections and surjections. A map $f: X \to Y$ is said to be *injective* (or *one-to-one*) and is called an *injection* when it assigns distinct values to distinct members of its domain:

$$x_1 \neq x_2 \implies f(x_1) \neq f(x_2) \qquad\qquad (x_1, x_2 \in X).$$

For example, of the two maps

$$f: \mathsf{R} \to \mathsf{R}, \qquad g: \mathsf{R} \to \mathsf{R}$$

$$x \mapsto x^3 \qquad\qquad x \mapsto x^2$$

the first is injective, but the second is not $[g(-1) = g(1)]$.

A map $f: X \to Y$ is said to be *surjective*, is said to map its domain X *onto* its codomain Y, and is called a *surjection* when its range equals its codomain:

$$f(X) = Y,$$

in other words, when

$$y \in Y \implies y = f(x) \quad \text{for some } x \in X.$$

In still other words, $f: X \to Y$ is surjective when $f^{-1}(y) \neq \varnothing$ for each $y \in Y$. Of the two maps

$$f: \mathsf{R} \to \{y \in \mathsf{R} \mid y \geq 0\}, \qquad f: \mathsf{R} \to \mathsf{R}$$

$$x \mapsto x^2 \qquad\qquad\qquad x \mapsto x^2$$

the first is surjective, but the second is not ($-1 \neq x^2$ for any $x \in \mathsf{R}$). Notice that whether a map is surjective depends in a crucial way on its codomain and not just on its domain and the rule applied to elements of that domain.

A map $f: X \to Y$ is said to be *bijective* and is called a *bijection* (or a *one-to-one correspondence between X and Y*) when it is both injective and surjective— in other words, when for each $y \in Y$ there is one and only one $x \in X$ such that $f(x) = y$. Thus a bijection effects a pairing of the elements of its domain one by one with the elements of its codomain.

The map

$$\{1, 2, 3, \ldots\} \to \{2, 4, 6, \ldots\}$$

$$n \mapsto 2n$$

is a one-to-one correspondence between the set of all positive integers and the set of all even positive integers. The maps

(*)
$$f: \mathsf{R} \to \mathsf{R}, \qquad g: \mathsf{R} \to \mathsf{R}$$
$$x \mapsto x^3 \qquad\qquad y \mapsto y^{1/3}$$

are both bijections from R onto R. In general, a bijection from a set onto itself is called a *permutation* of that set and may be regarded as a way of rearranging the elements of the set.

0.10 Proposition. Let $f: X \to Y$ and $g: Y \to Z$ be maps.

(1) If f and g are injective, then $g \circ f: X \to Z$ is injective.
(2) If f and g are surjective, then $g \circ f: X \to Z$ is surjective.
(3) If f and g are bijective, then $g \circ f: X \to Z$ is bijective.

Proof. (1) Assume f and g are injective. Let $x_1, x_2 \in X$ with $x_1 \neq x_2$. Then

$$f(x_1) \neq f(x_2)$$

since f is injective. Hence

$$g \circ f(x_1) = g(f(x_1)) \neq g(f(x_2)) = g \circ f(x_2)$$

since g is injective.

(2) Assume f and g are surjective. Then

$$g \circ f(X) = g(f(X)) = g(Y) = Z.$$

(3) follows from (1) and (2).   □

The symbol □ in the line above indicates the end of the proof.

Statements (1) and (2) above have partial converses: If $g \circ f$ is injective, then f is injective; if $g \circ f$ is surjective, then g is surjective. These converses are often applied in case $g \circ f$ is an identity map, which is bijective.

0.11 Inverse of a bijection. Suppose

$$f: X \to Y$$

is a bijection. Then a new map

$$f^{-1}: Y \to X,$$

called the *inverse of f*, is obtained by defining

$$f^{-1}(y) = \text{the unique } x \in X \text{ such that } f(x) = y$$

for each $y \in Y$. Thus

$$f(x) = y \iff f^{-1}(y) = x \qquad (x \in X, y \in Y).$$

Hence

$$f^{-1}(f(x)) = x \quad (x \in X), \quad f(f^{-1}(y)) = y \quad (y \in Y),$$

or more succinctly,

$$f^{-1} \circ f = i_X, \quad f \circ f^{-1} = i_Y,$$

where i_X and i_Y are the respective identity maps of X and Y.

In examples (*) of 0.9, the map g is the inverse of f. Moreover, f is the inverse of g, because in general for any bijection $f: X \rightarrow Y$ its inverse $f^{-1}: Y \rightarrow X$ is also a bijection and

$$(f^{-1})^{-1} = f.$$

If $f: X \rightarrow Y$ and $g: Y \rightarrow Z$ are bijections, then by 0.10 (3) their composite $g \circ f: X \rightarrow Z$ is also a bijection, with inverse

$$(g \circ f)^{-1} = f^{-1} \circ g^{-1}.$$

EXERCISES

12. Show that the following identities concerning the characteristic functions of subsets A and B in a set X hold for all $x \in X$:

(**a**) $\chi_{X \setminus A}(x) = 1 - \chi_A(x)$.

(**b**) $\chi_{A \cap B}(x) = \chi_A(x) \cdot \chi_B(x)$.

(**c**) $\chi_{A \cup B}(x) = \chi_A(x) + \chi_B(x) - \chi_{A \cap B}(x)$.

13. Verify the following properties of the absolute value function:

(**a**) $|x| \geq 0$.

(**b**) $|x| = 0 \iff x = 0$.

(**c**) $|\lambda x| = |\lambda| \cdot |x|$.

(**d**) $|x + y| \leq |x| + |y|$.

14. Let $\epsilon > 0$. Prove:

$$|x| < \epsilon \iff -\epsilon < x < \epsilon \qquad (x \in \mathbf{R}).$$

15. Let $f: \mathbf{R} \rightarrow \mathbf{R}$, $g: \mathbf{R} \rightarrow \mathbf{R}$ be defined by

$$f(x) = x + x^2, \quad g(x) = |x|.$$

Determine the ranges of f, g, $g \circ f$, and $f \circ g$.

16. If $f: X \rightarrow Y$ is a map, show that

$$f \circ i_X = f, \quad i_Y \circ f = f,$$

where i_X and i_Y are the identity maps of X and Y, respectively.

17. (**a**) Exhibit a map $f: \mathbf{R} \rightarrow \mathbf{R}$ and subsets A and B of $\mathbf{R}$ for which

$$f(A \cap B) \neq f(A) \cap f(B).$$

(**b**) If $A \subset X$ and $f: X \to Y$ is a map, prove that

$$f^{-1}(Y \setminus f(X \setminus A)) \subset A.$$

18. Let $p: X \times Y \to X$, $q: X \times Y \to Y$ be the projections. Is it necessarily true that $E = p(E) \times q(E)$ for any nonempty subset E of $X \times Y$?

19. Let $f: X \to Y$. Consider the maps

$$h: X \to \mathcal{P}(X), \qquad k: Y \to \mathcal{P}(Y).$$
$$x \mapsto \{x\} \qquad\qquad y \mapsto \{y\}$$

Construct a map $F: \mathcal{P}(X) \to \mathcal{P}(Y)$ making the following diagram commute:

$$
\begin{array}{ccc}
& f & \\
X & \longrightarrow & Y \\
h \downarrow & & \downarrow k \\
& F & \\
\mathcal{P}(X) & \longrightarrow & \mathcal{P}(Y)
\end{array}
$$

20. For real numbers α and β let $f: \mathsf{R} \to \mathsf{R}$ be the map $x \mapsto \alpha x + \beta$.

(**a**) For which α, β is f constant?
(**b**) For which α, β is f injective?
(**c**) For which α, β is f surjective?
(**d**) For which α, β is f bijective? For such α, β compute f^{-1}.

21. If $f: X \to Y$ is an injection and if $g_1, g_2: Z \to X$ are maps with $f \circ g_1 = f \circ g_2$, deduce that $g_1 = g_2$.

22. Let $f: X \to Y$ and $g: Y \to X$ be maps such that $g \circ f = i_X$, the identity map of X, and $f \circ g = i_Y$, the identity map of Y. Prove that f is bijective and $g = f^{-1}$.

23. For a set X, denote the set of all maps from X to $\{0, 1\}$ by 2^X. Note that the characteristic function $\chi_A \in 2^X$ for each $A \subset X$. Construct a one-to-one correspondence between $\mathcal{P}(X)$ and 2^X.

24. Let $\mathcal{C}$ be the set of all continuous functions $f: \mathsf{R} \to \mathsf{R}$. A map $I: \mathcal{C} \to \mathcal{C}$ is defined by

$$I(f)(x) = \int_0^x f(t) \, dt \qquad\qquad (f \in \mathcal{C}, \, x \in \mathsf{R}).$$

Is I injective? Is I surjective? (*Hint:* The fundamental theorem of calculus is useful here.)

3. FAMILIES AND PRODUCTS

A map may be thought of as labeling the various values of its range by the members of its domain at which it takes these values, and then the special language and notation of families are employed. Specifically, a map

$$x: I \to X$$

is called a *family in* X (or *of elements of* X) *indexed by* I, an element $i \in I$

is called an *index* of the family, and I is called the *index set* of the family; the value $x(i)$ of x at an index i is called the *ith coordinate of x* and is denoted by means of a subscript as

$$x_i,$$

and then x itself is denoted by

$$(x_i \mid i \in I).$$

(For technical reasons it is sometimes convenient to disregard the codomain of a map when it is considered a family.)

0.12 *N*-tuples and sequences. The families we shall encounter most often are indexed by subsets of the set $\mathsf{N} = \{0, 1, 2, \ldots\}$ of natural numbers. When for some natural number $n \geq 1$ a family $(x_i \mid i \in I)$ is indexed by

$$I = \{1, 2, \ldots, n\},$$

then $(x_i \mid i \in I)$ is called an *n-tuple* and is denoted by

$$(x_i \mid i = 1, 2, \ldots, n)$$

or more simply by

$$(x_1, x_2, \ldots, x_n).$$

A 2-tuple is a family $(x_i \mid i = 1, 2)$ assigning an object x_1 to 1 and an object x_2 to 2 and hence is conceptually the same thing as an ordered pair (x_1, x_2); indeed, we make no notational distinction between a 2-tuple and an ordered pair when we write (x_1, x_2). A 1-tuple $(x_i \mid i = 1)$ assigns an object x_1 to 1 and hence is essentially the same thing as the object x_1 itself.

A family $(x_i \mid i \in I)$ indexed by the set

$$I = \mathsf{N}$$

of all natural numbers is called a *sequence* and is denoted by

$$(x_i \mid i = 0, 1, \ldots)$$

or more simply by

$$(x_0, x_1, \ldots).$$

Families indexed by the set $I = \{1, 2, \ldots\}$ are called *sequences* too, and then the obvious notation is employed.

Suppose X is a set. If n is a positive integer, we define

$$X^n = \{(x_1, x_2, \ldots, x_n) \mid x_1 \in X, x_2 \in X, \ldots, x_n \in X\},$$

the set of all *n*-tuples of elements of X. Because a 2-tuple is essentially the same thing as an ordered pair, we shall make no distinction between X^2 and $X \times X$. Likewise, we shall make no distinction between X^1 and X.

An important instance of the preceding definition occurs when $X = \mathsf{R}$:

0.13 Euclidean spaces. If n is a positive integer, the set

$$\mathsf{R}^n = \{(x_1, x_2, \ldots, x_n) \mid x_1 \in \mathsf{R}, x_2 \in \mathsf{R}, \ldots, x_n \in \mathsf{R}\}$$

of all n-tuples of real numbers is n-*dimensional Euclidean space*; the n-tuple

$$0 = (0, 0, \ldots, 0)$$

all of whose coordinates are the real number 0 is the *origin* of R^n. In particular,

$$\mathsf{R}^1 = \mathsf{R}$$

is the *real line*, and

$$\mathsf{R}^2 = \mathsf{R} \times \mathsf{R}$$

is the (*Euclidean*) *plane*.

On R^n two algebraic operations of vector addition and multiplication by scalars are defined "coordinatewise" as follows. If

$$x = (x_1, x_2, \ldots, x_n) \in \mathsf{R}^n, \qquad y = (y_1, y_2, \ldots, y_n) \in \mathsf{R}^n,$$

then

$$x + y = (x_1 + y_1, x_2 + y_2, \ldots, x_n + y_n).$$

If

$$x = (x_1, x_2, \ldots, x_n) \in \mathsf{R}^n, \qquad \alpha \in \mathsf{R},$$

then

$$\alpha x = (\alpha x_1, \alpha x_2, \ldots, \alpha x_n).$$

If we set

$$-x = (-x_1, -x_2, \ldots, -x_n),$$

then the following properties say that those two operations make R^n into a (real) *vector space*:

$$(x + y) + z = x + (y + z)$$
$$0 + x = x = x + 0$$
$$x + (-x) = 0 = (-x) + x$$
$$x + y = y + x$$
$$(\alpha\beta)x = \alpha(\beta x)$$
$$(\alpha + \beta)x = \alpha x + \beta x$$
$$\alpha(x + y) = \alpha x + \alpha y$$
$$1x = x$$

Here x, y, z are arbitrary elements of R^n and α, β are arbitrary real numbers.

0.14 Line segments and lines. Let $x, y \in \mathsf{R}^n$. The *line segment joining* x *to* y is the subset

$$\{(1 - t)x + ty \mid 0 \le t \le 1\}$$

of R^n (note that this line segment reduces to the singleton $\{x\}$ in case $y = x$). If $x \ne y$, then the *line passing through* x *and* y is the subset

$$\{(1 - t)x + ty \mid t \in \mathsf{R}\}$$

of R^n containing that line segment. In R^3, for example, the line passing through two points $x = (x_1, x_2, x_3)$ and $y = (y_1, y_2, y_3)$ consists of all points $z = (z_1, z_2, z_3)$ whose coordinates satisfy the equations

$$\begin{cases} z_1 = (1 - t)x_1 + ty_1 \\ z_2 = (1 - t)x_2 + ty_2 \\ z_3 = (1 - t)x_3 + ty_3 \end{cases}$$

for some $t \in \mathsf{R}$, and these are the usual parametric equations for the line.

A subset K of R^n is said to be *convex* when it contains the line segment joining any two of its points—that is, when

$$x, y \in K \quad \Rightarrow \quad (1 - t)x + ty \in K \quad \text{for all } 0 \le t \le 1.$$

For example, the line segment joining two points of R^n is itself a convex set.

Again suppose X is an arbitrary set. For any index set I, we define

$$X^I = \{(x_i \mid i \in I) \mid x_i \in X \quad \text{for all } i \in I\},$$

the set of all families in X indexed by I. Taking in particular $I = P = \{1, 2, \ldots\}$, we obtain the set

$$X^P = \{(x_1, x_2, \ldots) \mid x_1 \in X, x_2 \in X, \ldots\}$$

of all sequences of elements of X.

Instead of families all of whose coordinates belong to the same set, we can consider families whose coordinates belong to (possibly) different sets.

0.15 Product of a family. Let $X_1, X_2, \ldots, X_n$ be sets where n is a positive integer. Then the *product of* $(X_1, X_2, \ldots, X_n)$ is the set

$$\overset{n}{\underset{i=1}{\mathsf{X}}} X_i = X_1 \times X_2 \times \cdots \times X_n$$

$$= \{(x_1, x_2, \ldots, x_n) \mid x_1 \in X_1, x_2 \in X_2, \ldots, x_n \in X_n\}$$

of all n-tuples whose first coordinate belongs to x_1, whose second coordinate belongs to x_2, and so on. Because we make no distinction between 2-tuples and ordered pairs, when $n = 2$ we have

$$\overset{2}{\underset{i=1}{\times}} X_i = X_1 \times X_2,$$

so that products in the present sense generalize products in the sense of 0.5. For $j = 1, 2, \ldots, n$, the *jth projection* is the map

$$p_j: X_1 \times X_2 \times \cdots \times X_j \times \cdots \times X_n \to X_j$$

$$(x_1, x_2, \ldots, x_j, \ldots, x_n) \mapsto x_j$$

sending each element of the product to its *j*th coordinate. In particular, when $n = 2$, we have two projections

$$p_1: X_1 \times X_2 \to X_1, \qquad p_2: X_1 \times X_2 \to X_2$$

$$(x_1, x_2) \mapsto x_1 \qquad\qquad (x_1, x_2) \mapsto x_2$$

which are the first and second projections considered in 0.7.

Now let $(X_1, X_2, \ldots)$ be a sequence of sets. The *product of* this sequence is the set

$$\overset{\infty}{\underset{i=1}{\times}} X_i = X_1 \times X_2 \times \cdots = \{ (x_1, x_2, \ldots) \mid x_1 \in X_1, x_2 \in X_2, \ldots \}$$

of all sequences whose *i*th coordinate belongs to the *i*th set X_i for each $i \in I$. As before, for each $j = 1, 2, \ldots$ we have the *jth projection*

$$p_j: X_1 \times X_2 \times \cdots \to X_j$$

$$(x_1, x_2, \ldots) \mapsto x_j.$$

Both kinds of products just considered can be denoted uniformly by

$$\underset{i \in I}{\times} X_i$$

where either $I = \{1, 2, \ldots, n\}$ for some n or else $I = \{1, 2, \ldots\}$. The same notation is used, with the obvious meaning, when the index set I is any subset of N. (Products indexed by arbitrary sets can be formed, too—see 3.41.)

Suppose there is a set X with

$$X_i = X \qquad\qquad (i \in I).$$

Then

$$\underset{i \in I}{\times} X_i = X^I.$$

In particular, for $I = \{1, 2, \ldots, n\}$ we have in this case

$$\overset{n}{\underset{i=1}{\times}} X_i = X^n.$$

For example,

$$R^n = R \times R \times \cdots \times R.$$

$$\underbrace{}$$

$$n \text{ copies}$$

0.16 Union and intersection of a family. If $(X_i \mid i \in I)$ is a family of sets, its *union* is the set

$$\bigcup_{i \in I} X_i = \{x \mid x \in X_i \text{ for some } i \in I\}$$

and, when $I \neq \varnothing$, its *intersection* is the set

$$\bigcap_{i \in I} X_i = \{x \mid x \in X_i \text{ for each } i \in I\}.$$

(The case $I = \varnothing$ is excluded when forming intersections, for it is vacuously true that every set x belongs to X_i for each $i \in \varnothing$, and to allow a "set of all sets" would lead to certain logical paradoxes.)

When $I = \{1, 2, \ldots, n\}$, the notations

$$\bigcup_{i=1}^{n} X_i, \qquad X_1 \cup X_2 \cup \cdots \cup X_n$$

are also used for $\bigcup_{i \in I} X_i$, and when $I = \{1, 2, \ldots\}$ the notations

$$\bigcup_{i=1}^{\infty} X_i, \qquad X_1 \cup X_2 \cup \cdots$$

are also used for $\bigcup_{i \in I} X_i$; similarly for intersections. Other notations such as

$$\bigcup_{i=2}^{5} X_i, \qquad X_0 \cup X_1 \cup X_2 \cup \cdots$$

should require no further explanation.

The union and intersection of two sets in the sense of 0.3 are special cases of the union and intersection of a family: if X_1 and X_2 are sets, then

$$X_1 \cup X_2 = \bigcup_{i=1}^{2} X_i, \qquad X_1 \cap X_2 = \bigcap_{i=1}^{2} X_i.$$

Moreover, the formulas involving union and intersection in 0.3 generalize to the case of arbitrary families. For example, we have the distributive laws

$$Y \cap \bigcup_{i \in I} X_i = \bigcup_{i \in I} (Y \cap X_i), \qquad Y \cup \bigcap_{i \in I} X_i = \bigcap_{i \in I} (Y \cup X_i).$$

A family $(X_i \mid i \in I)$ of sets is said to be *(pairwise) disjoint* when

$$i \neq j \implies X_i \cap X_j = \varnothing \qquad\qquad (i, j \in I).$$

Similarly, a collection $\mathfrak{a}$ of sets is said to be *(pairwise) disjoint* when

$$A \neq B \implies A \cap B = \varnothing \qquad\qquad (A, B \in \mathfrak{a}).$$

Observe that a family $(X_i \mid i \in I)$ need not be disjoint just because $\bigcap_{i \in I} X_i = \varnothing$; for example, the sequence $(\{n, n+1\} \mid n \in \mathbb{N})$ of subsets of $\mathbb{N}$ has empty intersection, yet $\{0, 1\} \cap \{1, 2\} \neq \varnothing$.

0.17 Proposition (De Morgan's laws). Let $(A_i \mid i \in I)$ be a family of sets with $I \neq \varnothing$, and let X be a set. Then

$$X \backslash \bigcup_{i \in I} A_i = \bigcap_{i \in I} (X \backslash A_i),$$

$$X \backslash \bigcap_{i \in I} A_i = \bigcup_{i \in I} (X \backslash A_i).$$

The formation of inverse images under a map $f \colon X \to Y$ "preserves" unions and intersections according to the formulas

$$f^{-1} \left(\bigcup_{i \in I} D_i \right) = \bigcup_{i \in I} f^{-1}(D_i),$$

$$f^{-1} \left(\bigcap_{i \in I} D_i \right) = \bigcap_{i \in I} f^{-1}(D_i).$$

For images, although we have the equality

$$f \left(\bigcup_{i \in I} A_i \right) = \bigcup_{i \in I} f(A_i),$$

we have, in general, only the inclusion

$$f \left(\bigcap_{i \in I} A_i \right) \subset \bigcap_{i \in I} f(A_i).$$

0.18 Union and intersection of a collection. It is sometimes convenient to form the union or intersection of a number of sets even when these sets have not been explicitly indexed as the values of a family. Thus if $\mathcal{C}$ is a collection of sets, the *union of* $\mathcal{C}$ is the set

$$\bigcup \mathcal{C} = \{x \mid x \in A \quad \text{for some } A \in \mathcal{C}\},$$

and when $\mathcal{C}$ is nonempty the *intersection of* $\mathcal{C}$ is the set

$$\bigcap \mathcal{C} = \{x \mid x \in A \quad \text{for every } A \in \mathcal{C}\}.$$

If we do want to index the sets belonging to $\mathcal{C}$, the simplest way is to index each $A \in \mathcal{C}$ by itself, so that we can also write

$$\bigcup_{A \in \mathcal{C}} A = \bigcup \mathcal{C}, \qquad \bigcap_{A \in \mathcal{C}} A = \bigcap \mathcal{C}.$$

As an example let $\mathcal{C}$ be the collection

$$\mathcal{C} = \left\{ \left] -\frac{1}{n}, 1 - \frac{1}{n} \right[\, \middle| \, n = 2, 3, \ldots \right\},$$

where

$$\left]-\frac{1}{n},\, 1-\frac{1}{n}\right[= \left\{x \in \mathsf{R}\,\middle|\, -\frac{1}{n} < x < 1-\frac{1}{n}\right\} \qquad (n = 2, 3, \ldots).$$

Then

$$\bigcup \mathcal{Q} = \bigcup_{n=2}^{\infty}\left]-\frac{1}{n},\, 1-\frac{1}{n}\right[= \left]-\frac{1}{2},\, 1\right[$$

$$= \{x \in \mathsf{R}\,|\, -\tfrac{1}{2} < x < 1\},$$

$$\bigcap \mathcal{Q} = \bigcap_{n=2}^{\infty}\left]-\frac{1}{n},\, 1-\frac{1}{n}\right[= \{x \in \mathsf{R}\,|\, 0 \le x < \tfrac{1}{2}\}.$$

As another example, if

$$\mathcal{Q} = \{A, B\},$$

then

$$\bigcup \mathcal{Q} = \bigcup \{A, B\} = A \cup B, \qquad \bigcap \mathcal{Q} = \bigcap \{A, B\} = A \cap B.$$

As a final example, if X is any set, then

$$\bigcup \mathcal{P}(X) = X, \qquad \bigcap \mathcal{P}(X) = \varnothing.$$

EXERCISES

25. (a) Find a sequence $(X_n \mid n = 0, 1, 2, \ldots)$ of sets such that X_{n+1} is a proper subset of X_n for each n. Then compute $\bigcup_{n=0}^{\infty} X_n$ and $\bigcap_{n=0}^{\infty} X_n$.
(b) Find a sequence $(Y_n \mid n = 0, 1, 2, \ldots)$ of sets such that $Y_n \in Y_{n+1}$ for each n. Then compute $\bigcup_{n=0}^{\infty} Y_n$ and $\bigcap_{n=0}^{\infty} Y_n$.

26. Let $x = (1, 2, 3, 4)$, $y = (4, 3, 2, 1) \in \mathsf{R}^4$ and let L be the line in R^4 passing through x and y.

(a) Determine two points other than x and y belonging to L.
(b) Does the origin of R^4 belong to L?

27. Let $K = \{x \in \mathsf{R}^n \colon |x_1| + |x_2| + \cdots + |x_n| \le 1\}$.

(a) Draw pictures of K for the cases $n = 1, 2$, and 3.
(b) Show that K is convex for arbitrary $n = 1, 2, \ldots$.

28. Prove or disprove:

(a) If $(K_i \mid i \in I)$ is any family of convex subsets of R^n, then $\bigcap_{i \in I} K_i$ is convex.
(b) If $(K_i \mid i \in I)$ is any family of convex subsets of R^n, then $\bigcup_{i \in I} K_i$ is convex.

29. Let $X_1 = \{0, 1\}$, $X_2 = \{1, 2\}$, $X_3 = \{2, 3\}$, $X_n = \{3\}$ for $n \ge 4$. Compute:

(a) $\bigcup_{n=1}^{3} X_n$.   (d) $\bigcup_{n=1}^{\infty} X_n$.

(b) $\bigcap_{n=1}^{3} X_n$.   (e) $\bigcap_{n=1}^{\infty} X_n$.

(c) $\bigtimes_{n=1}^{3} X_n$.   (f) $\bigtimes_{n=1}^{\infty} X_n$.

30. (**a**) Prove: If $(X_i \mid i \in I)$ and $(Y_i \mid i \in I)$ are families of sets indexed by the same set I, then

$$(\bigcap_{i \in I} X_i) \times (\bigcap_{i \in I} Y_i) = \bigcap_{i \in I} (X_i \times Y_i).$$

(**b**) Does the analog of (a) for union instead of intersection also hold?

31. Let $(X_1, X_2, \ldots, X_n)$ be an n-tuple of sets. For each $1 \le j \le n$ let $p_j \colon \mathsf{X}_{i=1}^{n} X_i \to X_j$ be the jth projection.

(**a**) If $1 \le j \le n$ and $A \subset X_j$, express the subset $p_j^{-1}(A)$ of $\mathsf{X}_{i=1}^{n} X_i$ as a product.

(**b**) If $A_j \subset X_j$ for each $j = 1, \ldots, n$, express the subset $\bigcap_{j=1}^{n} p_j^{-1}(A_j)$ of $\mathsf{X}_{i=1}^{n} X_i$ as a product.

32. If $\mathcal{C}$ is a collection of sets and $X \in \mathcal{C}$, show that $\bigcap \mathcal{C} \subset X \subset \bigcup \mathcal{C}$.

33. Find collections $\mathcal{C}$ and $\mathcal{B}$ of sets for which

$$(\bigcap \mathcal{C}) \cap (\bigcap \mathcal{B}) \ne \bigcap (\mathcal{C} \cap \mathcal{B}).$$

34. Let $\mathcal{C}$ and $\mathcal{B}$ be collections of sets with $\mathcal{C} \subset \mathcal{B}$. Prove:

$$\bigcup \mathcal{C} \subset \bigcup \mathcal{B}, \qquad \bigcap \mathcal{B} \subset \bigcap \mathcal{C}.$$

4. COUNTABILITY

This section concerns sets having so small a number of elements that their elements can be "counted". The underlying idea below is to regard two sets as having the same "number" of elements when the elements of one of the sets can be paired up with the elements of the other—in other words, when there is a one-to-one correspondence between the two sets.

0.19 Definition. A set X is said to be *finite* when either $X = \varnothing$ or else there exists a bijection from the set $\{1, 2, \ldots, n\}$ onto X for some positive integer n, and X is said to be *infinite* when it is not finite.

Obviously a set will be finite (respectively, infinite) if there is a bijection from it onto some finite (respectively, infinite) set.

It is easy to exhibit finite sets: any singleton $\{x\}$ is finite. It is somewhat harder to establish the existence of infinite sets.

0.20 Example. The set N of all natural numbers is infinite. In fact, just suppose N is finite. Since N is nonempty, there exists a bijection

$$f \colon \{1, 2, \ldots, n\} \to \mathsf{N}$$

for some positive integer n. Define

$$g \colon \{1, 2, \ldots, n + 1\} \to \mathsf{N}$$

by

$$g(i) = \begin{cases} f(i) + 1 & \text{if } i < n + 1, \\ \\ 0 & \text{if } i = n + 1. \end{cases}$$

Then g is bijective, so the composite

$$f^{-1} \circ g : \{1, 2, \ldots, n + 1\} \to \{1, 2, \ldots, n\}$$

is a bijection.

We use induction on n to show, on the contrary, that for each $n \geq 1$ there does not exist any bijection from $\{1, 2, \ldots, n + 1\}$ to $\{1, 2, \ldots, n\}$. Certainly there cannot exist a bijection from $\{1, 2\}$ to $\{1\}$. Now let $n \geq 1$ and assume there exists no bijection from $\{1, 2, \ldots, n + 1\}$ to $\{1, 2, \ldots, n\}$. From this inductive hypothesis we must deduce that there exists no bijection from $\{1, 2, \ldots, n + 1, n + 2\}$ to $\{1, 2, \ldots, n, n + 1\}$. Just suppose there did exist some bijection

$$h : \{1, 2, \ldots, n + 1, n + 2\} \to \{1, 2, \ldots, n, n + 1\}.$$

Then the modified map

$$h' : \{1, 2, \ldots, n + 1, n + 2\} \to \{1, 2, \ldots, n, n + 1\}$$

given by

$$h'(i) = \begin{cases} h(i) & \text{if } i \neq n + 2, h(i) \neq n + 1, \\ \\ h(n + 2) & \text{if } h(i) = n + 1, \\ \\ n + 1 & \text{if } i = n + 2 \end{cases}$$

would also be a bijection, but with

$$h'(n + 2) = n + 1.$$

Hence the map obtained from h' by restricting its domain to $\{1, 2, \ldots, n + 1\}$ and its codomain to $\{1, 2, \ldots, n\}$ would be a bijection, contrary to the inductive assumption that no such bijection exists.

The next several results give methods for obtaining finite sets from other finite sets. The first is a technical result that will be subsumed in 0.23.

0.21 Lemma. Let A and B be finite sets with A disjoint from B. Then $A \cup B$ is finite.

Proof. There is nothing to prove if $A = \varnothing$ or $B = \varnothing$, so we assume neither A nor B is empty. There exist positive integers m and n and bijections

$$f : \{1, \ldots, m\} \to A, \qquad g : \{1, \ldots, n\} \to B.$$

Then the map

$$h\colon \{1, \ldots, m, m+1, \ldots, m+n\} \to A \cup B$$

given by

$$h(i) = \begin{cases} f(i) & \text{if } 1 \leq i \leq m, \\ g(i-m) & \text{if } m+1 \leq i \leq m+n \end{cases}$$

is a bijection. ☐

0.22 Proposition. Any subset of a finite set is finite.

Proof. It suffices to show that for each $n \geq 1$ any subset of $\{1, \ldots, n\}$ is finite. We show this by induction. The only subsets of $\{1\}$ are $\varnothing$ and $\{1\}$, and both are finite. Assume that each subset of $\{1, \ldots, n\}$ is finite. Let $A \subset \{1, \ldots, n, n+1\}$. If $A \subset \{1, \ldots, n\}$, then A is finite by the inductive assumption. Suppose now $A \not\subset \{1, \ldots, n\}$. Then $n+1 \in A$. By the inductive hypothesis the subset $A \setminus \{n+1\}$ of $\{1, \ldots, n\}$ is finite, and $\{n+1\}$ is finite. From 0.21 it follows that

$$A = (A \setminus \{n+1\}) \cup \{n+1\}$$

is finite. ☐

According to this proposition, a set is infinite if it contains an infinite set. In view of 0.20, then, the sets Z, Q, and R are all infinite.

Below we shall speak of a *finite family*, meaning a family indexed by a finite set. (Later we will need to form the intersection of a finite family of sets, and then we shall understand that the index set is not only finite, but nonempty as well.)

0.23 Proposition. The union of a finite family of finite sets is finite.

Proof. First we note that the union of any two (not necessarily disjoint) finite sets A and B is finite, for

$$A \cup B = (A \setminus B) \cup (B \setminus A) \cup (A \cap B),$$

the pairwise disjoint sets $A \setminus B$, $B \setminus A$, $A \cap B$ are finite by 0.22, and so $A \cup B$ is finite by 0.21.

Now let $(X_i \mid i \in I)$ be a finite family of finite sets. If $I = \varnothing$, then $\bigcup_{i \in I} X_i = \varnothing$. Suppose $I \neq \varnothing$. Then there is bijection $\sigma\colon \{1, \ldots, n\} \to I$ for some $n \geq 1$. The family $(A_j \mid j = 1, \ldots, n)$ defined by

$$A_j = X_{\sigma(j)} \qquad\qquad (j = 1, \ldots, n)$$

is indexed by $\{1, \ldots, n\}$ and has the union

$$\bigcup_{j=1}^{n} A_j = \bigcup_{i \in I} X_i.$$

In view of what we have just shown, it suffices to prove that for each $n = 1, 2, \ldots$ each family $(A_j \,|\, j = 1, \ldots, n)$ of finite sets has a finite union. We use induction on n. The case $n = 1$ is obvious, and the case $n = 2$ follows from the first paragraph. The inductive step follows from the formula

$$\bigcup_{j=1}^{n+1} A_j = \bigcup_{j=1}^{n} A_j \cup A_{n+1}$$

with the aid of the first paragraph.   ☐

0.24 Proposition. The product of a finite family of finite sets is finite.

Proof. First we show that the product of two finite sets A and B is finite. We have

$$A \times B = \bigcup_{a \in A} (\{a\} \times B).$$

For each $a \in A$ the map

$$\{a\} \times B \to B$$

$$(a, b) \mapsto b$$

is bijective, so $\{a\} \times B$ is finite because B is. It follows from 0.23 that $A \times B$ is finite.

As in 0.23, to prove the proposition it suffices to show that for each $n = 1, 2, \ldots$, each family $(X_i \,|\, i = 1, \ldots, n)$ of finite sets has a finite product. The case $n = 1$ is obvious, and the case $n = 2$ follows from what we just proved. To execute the necessary inductive step use the fact that the map

$$\mathop{\mathsf{X}}_{i=1}^{n+1} X_i \to \left(\mathop{\mathsf{X}}_{i=1}^{n} X_i \right) \times X_{n+1}$$

$$(x_1, \ldots, x_n, x_{n+1}) \mapsto ((x_1, \ldots, x_n), x_{n+1})$$

is a bijection.   ☐

For our final result concerning finite sets we shall need the *well-ordering principle for* N: Any nonempty subset of N has a least member. (This principle is logically equivalent to the principle of mathematical induction.)

0.25 Proposition. Let X be a finite set and let $f \colon X \to Y$ be a surjection. Then Y is finite.

Proof. If $X = \varnothing$, then $Y = \varnothing$, so we suppose $X \neq \varnothing$. Without loss of generality we may assume

$$X = \{1, \ldots, n\}$$

for some positive integer n. Define a map

$$g\colon Y \to X$$

as follows. If $y \in Y$, then $f^{-1}(y) \neq \varnothing$, and we take $g(y)$ to be the least element of $f^{-1}(y)$. Then g is an injection, the set $g(Y)$ is finite by 0.22, and hence Y is finite.   □

We now distinguish between two kinds of infinite sets.

0.26 Definition. A set X is *denumerable* if there exists some bijection

$$f\colon \mathsf{N} \to X,$$

countable if it is finite or denumerable, and *uncountable* if it is not countable. (*Caution:* Some authors use 'countable' to mean what we have called 'denumerable'.)

Because N is infinite, each denumerable set is infinite. Hence a set is infinite if and only if it is either denumerable or else uncountable. Of course, each set is either countable or else uncountable.

The set $\{1, 2, \ldots\}$ of all positive integers is denumerable because the map $n \mapsto n + 1$ from N to $\{1, 2, \ldots\}$ is a bijection. Similarly, the set $\{2, 4, 6, \ldots\}$ of all even positive integers is denumerable.

A set will obviously be denumerable (respectively, countable, uncountable) if there is a bijection from it to some denumerable (respectively, countable, uncountable) set.

A bijection $f\colon \mathsf{N} \to X$ is nothing but a sequence $(x_n \mid n \in \mathsf{N})$ of elements of X such that

$$X = \{x_n \mid n \in \mathsf{N}\}, \qquad x_i \neq x_j \qquad (i \neq j).$$

Hence a set X is denumerable precisely when such a sequence exists.

0.27 Proposition. An infinite subset of a countable set is denumerable.

Proof. Let A be an infinite subset of a countable set X. By 0.22 the set X must be denumerable, and without loss of generality we may assume $X = \mathsf{N}$. Let x_0 be the least element of A. Since the subset $A \backslash \{x_0\}$ of N is infinite, it is nonempty; let x_1 be the least element of $A \backslash \{x_0\}$. Since $A \backslash \{x_0, x_1\}$ is also infinite, it is nonempty; let x_2 be the least element of $A \backslash \{x_0, x_1\}$. In general, once $x_0, \ldots, x_n$ have been constructed let x_{n+1} be the least element of $A \backslash \{x_0, \ldots, x_n\}$.

By construction $x_i \neq x_j$ whenever $i \neq j$. An easy induction on m establishes

$$m \in A \quad \Rightarrow \quad m = x_n \ \text{ for some } n \in \mathsf{N}.$$

Hence $A = \{x_n \mid n \in \mathsf{N}\}$.   □

An obvious corollary is that each subset of a countable set is countable. By 0.20 and 0.22, a set is infinite if it contains some denumerable subset. Conversely we have the following proposition.

0.28 Proposition. Let X be an infinite set. Then there exists a denumerable subset of X.

Proof. Arbitrarily choose an element $x_0 \in X$. The set $X \setminus \{x_0\}$ is infinite and therefore nonempty; arbitrarily choose an $x_1 \in X \setminus \{x_0\}$. Next choose an $x_2 \in X \setminus \{x_0, x_1\}$. Continuing in this way we obtain a sequence $(x_n \mid n \in \mathsf{N})$ in X with $x_i \neq x_j$ whenever $i \neq j$. Then $D = \{x_n \mid n \in \mathsf{N}\}$ is the desired denumerable subset of X.   □

As convincing as this argument might seem, it is not complete; a guarantee is needed that *all* the necessary choices can be made "simultaneously". This guarantee is provided by the *axiom of choice*, one version of which is: *If $\mathcal{Q}$ is any collection of nonempty sets, then there exists some map c having domain $\mathcal{Q}$ such that $c(A) \in A$ for each $A \in \mathcal{Q}$.* Such a "choice function" c should be thought of as a way of simultaneously choosing an element from each member of $\mathcal{Q}$. (A rigorous proof of 0.28 is given on pages 235–236 of Eisenberg [9].)

0.29 Proposition. Let X be a countable set and let $f \colon X \to Y$ be a surjection. Then Y is countable.

Proof. If X is finite, then by 0.25 the set Y is finite and hence countable. Suppose now that X is denumerable. Without loss of generality we may assume that $X = \mathsf{N}$. For each $y \in Y$ the subset $f^{-1}(y)$ of N is nonempty, and we let $g(y)$ be the least member of $f^{-1}(y)$. The map $g \colon Y \to \mathsf{N}$ so defined is clearly injective. The set $g(Y)$ is countable, so Y is countable.   □

From 0.29 it follows that a nonempty set Y is countable if and only if there is a sequence $(x_n \mid n \in \mathsf{N})$ in Y with $Y = \{x_n \mid n \in \mathsf{N}\}$—in other words, the elements of Y can be arranged in a sequence.

0.30 Lemma. The set $\mathsf{N} \times \mathsf{N}$ is denumerable.

Proof. Because it contains the infinite set $\mathsf{N} \times \{0\}$, the set $\mathsf{N} \times \mathsf{N}$ is infinite. To show that this set is countable as well, we need only construct an injection $\mathsf{N} \times \mathsf{N} \to \mathsf{N}$. We claim that

$$\mathsf{N} \times \mathsf{N} \to \mathsf{N}$$

$$(m, n) \mapsto 2^m 3^n$$

is such an injection. Suppose $(m, n), (i, j) \in \mathsf{N} \times \mathsf{N}$ with

$$2^m 3^n = 2^i 3^j.$$

We show that $m = i$ and $n = j$. If $m \neq i$, say $m > i$, then

$$2^{m-i}3^n = 3^j$$

which is impossible since $2^{m-i}3^n$ is even whereas 3^j is odd. Hence $m = i$. Then $2^m = 2^i$, $3^n = 3^j$, and consequently $n = j$. □

When we refer to a countable family below, we shall mean a family indexed by a countable set.

0.31 Theorem. The union of a countable family of countable sets is countable.

Proof. Let $(X_i \mid i \in I)$ be a countable family of countable sets. If $I = \varnothing$, there is nothing to prove, so we assume $I \neq \varnothing$. Without loss of generality we may assume $I \subset \mathsf{N}$. In fact, we may assume that $I = \mathsf{N}$, for if we choose some $j \in I$ and define

$$X_n = X_j \qquad\qquad (n \in \mathsf{N} \setminus I),$$

then

$$\bigcup_{n \in \mathsf{N}} X_n = \bigcup_{i \in I} X_i.$$

For each $n \in \mathsf{N}$ there is a surjection

$$f_n : \mathsf{N} \to X_n.$$

Define

$$f : \mathsf{N} \times \mathsf{N} \to \bigcup_{n \in \mathsf{N}} X_n$$

by

$$f(m, n) = f_n(m).$$

Clearly f is surjective. That the union of $(X_n \mid n \in \mathsf{N})$ is countable now follows from 0.29 and 0.30. □

A particular consequence of 0.31 is that the set Z of all integers is denumerable: The set Z is infinite because it contains N. It is countable because

$$\mathsf{Z} = \mathsf{N} \cup \{-n \mid n = 1, 2, \ldots\}.$$

0.32 Corollary. The product of a finite family of countable sets is countable.

Proof. The product of two countable sets A and B is countable because

$$A \times B = \bigcup_{a \in A} (\{a\} \times B).$$

The general result now follows by induction. □

In particular, if X is a countable set, then X^n is countable for each $n = 1, 2, \ldots$.

From 0.32 we can deduce that the set Q of all rational numbers is denumerable: The set Q is infinite because it contains N. It is countable because the map

$$Z \times (Z \backslash \{0\}) \to Q$$

$$(m, n) \mapsto \frac{m}{n}$$

is surjective and the sets Z and $Z \backslash \{0\}$ are countable.

The product of an infinite family of countable sets need not be countable:

0.33 Example. Let $P = \{1, 2, \ldots\}$. Then the set $\{0, 1\}^P$ of all sequences of 0s and 1s is uncountable. In fact, suppose this set is countable. Then

$$\{0, 1\}^P = \{x_n \mid n = 1, 2, \ldots\}$$

for some sequence $(x_n \mid n = 1, 2, \ldots)$. For each n the member x_n of $\{0, 1\}^P$ is itself a sequence

$$x_n = (x_{ni} \mid i = 1, 2, \ldots).$$

We now use Georg Cantor's original *diagonal argument* to obtain a sequence y of 0s and 1s different from x_n for each n, thereby deriving a contradiction. Consider the array shown below, in which the values of x_n are written on the nth line.

$$
\begin{array}{cccccc}
x_{11} & x_{12} & x_{13} & \cdots & x_{1n} & \cdots \\
\\
x_{21} & x_{22} & x_{23} & \cdots & x_{2n} & \cdots \\
\\
x_{31} & x_{32} & x_{33} & \cdots & x_{3n} & \cdots \\
\\
\vdots & \vdots & \vdots & & \vdots \\
\\
x_{n1} & x_{n2} & x_{n3} & \cdots & x_{nn} & \cdots \\
\\
\vdots & \vdots & \vdots & & \vdots
\end{array}
$$

Form the sequence $(x_{11}, x_{22}, \ldots)$ by taking the elements on the diagonal of this array. Define a sequence

$$y = (y_1, y_2, \ldots)$$

by

$$y_n = \begin{cases} 1 & \text{if } x_{nn} = 0, \\ \\ 0 & \text{if } x_{nn} = 1. \end{cases}$$

If $n \geq 1$, then $y \neq x_n$ because $y_n \neq x_{nn}$. $\quad\square$

Readers familiar with binary (base 2) expansions of real numbers will be able to deduce from 0.33 that the set R of all real numbers is uncountable. We give a different proof of this fact in the next section (see 0.46).

EXERCISES

35. A certain collection $\mathcal{C}$ of sets has the property that $A \cap B \in \mathcal{C}$ whenever $A \in \mathcal{C}$ and $B \in \mathcal{C}$. Show that the intersection of any finite family of members of $\mathcal{C}$ is a member of $\mathcal{C}$.

36. Let $\mathfrak{I} = \{A \mid A \subset R, R \setminus A \text{ finite}\}$. Prove:

(**a**) If $(A_i \mid i \in I)$ is a finite family with $A_i \in \mathfrak{I}$ for each $i \in I$, then $\bigcap_{i \in I} A_i \in \mathfrak{I}$.
(**b**) If $(A_i \mid i \in I)$ is any family with $A_i \in \mathfrak{I}$ for each $i \in I$, then $\bigcup_{i \in I} A_i \in \mathfrak{I}$.

37. If X is a finite set, prove that the power set $\mathcal{P}(X)$ of X is also finite.

38. Let $f \colon X \to X$ be a map from a finite set X into X. Prove that f is bijective if and only if f is injective, or equivalently, if and only if f is surjective.

39. Verify that each of the following sets is denumerable:

(**a**) $N \setminus \{n\}$, where $n \in N$.
(**b**) $N \setminus \{m, n\}$, where $m, n \in N$ with $m \neq n$.
(**c**) $N \setminus F$, where F is any finite subset of N.

40. Show that each of the following sets is infinite and determine whether each is denumerable or uncountable:

(**a**) The set of all nonvertical lines in R^2 having rational slopes.
(**b**) $\{x \in R^2 \mid x_2 \in Q\}$.
(**c**) $\{x \in R \mid x = m/2^n \text{ for some } m \in Z, n \in N\}$.

41. (**a**) Must the union of a denumerable family of denumerable sets be denumerable?
(**b**) Must the union of a nonempty countable family of denumerable sets be denumerable?

42. A real number is said to be *algebraic* if it is a root of some polynomial

$$(*) \qquad\qquad p = a_0 x^n + a_1 x^{n-1} + \cdots + a_{n-1}x + a_n$$

having its coefficients $a_0, a_1, \ldots, a_n$ all integers.

(**a**) Show that each rational number is algebraic.
(**b**) Exhibit three algebraic numbers that are not rational.
(**c**) Prove that the set A of all algebraic real numbers is denumerable. [*Hint:* You may assume that a polynomial of degree n has at most n real roots. If p is a polynomial given by $(*)$, define the "height" $h(p)$ by

$$h(p) = n + |a_0| + |a_1| + \cdots + |a_n|.$$

How many polynomials p are there with $h(p) = m$?]
(**d**) Deduce that the set $R \setminus A$ of all *transcendental numbers* is uncountable. (This establishes, in particular, the existence of transcendental numbers, but it does not provide any specific examples of such numbers. Two transcendental numbers are π and e.)

43. Prove:

> (**a**) The collection of all finite subsets of N is denumerable.
> (**b**) The collection of all denumerable subsets of N is uncountable.

5. ORDER-COMPLETENESS OF THE REAL NUMBERS

In addition to its operations of addition and multiplication which give it its algebraic structure, the set R of real numbers has its usual ordering relation (0.5) which makes it a totally ordered set.

0.34 Definition. A *totally ordered set* is a set X together with a relation $\leq$ in X to X, called a *total ordering of* X, which has the four properties:

> (1) Reflexivity: For each $x \in X$,
> $$x \leq x.$$
>
> (2) Antisymmetry: For every $x, y \in X$,
> $$x \leq y \quad \text{and} \quad y \leq x \ \Rightarrow \ x = y.$$
>
> (3) Transitivity: For every $x, y, z \in X$,
> $$x \leq y \quad \text{and} \quad y \leq z \ \Rightarrow \ x \leq z.$$
>
> (4) Comparability: If $x, y \in X$, then
> $$x \leq y \quad \text{or} \quad y \leq x.$$

Any subset Y of a totally ordered set X becomes a totally ordered set in its own right by restriction of the total ordering of X just to elements of Y. In this way N, Q, and Z are totally ordered sets. An interesting method for totally ordering the plane R $\times$ R is given in Exercise 46.

0.35 Definition. Let E be a subset of a totally ordered set. A *greatest* (respectively, *least*) *element of* E is an $x \in E$ such that

$$y \in E \ \Rightarrow \ y \leq x \qquad (\text{respectively}, y \in E \ \Rightarrow \ x \leq y).$$

A nonempty set E need not have a greatest or a least element; consider $E = \{x \in R \mid 0 < x < 1\}$, for example. If E does have a greatest (respectively, least) element, however, antisymmetry guarantees it has exactly one such, which is then also called the *maximum* (respectively, *minimum*) of E and is denoted by

$$\max E \qquad (\text{respectively}, \min E).$$

Notations such as

$$\max_{1 \leq i \leq n} x_i, \qquad \min_{i \in I} x_i,$$

used when the elements of E are indexed, should be self-explanatory.

An easy induction shows that a nonempty finite subset of a totally ordered set always has both a greatest and a least element.

If x and y are elements of a totally ordered set, we write

$$x < y$$

to mean

$$x \leq y \quad \text{and} \quad x \neq y,$$

so that

$$x \leq y \iff x < y \quad \text{or} \quad x = y.$$

Notations such as $y > x$ and $y \geq x$ are used for arbitrary totally ordered sets the same way they are used for the set R.

0.36 Definition. An *interval in* a totally ordered set X is a subset J of X with the property

$$x, y \in J \quad \text{and} \quad z \in X \quad \text{and} \quad x < z < y \implies z \in J,$$

in other words, any element of X lying "between" two elements of J must belong to J.

According to this definition, both $\varnothing$ and X are intervals in X. For each $a \in X$ the *open rays*

$$]a, \rightarrow[\,= \{x \in X \mid x > a\}, \qquad]\leftarrow, a[\,= \{x \in X \mid x < a\}$$

and the *closed rays*

$$[a, \rightarrow[\,= \{x \in X \mid x \geq a\}, \qquad]\leftarrow, a] = \{x \in X \mid x \leq a\}$$

are intervals in X. Also, for $a, b \in X$ with $a \leq b$ the *open interval*

$$]a, b[\,= \{x \in X \mid a < x < b\},$$

the *closed interval*

$$[a, b] = \{x \in X \mid a \leq x \leq b\},$$

and the *half-open intervals*

$$]a, b] = \{x \in X \mid a < x \leq b\}, \qquad [a, b[\,= \{x \in X \mid a \leq x < b\}$$

are intervals in X. (Notice that we use a backwards square bracket where the reader has probably used a parenthesis to denote omission of an endpoint.)

An interval in an arbitrary totally ordered set need not have one of the forms just listed. In Q, for example, the set

$$\{x \in \mathsf{Q} \mid x > 0, x^2 < 2\}$$

s an interval that is not of any of those forms. We shall shortly prove, however, that any interval in R does have one of those forms (see 0.48).

Although some of the generalities below make sense for arbitrary totally ordered sets, we shall restrict our attention to the totally ordered set of all real numbers.

0.37 Definition. Let $A \subset \mathsf{R}$. An *upper bound of A* (*in* R) is a real number b such that

$$x \in A \quad \Rightarrow \quad x \leq b.$$

A *lower bound of A* (*in* R) is a real number b such that

$$x \in A \quad \Rightarrow \quad b \leq x.$$

A subset A of R can have neither an upper nor a lower bound—for example, $A = \mathsf{R}$. Moreover, if A does have an upper (respectively, a lower) bound b, it will not be unique, because every real number y with $y \geq b$ (respectively, $y \leq b$) will also be an upper (respectively, a lower) bound of A.

0.38 Definition. Let $A \subset \mathsf{R}$. A real number is called a *supremum*, or *least upper bound*, *of A* if it is a least element of the set of all upper bounds of A, and an *infimum*, or *greatest lower bound*, *of A* if it is a greatest element of the set of all lower bounds.

In order that a subset A of R have a supremum (respectively, infimum) it is evidently necessary that A have at least one upper (respectively, lower) bound. If A does have a supremum (respectively, an infimum), it will be unique and is then denoted by

$$\sup A \qquad (\text{respectively, } \inf A).$$

According to this definition,

$$b = \sup A$$

precisely when

$$x \in A \quad \Rightarrow \quad x \leq b, \qquad b' < b \quad \Rightarrow \quad b' < y \quad \text{for some } y \in A.$$

Similarly,

$$b = \inf A$$

precisely when

$$x \in A \quad \Rightarrow \quad b \leq x, \qquad b < b' \quad \Rightarrow \quad y < b' \quad \text{for some } y \in A.$$

An infimum of an interval J in R is called a *left endpoint of J*, and a supremum of J is called a *right endpoint of J*. Thus a real number a is a left endpoint of each interval of the form $]a, \rightarrow[$, $[a, \rightarrow[$, $]a, b[$, $[a, b]$, $[a, b[$, or $]a, b]$; a real number b is a right endpoint of each interval of the form $]\leftarrow, b[$, $]\leftarrow, b]$, $]a, b[$, $[a, b]$, $[a, b[$, or $]a, b]$. Of course, an interval in R need not have a left endpoint or a right endpoint; when it does, the endpoint need not belong to the interval. Thus *the supremum or infimum of a set need not belong to the set.*

When a set A has a greatest (respectively, least) element b, then $b = \sup A$ (respectively, $b = \inf A$). Hence suprema and infima are generalizations of greatest and least elements, respectively.

We shall accept the following statement as an article of faith.

0.39 Axiom (order-completeness of R). Each nonempty subset of R that has an upper bound has a supremum.

This axiom, which distinguishes R from other number systems such as Q having both an algebraic and an order structure, asserts that there are no "holes" in the real line. It guarantees, for example, the existence of a square root of each real number $a \geq 0$ (if $b = \sup \{x \in \mathsf{R} \mid x \geq 0,\ x^2 \leq a\}$, then $b^2 = a$). We proceed to deduce several of its more important consequences.

0.40 Proposition. Each nonempty subset of R that has a lower bound has an infimum.

Proof. Let $\varnothing \neq B \subset \mathsf{R}$ and let B have a lower bound $m \in \mathsf{R}$. Define

$$A = \{-x \mid x \in B\}.$$

Then $A \neq \varnothing$ and $-m$ is an upper bound of A. By Axiom 0.39, A has a supremum b. We claim

$$-b = \inf B.$$

Clearly $-b$ is a lower bound of B. If b' is any lower bound of B, then $-b'$ is an upper bound of A, $b \leq -b'$ because b is the least upper bound of A, and hence $b' \leq -b$.   □

0.41 Theorem (Archimedean property). Let x and ϵ be real numbers with $\epsilon > 0$. Then there exists a positive integer n such that

$$n\epsilon > x.$$

Proof. Suppose, to the contrary, that

$$n\epsilon \leq x \qquad\qquad (n = 1, 2, \ldots).$$

Then

$$n \leq \frac{x}{\epsilon} \qquad\qquad (n = 1, 2, \ldots),$$

so x/ϵ is an upper bound of the set

$$P = \{1, 2, \ldots\}$$

of all positive integers. By order-completeness we may form

$$b = \sup P.$$

Then

$$n \leq b \qquad\qquad (n \in P)$$

so also

$$n + 1 \leq b \qquad\qquad (n \in P).$$

Hence

$$n \leq b - 1 \qquad\qquad (n \in P),$$

in other words, $b - 1$ is an upper bound of P. This is impossible because b is the *least* upper bound of P. □

The Archimedean property says that by starting at the origin and laying end to end sufficiently many line segments of fixed length ϵ, no matter how small ϵ may be, we can get beyond any given point on the real line (see Fig. 0.1).

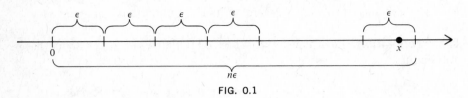

FIG. 0.1

0.42 Corollary. Let $y \in R$ with

$$0 \leq y \leq \frac{1}{n} \qquad\qquad (n = 1, 2, \ldots).$$

Then $y = 0$.

Proof. By hypothesis $ny \leq 1$ for each $n = 1, 2, \ldots$. If $y > 0$, then by taking $\epsilon = y$ and $x = 1$ in 0.41 we see, however, that $ny > 1$ for some n. □

Since $2^n > n$, we have also

$$0 \leq y \leq \frac{1}{2^n} \qquad (n = 1, 2, \ldots) \quad \Rightarrow \quad y = 0.$$

0.43 Corollary (order-density of Q in R). Each open interval in R contains a rational number.

Proof. Any open interval $]a, b[$ in R can be put into the form $]x - \epsilon, x + \epsilon[$ by taking $x = \frac{1}{2}(a + b)$ and $\epsilon = \frac{1}{2}(b - a)$. Hence we must show

(*) $\qquad x \in R, \epsilon > 0 \quad \Rightarrow \quad |x - q| < \epsilon$ for some $q \in Q$.

It is enough to consider the case $x \geq 0$; indeed, if $x < 0$ and if $q \in Q$ with $|(-x) - q| < \epsilon$, then $|x - (-q)| < \epsilon$ and $-q \in Q$.

Let $x \geq 0$ and $\epsilon > 0$. By 0.41 there exists a positive integer n with

$$\frac{1}{n} < \epsilon.$$

Also, by 0.41 there exists a positive integer k with $k > nx$; let m be the least such k. Then $m - 1$ is not such a k, that is,

$$m - 1 \leq nx,$$

so

$$\frac{m}{n} - \frac{1}{n} \leq x < \frac{m}{n}.$$

Hence the rational number $q = m/n$ satisfies $|x - q| < \epsilon$. $\quad$ []

Property (*) above says that any real number can be approximated as closely as we wish by rational numbers.

In the next theorem we refer to a decreasing sequence of closed intervals. In general, a sequence $(X_n \mid n \in \mathsf{N})$ of sets is said to be *decreasing* when

$$X_0 \supset X_1 \supset \cdots \supset X_n \supset X_{n+1} \supset \cdots$$

and *increasing* when

$$X_0 \subset X_1 \subset \cdots \subset X_n \subset X_{n+1} \subset \cdots.$$

Note that a sequence $(X_n \mid n \in \mathsf{N})$ of sets will be both decreasing and increasing in case $X_0 = X_1 = X_2 = \cdots$.

0.44 Theorem (nested interval property). Let $([a_n, b_n] \mid n \in \mathsf{N})$ be a decreasing sequence of closed intervals in R with

$$\lim_{n \to \infty}(b_n - a_n) = 0.$$

Then there exists a unique point

$$x \in \bigcap_{n=0}^{\infty} [a_n, b_n].$$

Proof. Since $[a_n, b_n] \subset [a_0, b_0]$,

$$a_n \leq b_0 \qquad\qquad (n \in \mathsf{N}).$$

Thus b_0 is an upper bound of the nonempty set $\{a_n \mid n \in \mathsf{N}\}$. By Axiom 0.39 we may form

$$x = \sup \{a_n \mid n \in \mathsf{N}\}.$$

We show that $x \in [a_n, b_n]$ for every n. Since x is an upper bound of $\{a_n \mid n \in \mathsf{N}\}$, already $a_n \leq x$ for each n. Just suppose $b_m < x$ for some m. Now

$$n \leq m \quad \Rightarrow \quad [a_n, b_n] \supset [a_m, b_m] \quad \Rightarrow \quad a_n \leq a_m \leq b_m,$$

and

$$n > m \quad \Rightarrow \quad [a_n, b_n] \subset [a_m, b_m] \quad \Rightarrow \quad a_n \leq b_n \leq b_m.$$

Hence b_m is also an upper bound of $\{a_n \mid n \in \mathsf{N}\}$, and this is impossible because x is the least upper bound of this set.

Suppose $y \in [a_n, b_n]$ for every n but $y \neq x$. Since $\lim_{n \to \infty} (b_n - a_n) = 0$, there exists an n with

$$b_n - a_n < |x - y|.$$

However,

$$|x - y| \leq b_n - a_n$$

because $x \in [a_n, b_n]$ and $y \in [a_n, b_n]$. $\square$

We have taken the order-completeness of R as an axiom and from it have deduced the more intuitively appealing Archimedean property and nested interval property. Conversely, order-completeness can be deduced from these two properties together.

0.45 Theorem. Each closed interval $[a, b]$ in R with $a < b$ is uncountable.

Proof. Suppose to the contrary that $[a, b]$ is countable. Then

$$[a, b] = \{x_n \mid n \in \mathsf{N}\}$$

for some sequence $(x_n \mid n \in \mathsf{N})$. Choose a closed interval $[a_0, b_0] \subset [a, b]$ with

$$x_0 \notin [a_0, b_0], \qquad b_0 - a_0 < \tfrac{1}{2}(b - a).$$

Next, choose a closed interval $[a_1, b_1] \subset [a_0, b_0]$ with

$$x_1 \notin [a_1, b_1], \qquad b_1 - a_1 < \tfrac{1}{2}(b_0 - a_0).$$

Continuing in this way, we obtain a decreasing sequence $([a_n, b_n] \mid n \in \mathsf{N})$ of closed intervals with

$$x_n \notin [a_n, b_n], \qquad b_n - a_n < \frac{1}{2^{n+1}} (b - a).$$

Clearly $\lim_{n \to \infty} (b_n - a_n) = 0$. By the nested interval property there is some

$$x \in \bigcap_{n=0}^{\infty} [a_n, b_n].$$

In particular, $x \in [a_0, b_0] \subset [a, b]$, so $x = x_n$ for some n. Then $x_n = x \in [a_n, b_n]$, contrary to the choice of x_n. $\square$

0.46 Corollary. The set R of all real numbers is uncountable.

0.47 Corollary. Each open interval in R contains an irrational number.

Proof. Consider an open interval $]a, b[$, where $a < b$. Choose $c < d$ with $[c, d] \subset]a, b[$. The set $[c, d]$ is uncountable by 0.45, so $]a, b[$ is uncountable also. Now

$$]a, b[\ = \ (\mathbb{Q} \cap \ ]a, b[) \cup ((\mathbb{R}\backslash\mathbb{Q}) \cap \ ]a, b[)$$

and $\mathbb{Q} \cap \ ]a, b[$ is countable, so $(\mathbb{R}\backslash\mathbb{Q}) \cap \ ]a, b[$ is uncountable. In particular, $(\mathbb{R}\backslash\mathbb{Q}) \cap \ ]a, b[$ is nonempty. $\square$

To conclude our treatment of order-completeness we determine all intervals in $\mathbb{R}$.

0.48 Theorem. A subset of $\mathbb{R}$ is an interval in $\mathbb{R}$ if and only if it is the empty set, an open ray, a closed ray, an open interval, a closed interval, a half-open interval, or $\mathbb{R}$ itself.

Proof. We have already noted that the empty set, open rays, and so on, are intervals in any totally ordered set. Conversely, let J be an interval in $\mathbb{R}$. Suppose $J \neq \varnothing$. We distinguish several cases.

Case (1). J has both a lower bound and an upper bound. Then order-completeness allows us to form

$$a = \inf J, \qquad b = \sup J.$$

Clearly

$$x < a \quad \text{or} \quad x > b \ \Rightarrow \ x \notin J,$$

so

$$J \subset [a, b].$$

Also,

$$]a, b[\ \subset J,$$

for if $a < x < b$, then x is neither a lower nor an upper bound of J, there exist $c, d \in J$ with $c < x < d$, and $x \in J$ because J is an interval. In this case we conclude that $J = [a, b]$ if $a \in J$ and $b \in J$, $J = [a, b[$ if $a \in J$ and $b \notin J$, $J = \ ]a, b]$ if $a \notin J$ and $b \in J$, and $J = \ ]a, b[$ if $a \notin J$ and $b \notin J$.

Case (2). J has a lower bound but not an upper bound. Let

$$a = \inf J.$$

Clearly

$$x < a \ \Rightarrow \ x \notin J$$

so

$$J \subset [a, \rightarrow[.$$

Also,

$$]a, \rightarrow[\ \subset J,$$

for if $x > a$, then x is not a lower bound of J, there exists $c \in J$ with $c < x$, there exists $d \in J$ with $x < d$ because x is not an upper bound of J, and $x \in J$

because J is an interval. In this case, $J = [a, \rightarrow[$ if $a \in J$, and $J =]a, \rightarrow[$ if $a \notin J$.

Case (3). J has an upper bound but not a lower bound. Let

$$b = \sup J.$$

As in Case (2); $J =]\leftarrow, b]$ if $b \in J$, and $J =]\leftarrow, b[$ if $b \notin J$.

Case (4). J has neither an upper nor a lower bound. Then an arbitrary real number x is neither a lower nor an upper bound of J, there exist $c, d \in J$ with $c < x < d$, and $x \in J$ because J is an interval. Hence $J = \mathsf{R}$ in this case.   ☐

EXERCISES

44. Let $\leq$ be a total ordering of a set X. Define a new relation $\leq'$ in X to X by the rule

$$x \leq' y \implies y \leq x \qquad\qquad (x, y \in X).$$

(**a**) Verify that $\leq'$ is also a total ordering of X.
(**b**) Show that a least element x of a set $E \subset X$ for $\leq$ is a greatest element of E for $\leq'$.

45. In each case below show that the given relation $\leq$ in X to X has all the properties of a total ordering *except* comparability (such a relation is called a *partial ordering* of the set X).

(**a**) $X = \mathsf{N}\backslash\{0\}$; for $x, y \in X$, $x \leq y$ means that x divides y, that is, $y = kx$ for some $k \in X$.
(**b**) X is the set of all functions $f: \mathsf{R} \to \mathsf{R}$; for $f, g \in X, f \leq g$ means that $f(t) \leq g(t)$ for all $t \in \mathsf{R}$.

46. On $X = \mathsf{R} \times \mathsf{R}$, define a relation $\leq$ as follows: For $x = (x_1, x_2), y = (y_1, y_2) \in X$,

$$x \leq y \iff (x_1 = y_1 \text{ and } x_2 \leq y_2) \quad \text{or} \quad (x_1 < y_1).$$

(**a**) Verify that $\leq$ is a total ordering of X.
(**b**) If $a \in X$, describe geometrically the rays $[a, \rightarrow[, \ ]a, \rightarrow[, \ ]\leftarrow, a]$, and $]\leftarrow, a[$.
(**c**) If $a, b \in X$ with $a < b$, describe geometrically the intervals $[a, b], [a, b[,$ $]a, b]$, and $]a, b[$.

47. Let A and B be nonempty sets of positive real numbers. Denote by $A + B$ the set $\{a + b \mid a \in A, b \in B\}$.

(**a**) Suppose each of A and B has a maximum. Show that $A + B$ does also and $\max (A + B) \leq \max A + \max B$.
(**b**) Suppose each of A and B has a supremum. Show that $A + B$ does also and $\sup (A + B) \leq \sup A + \sup B$.
(**c**) Can the inequalities in (a) and (b) be strengthened to equalities?

48. Reprove Proposition 0.40 as follows: Let $\varnothing \neq B \subset \mathsf{R}$ and let B have a lower bound. Define L to be the set of all lower bounds of B. Show that L has a supremum and then that sup $L = \inf B$.

49. Let E be a nonempty set of integers.

(**a**) Suppose E has an upper bound in R. Use order-completeness to show that E has a greatest element.

(**b**) Suppose E has an upper bound b that is an integer. Without using order-completeness, show that E has a greatest element. (*Hint:* Apply the well-ordering principle for N to the set $\{b - x \mid x \in E\}$.)

50. Let $x \in$ R. Establish the existence of some integer n such that $n \le x < n + 1$. Then prove that such an integer n is necessarily unique.

51. Let $x \in$ R and $\epsilon > 0$. Show that there exist infinitely many rational numbers q with $|x - q| < \epsilon$.

52. (**a**) Exhibit a denumerable collection $\mathfrak{D}$ of pairwise disjoint intervals in R such that $\bigcup \mathfrak{D} =$ R.

(**b**) Prove that there does not exist any uncountable collection $\mathfrak{A}$ of pairwise disjoint open intervals in R. (*Hint:* Use the order-density of Q in R.)

53. A *Dedekind cut* is a pair (A, B) of subsets of Q such that $A \ne \varnothing$, $B \ne \varnothing$, $A \cap B = \varnothing$, $A \cup B =$ Q, and $a < b$ for each $a \in A$ and $b \in B$. For example, the pair (A, B) given by

$$A = \{a \in \text{Q} \mid a \le \sqrt{2}\}, \qquad B = \{b \in \text{Q} \mid \sqrt{2} \le b\}$$

is a Dedekind cut. Prove that if (A, B) is any Dedekind cut, then there exists a unique real number x with either

$$A = \{a \in \text{Q} \mid a \le x\}, \qquad B = \{b \in \text{Q} \mid x < b\}$$

or

$$A = \{a \in \text{Q} \mid a < x\}, \qquad B = \{b \in \text{Q} \mid x \le b\}.$$

[*Note:* The number x will not always be rational; when x is irrational, the two possibilities for (A, B) indicated above are really one and the same.]

54. Let $([c_n, d_n] \mid n \in$ N$)$ be a decreasing sequence of closed intervals in R. Show that

$$\bigcap_{n=0}^{\infty} [c_n, d_n] \ne \varnothing.$$

[*Hint:* Apply the nested interval property to an appropriate sequence $([a_n, b_n] \mid n \in$ N$)$ with $[a_n, b_n] \subset [c_n, d_n]$ for each n.]

55. Does the nested interval property remain true if the intervals are open instead of closed?

6. EQUIVALENCE RELATIONS

Suppose the objects in a given collection are classified by some rule that declares certain of the objects to be alike and others unlike. We may represent the given collection by a set X and the classifying rule by a relation $\sim$ in X to X, with $x \sim y$ being interpreted to mean that x is like y under the rule. Now the connotations of 'like' suggest that each x is like itself, that y is like x if x is like y, and that x is like z if x is like y and y is like z. Hence the relation $\sim$ representing the classifying rule should be an "equivalence relation" in the following sense.

0.49 Definition. Let X be a set. An *equivalence relation on X* is a relation $\sim$ in X to X that is reflexive:

$$x \in X \quad \Rightarrow \quad x \sim x,$$

symmetric:

$$x, y \in X \quad \text{and} \quad x \sim y \quad \Rightarrow \quad y \sim x,$$

and transitive:

$$x, y, z \in X \quad \text{and} \quad x \sim y \quad \text{and} \quad y \sim z \quad \Rightarrow \quad x \sim z.$$

0.50 Examples

(1) Fix a positive integer m. Let $\sim$ be the relation in Z to Z such that

$$n \sim k \quad \Leftrightarrow \quad n - k = dm \quad \text{for some } d \in \mathsf{Z}.$$

Equivalently, $n \sim k$ if and only if n and k leave the same remainder when divided by m. Then $\sim$ is an equivalence relation, which is usually denoted $\equiv \pmod{m}$ and is called *congruence modulo m*.

Take $m = 2$ above. Then

$$n \sim k \quad \Leftrightarrow \quad n \equiv k \pmod 2$$
$$\Leftrightarrow \quad n - k \text{ is even}$$
$$\Leftrightarrow \quad n \text{ and } k \text{ are both even or both odd.}$$

(2) Let X be the set of all triangles in the plane. Define $\sim$ by

$$x \sim y \quad \Leftrightarrow \quad x \text{ is congruent to } y.$$

Then $\sim$ is an equivalence relation on X.

(3) Let X be any set. The relation $\sim$ in X to X defined by

$$x \sim y \quad \Leftrightarrow \quad x = y$$

is an equivalence relation on X that classifies no object as like another distinct from itself.

(4) Let $\mathcal{A}$ be a collection of sets. For $A, B \in \mathcal{A}$, let

$$A \sim B \quad \Leftrightarrow \quad \text{there exists a bijection } f \colon A \to B.$$

Then $\sim$ is an equivalence relation on $\mathcal{A}$ that classifies two sets belonging to

α as alike when they have the same "number" of elements. If $N \in \alpha$, then for $A \in \alpha$ we have

$$A \sim N \quad \Leftrightarrow \quad A \text{ is denumerable.}$$

Given an equivalence relation on a set X, we may group together with each $x \in X$ all those elements of X (including x) that are like x.

0.51 Definition. Let $\sim$ be an equivalence relation on a set X. For each $x \in X$, the *equivalence class of x under* $\sim$ is the subset

$$x/\sim \; = \{y \in X \mid x \sim y\}$$

of X, and each $y \in x/\sim$ is a *representative of* this equivalence class. The *quotient of X under* $\sim$ is the set

$$X/\sim \; = \{x/\sim : x \in X\}$$

of all these equivalence classes, and the *quotient map induced by* $\sim$ is the surjection

$$p \colon X \to X/\sim$$

$$x \mapsto x/\sim$$

which sends each element of X to its equivalence class under $\sim$.

In Example 0.50(1), if $m = 2$, then

$$n/\sim \; = \begin{cases} E & \text{if } n \text{ is even,} \\ O & \text{if } n \text{ is odd,} \end{cases}$$

where $E = \{\ldots, -4, -2, 0, 2, 4, \ldots\}$ and $O = \{\ldots, -3, -1, 1, 3, \ldots\}$, so $\mathbb{Z}/\sim \; = \{E, O\}$. In Example (3), $x/\sim \; = \{x\}$ for each $x \in X$, and $X/\sim \; = \{\{x\} \mid x \in X\}$.

In the terms of Definition 0.51 we can say in several different ways when two objects are alike under an equivalence relation.

0.52 Lemma. Let $\sim$ be an equivalence relation on a set X and let $x, y \in X$. Then the following statements are equivalent:

(1) $x \sim y$.
(2) $x/\sim \; = y/\sim$.
(3) $p(x) = p(y)$, where $p \colon X \to X/\sim$ is the quotient map.
(4) $x/\sim$ intersects $y/\sim$.

Proof. (1) $\Rightarrow$ (2). Assume (1). If $z \in x/\sim$, then $x \sim z$, by symmetry $z \sim x$, by transitivity $z \sim y$, by symmetry $y \sim z$, and so $z \in y/\sim$. Hence $x/\sim \; \subset y/\sim$. Similarly, $y/\sim \; \subset x/\sim$.

(2) $\Rightarrow$ (1). Assume (2). Then $y \in y/\sim$ because $y \sim y$, so $y \in x/\sim$, that is, $x \sim y$.

(2) $\Leftrightarrow$ (3) by definition of p, and (2) $\Rightarrow$ (4) because $x \in x/\sim$.

(4) $\Rightarrow$ (1). Assume (4). Choose

$$z \in (x/\sim) \cap (y/\sim).$$

Then

$$x \sim z, \qquad y \sim z,$$

so by symmetry and transitivity $x \sim y$. $\square$

This lemma tells us that any two equivalence classes are either identical or else disjoint.

0.53 Definition. A collection $\mathcal{Q}$ of nonempty subsets of a set X is called a *partition of* X when each point of X belongs to some member of $\mathcal{Q}$ (that is, $X = \bigcup \mathcal{Q}$) and the collection $\mathcal{Q}$ is disjoint, in other words, when each $x \in X$ belongs to exactly one member of A.

0.54 Theorem. Let X be a set.

(1) If $\sim$ is an equivalence relation on X, then the quotient set $X/\sim$ is a partition of X.

(2) If $\mathcal{Q}$ is a partition of X, then the relation $\sim$ in X to X given by

$$x \sim y \quad \Leftrightarrow \quad x \in A \quad \text{and} \quad y \in A \quad \text{for some } A \in \mathcal{Q}$$

is an equivalence relation on X, and said to be *induced by* $\mathcal{Q}$.

Proof. (1) follows from 0.52 and the fact that $x \in x/\sim \in X/\sim$ for each $x \in X$.

(2) Let $\mathcal{Q}$ be a partition of X and define $\sim$ as indicated. Suppose $x, y, z \in X$ with $x \sim y$ and $y \sim z$. Then there exist $A, B \in \mathcal{Q}$ with

$$x, y \in A, \qquad y, z \in B.$$

Since $y \in A \cap B$, we have $A = B$. Then $x, z \in A$, so $x \sim z$. This proves transitivity. The proofs of reflexivity and symmetry are even easier. $\square$

A partition $\mathcal{Q}$ of a set X may be thought of as a way of classifying the objects of X by dividing them into different classes. Then the equivalence relation induced by $\mathcal{Q}$ considers two objects alike precisely when they belong to the same class. On the intuitive level it is clear that equivalence relations and partitions are two different mathematical constructs expressing a single concept of classification. On the formal level the same thing is true, for the map

$$\sim \mapsto X/\sim$$

assigning to each equivalence relation on X the quotient under $\sim$ is a one-to-one correspondence between the collection of all equivalence relations on X and the collection of all partitions of X (see Exercise 63).

EXERCISES

56. In each case below decide whether the specified relation $\sim$ is an equivalence relation on the given set X.

(a) $X = \mathsf{R}$; for $x, y \in X$, $x \sim y \iff y - x \in \mathsf{Z}$.

(b) $X = \mathsf{R}$; for $x, y \in X$, $x \sim y \iff y - x \in \mathsf{Q}$.

(c) $X = \mathsf{Z} \setminus \{0\}$; for $x, y \in X$, $x \sim y \iff y = kx$ for some $k \in X$.

(d) $X = \mathcal{P}(\mathsf{R})$; for $A, B \in X$, $A \sim B \iff A \cap B \neq \varnothing$.

57. In each case below verify that the specified relation $\sim$ is an equivalence relation on the given set X and determine the equivalence class of each $x \in X$.

(a) $X = \mathsf{Z} \times (\mathsf{Z} \setminus \{0\})$; for $(m, n) \in X$ and $(i, j) \in X$, $(m, n) \sim (i, j)$ if and only if $mj = ni$.

(b) $X = [0, 1] \times [0, 1]$; for $(t, s) \in X$ and $(u, v) \in X$, $(t, s) \sim (u, v)$ if and only if $(t, s) = (u, v)$ or else $t = u$ and $s = 1 - v$.

(c) X is the set of all differentiable functions $f\colon \mathsf{R} \to \mathsf{R}$; for $f \in X$ and $g \in X$, $f \sim g$ if and only if $f'(t) = g'(t)$ for all $t \in \mathsf{R}$, where the prime denotes the derivative.

58. Let A be a nonempty proper subset of a set X. For $x, y \in X$, let

$$x \sim y \iff x = y \quad \text{or} \quad (x \in A \text{ and } y \in A).$$

(a) Verify that $\sim$ is an equivalence relation on X.

(b) Show that $X/\!\sim\, = \{A\} \cup (X \setminus A)$.

(c) Let $p\colon X \to X/\!\sim$ be the quotient map. Show that its restriction

$$f\colon X \setminus A \to (X/\!\sim) \setminus \{A\}$$

$$x \mapsto p(x)$$

is a bijection.

59. The *equivalence kernel* of a map $f\colon X \to Y$ is the relation $\sim$ in X to X given by

$$x \sim y \iff f(x) = f(y) \qquad\qquad (x, y \in X).$$

(a) Show that the equivalence kernel of a map $f\colon X \to Y$ is an equivalence relation on X.

(b) Determine explicitly the equivalence kernel of the map $f\colon \mathsf{R} \to \mathsf{R}$ given by the formula $f(x) = \cos 2\pi x$.

(c) Prove that an arbitrary equivalence relation on a set X is the equivalence kernel of some map $f\colon X \to Y$.

60. In each case below, either construct a partition $\mathcal{Q}$ of the plane R^2 with the stated properties or else tell why such a partition does not exist.

(a) The partition $\mathcal{Q}$ is denumerable and each $A \in \mathcal{Q}$ is denumerable.

(**b**) The partition $\mathcal{C}$ is denumerable and each $A \in \mathcal{C}$ is uncountable.

(**c**) The partition $\mathcal{C}$ is uncountable and each $A \in \mathcal{C}$ is denumerable.

61. Let $\sim$ and $\simeq$ be two equivalence relations on the same set X such that $x \sim y$ implies $x \simeq y$ for all $x, y \in X$. Denote by $\mathcal{C}$ and $\mathcal{B}$ the quotient sets $X/\sim$ and $X/\simeq$, respectively. How is $\mathcal{C}$ related to $\mathcal{B}$?

62. For each line L in the plane R^2 passing through the origin $(0, 0)$, call the set $L \setminus \{(0, 0)\}$ a "deleted line". Verify that the collection $\mathcal{C}$ of all deleted lines is a partition of $\mathsf{R}^2 \setminus \{(0, 0)\}$. Then describe algebraically the equivalence relation $\sim$ on $\mathsf{R}^2 \setminus \{(0, 0)\}$ induced by $\mathcal{C}$.

63. Let X be a nonempty set. For each partition $\mathcal{C}$ of X, denote by $X/\mathcal{C}$ the equivalence relation on X induced by $\mathcal{C}$, so that for $x \in X$ and $y \in X$, $x(X/\mathcal{C})y \iff x \in A$ and $y \in A$ for some $A \in \mathcal{C}$.

(**a**) Let $\mathcal{C}$ be a partition of X. Then $X/\mathcal{C}$ is an equivalence relation $\sim$ on X whose quotient set $X/\sim = X/(X/\mathcal{C})$ is therefore a partition of X. Prove that $X/\sim = \mathcal{C}$, in other words,

$$X/(X/\mathcal{C}) = \mathcal{C}.$$

(**b**) Let $\sim$ be any equivalence relation on X. Then the quotient set $X/\sim$ is a partition $\mathcal{C}$ of X which therefore induces the equivalence relation $X/\mathcal{C} = X/(X/\sim)$ on X. Prove that $X/\mathcal{C} = \sim$, in other words,

$$X/(X/\sim) = \sim.$$

(**c**) Use (a) and (b) to show that the rule

$$\sim \mapsto X/\sim$$

defines a one-to-one correspondence between the collection η of all equivalence relations on X and the collection π of all partitions of X.

1

Metric Spaces

As a prelude to our study of topology proper, in this chapter we discuss "metric spaces"—sets of points on which there are given numerical measures of distances between these points. Our first aim here is to provide a number of important examples that will recur throughout the entire text. Our second aim is to abstract from the familiar setting of one-, two-, and three-dimensional Euclidean spaces notions of continuous functions and convergent sequences meaningful in any metric space. Our third aim is to demonstrate that many concepts associated with metric spaces, and in particular continuity, depend not on the particular measures of distance involved, but only on the "open" sets they define. Thus we prepare the way for the more abstract concept of a "topological space" in which there need be no notion of distance at all.

The final section of the chapter concerns completeness, a notion not amenable to generalization to arbitrary topological spaces. Indeed, this metric property will not reappear until Section 2 of Chapter 4, where it is related to the topological property of compactness. Yet completeness is hardly insignificant; the fact that a certain metric space whose "points" are functions is complete will allow us to demonstrate that "most" continuous real-valued functions on an interval have derivatives at no point.

1. METRICS

Consider a real-valued function $f: \mathsf{R}^n \to \mathsf{R}$ defined on n-dimensional Euclidean space R^n consisting of all n-tuples $x = (x_1, x_2, \ldots, x_n)$ of real numbers. For example, when $n = 1$, R^n is the real line, and f might be the function of a single variable given by $f(x) = x \sin x$. When $n = 2$, R^n is the plane, and f might be the function of two variables given by

$$f(x_1, x_2) = (x_1^2 + x_2^2)/(1 + x_1^2 + x_2^2).$$

When $n = 3$, R^n is 3-space, and f might be the function of three variables given by

$$f(x_1, x_2, x_3) = x_1 x_2 \sin |1 + x_3|.$$

From calculus the reader is doubtless familiar (at least for $n = 1$, 2, or 3) with the following informal notion of continuity of the function f at a point $x \in \mathsf{R}^n$:

> $f(u)$ will be as close to $f(x)$ as one wishes provided that $u \in \mathsf{R}^n$ is close enough to x.

More precisely, f is continuous at x when:

> Given any $\epsilon > 0$, there is a corresponding $\delta > 0$ such that $f(u)$ is at a distance less than ϵ from $f(x)$ whenever $u \in \mathsf{R}^n$ is at a distance less than δ from x.

As it stands the preceding statement is not quite an adequate definition of continuity, for it does not prescribe how distances between points in R or in R^n are to be measured. It is to numerical measures of distance in Euclidean and other "spaces" that this section is devoted.

Formulas for distance in R^n are familiar enough for dimensions $n = 1$, 2, and 3. The distance from a point x to a point y in R^1 is

$$d(x, y) = |x - y|.$$

The distance from a point $x = (x_1, x_2)$ to a point $y = (y_1, y_2)$ in R^2 is

$$d(x, y) = [(x_1 - y_1)^2 + (x_2 - y_2)^2]^{1/2} = \left[\sum_{i=1}^{2} (x_i - y_i)^2 \right]^{1/2}.$$

The distance from a point $x = (x_1, x_2, x_3)$ to a point $y = (y_1, y_2, y_3)$ in R^3 is

$$d(x, y) = [(x_1 - y_1)^2 + (x_2 - y_2)^2 + (x_3 - y_3)^2]^{1/2} = \left[\sum_{i=1}^{3} (x_i - y_i)^2 \right]^{1/2}.$$

The distance formula for dimension 1 may be made to resemble the formulas for dimensions 2 and 3; recalling that $[a^2]^{1/2} = |a|$ for any real number a, for points $x = x_1$ and $y = y_1$ in R^1 we may write

$$d(x, y) = \left[\sum_{i=1}^{1} (x_i - y_i)^2 \right]^{1/2}.$$

Distance in R^n for arbitrary dimension n may now be defined so as to fit the same "square root of the sum of the squares of the differences of the coordinates" pattern.

1.1 Definition. Let n be a positive integer. For any points $x = (x_1, \ldots, x_n)$ and $y = (y_1, \ldots, y_n)$ in R^n, the *Euclidean distance from x to y* is defined to be the real number

$$[(x_1 - y_1)^2 + \cdots + (x_n - y_n)^2]^{1/2} = \left[\sum_{i=1}^{n} (x_i - y_i)^2\right]^{1/2}.$$

The assignment of the Euclidean distance from x to y to each ordered pair (x, y) of points in R^n is a function

$$d \colon \mathsf{R}^n \times \mathsf{R}^n \to \mathsf{R}$$

called the *Euclidean metric on R^n*.

This definition does make sense, for the quantity in square brackets is nonnegative and hence has a unique (nonnegative) square root.

We must emphasize that the formula

$$d(x, y) = \left[\sum_{i=1}^{n} (x_i - y_i)^2\right]^{1/2}$$

for the Euclidean distance from x to y is our *definition* of this distance, not something we proved. Of course, for dimensions $n = 2$ and 3 this formula is often proved in elementary mathematics courses under certain assumptions about distance (compare Exercise 4).

We are going to derive several properties of Euclidean distance from our definition by exploiting the algebraic structure that R^n has a vector space (0.13). We begin by observing that the distance from one point to another in R^n can always be expressed as the distance from a point to the origin $0 = (0, \ldots, 0)$ in R^n. Indeed, given $x = (x_1, \ldots, x_n)$ and $y = (y_1, \ldots, y_n)$ in R^n, we have $x - y = (x_1 - y_1, \ldots, x_n - y_n)$, so

$$d(x, y) = d(x - y, 0).$$

In view of this observation it is natural to consider for each $x = (x_1, \ldots, x_n) \in \mathsf{R}^n$ the *Euclidean norm of x*, the real number defined by

$$\|x\| = d(x, 0),$$

or in terms of coordinates,

$$\|x\| = \left[\sum_{i=1}^{n} x_i^2\right]^{1/2}.$$

In dimensions $n = 2$ and $n = 3$, $\|x\|$ is the usual "length" of the vector x. In dimension $n = 1$, $\|x\| = |x|$. Then the following lemma generalizes familiar properties of absolute value.

1.2 Lemma. For all $x \in \mathbf{R}^n$, $y \in \mathbf{R}^n$, and $\lambda \in \mathbf{R}$:

(1) $||x|| \geq 0$.

(2) $||x|| = 0$ if and only if $x = 0$.

(3) $||\lambda x|| = |\lambda| \cdot ||x||$.

(4) $||x + y|| \leq ||x|| + ||y||$.

Proof. (3) Write $x = (x_1, \ldots, x_n)$. Then $\lambda x = (\lambda x_1, \ldots, \lambda x_n)$, so

$$||\lambda x|| = \left[\sum_{i=1}^{n} (\lambda x_i)^2 \right]^{1/2} = \left[\lambda^2 \sum_{i=1}^{n} x_i^2 \right]^{1/2}$$

$$= |\lambda| \cdot \left[\sum_{i=1}^{n} x_i^2 \right]^{1/2} = |\lambda| \cdot ||x||.$$

(4) By (1), both $||x + y||$ and $||x|| + ||y||$ are nonnegative. Hence we need only show

$$||x + y||^2 \leq (||x|| + ||y||)^2.$$

Now

$$(||x|| + ||y||)^2 = ||x||^2 + 2 ||x|| \cdot ||y|| + ||y||^2$$

and

$$||x + y||^2 = \sum_{i=1}^{n} (x + y)_i{}^2 = \sum_{i=1}^{n} (x_i + y_i)^2$$

$$= \sum_{i=1}^{n} (x_i^2 + 2x_i y_i + y_i^2)$$

$$= \sum_{i=1}^{n} x_i^2 + 2 \sum_{i=1}^{n} x_i y_i + \sum_{i=1}^{n} y_i^2$$

$$= ||x||^2 + 2 \sum_{i=1}^{n} x_i y_i + ||y||^2.$$

It remains only to show

$$\sum_{i=1}^{n} x_i y_i \leq ||x|| \cdot ||y||.$$

This will follow from the next lemma.

1.3 Lemma (Cauchy-Schwarz inequality). Let $x = (x_1, \ldots, x_n) \in \mathbf{R}^n$ and $y = (y_1, \ldots, y_n) \in \mathbf{R}^n$. Then

$$\left| \sum_{i=1}^{n} x_i y_i \right| \leq ||x|| \cdot ||y||.$$

Proof. Set $\alpha = \sum_{i=1}^{n} x_i y_i$ and consider the quadratic polynomial

$$p(\lambda) = ||x||^2 \lambda^2 + 2\alpha\lambda + ||y||^2$$

in λ. It will suffice to show that its discriminant

$$\delta = (2\alpha)^2 - 4\,||x||^2 \cdot ||y||^2$$

is ≤ 0, for then $\alpha^2 \leq ||x||^2 \cdot ||y||^2$ and hence $|\alpha| \leq ||x|| \cdot ||y||$.

For each real number λ,

$$p(\lambda) = \lambda^2 \sum_{i=1}^{n} x_i^2 + 2\lambda \sum_{i=1}^{n} x_i y_i + \sum_{i=1}^{n} y_i^2$$

$$= \sum_{i=1}^{n} (\lambda x_i + y_i)^2$$

$$\geq 0.$$

If the discriminant $\delta > 0$, then $p(\lambda)$ would have exactly two distinct real roots and consequently would assume negative as well as nonnegative values. Hence $\delta \leq 0$, as required.   □

Since $d(x, y) = d(x - y, 0)$, then

$$d(x, y) = ||x - y||$$

for all x, $y \in \mathbf{R}^n$. Using this expression for the Euclidean metric in terms of the Euclidean norm, from Lemma 1.2 we can at last derive the properties of the Euclidean metric that will be important in the sequel.

1.4 Proposition. Let d be the Euclidean metric on $\mathbf{R}^n$. Then for all x, y, $z \in \mathbf{R}^n$:

(1) $d(x, y) \geq 0$.
(2) $d(x, y) = 0$ if and only if $x = y$.
(3) $d(x, y) = d(y, x)$.
(4) $d(x, z) \leq d(x, y) + d(y, z)$.

Proof. (2) Since $d(x, y) = ||x - y||$, then by 1.2(2), $d(x, y) = 0$ if and only if $x - y = 0$, that is, $x = y$.

(3) By 1.2(3),

$$d(x, y) = ||x - y|| = ||(-1)(y - x)|| = |-1| \cdot ||y - x|| = d(y, x).$$

(4) By 1.2(4),

$$d(x, z) = ||x - z|| = ||(x - y) + (y - z)|| \leq ||x - y|| + ||y - z||$$

$$= d(x, y) + d(y, z).   □$$

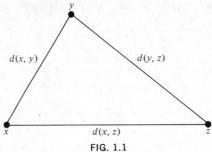

FIG. 1.1

Property 1.4(2) says that the distance from one point to another is zero precisely when the two points are actually the same. Property (3) says that the distance from one point to a second is the same as the distance from the second to the first. Property (4), called the *triangle inequality*, says that the length of one side of a triangle is at most the sum of the lengths of the other two sides (see Fig. 1.1). These properties seem geometrically obvious, so was all the work to prove them really necessary? Yes, because our definition of Euclidean distance was purely algebraic.

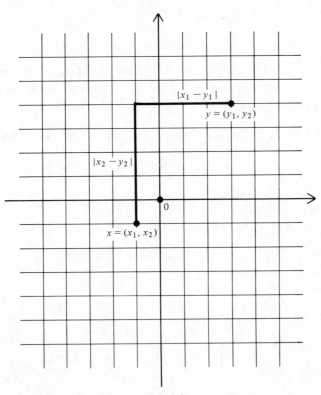

FIG. 1.2

Let us return for a moment to the plane R^2. A crow flying in a straight line from one point to another would want to use the Euclidean metric to compute how far he travels. The driver of a taxicab cruising a city's streets that are laid out in a rectangular grid would not. Instead, he would total the distance he travels on north–south streets and the distance he travels on east–west streets. (Just as we implicitly assumed the portion of the earth's surface the crow flies over is approximately planar, so we now assume the streets the taxicab cruises are not one way.) By superimposing a rectangular coordinate system on the street grid with axes parallel to the streets (and origin at City Hall), we see that the taxicab driver would compute the distance from $x = (x_1, x_2)$ to $y = (y_1, y_2)$ to be

$$|x_1 - y_1| + |x_2 - y_2|$$

(see Fig. 1.2).

This discussion suggests a new definition of distance in R^n.

1.5 Definition. For points $x = (x_1, \ldots, x_n)$ and $y = (y_1, \ldots, y_n)$ in R^n, the *taxicab distance from x to y* is defined to be the real number

$$\sum_{i=1}^{n} |x_i - y_i|.$$

The function

$$d_1 : R^n \times R^n \to R$$

assigning to each ordered pair (x, y) of points in R^n the taxicab distance from x to y is called the *taxicab metric on R^n*.

The taxicab metric d_1 has the very same properties listed in 1.4 for the Euclidean metric, namely, for all $x, y, z \in R^n$:

(1) $d_1(x, y) \geq 0$.
(2) $d_1(x, y) = 0$ if and only if $x = y$.
(3) $d_1(x, y) = d_1(y, x)$.
(4) $d_1(x, z) \leq d_1(x, y) + d_1(y, z)$.

The proofs of (1)–(4) here are all quite easy. For example,

$$|x_i - z_i| = |(x_i - y_i) + (y_i - z_i)| \leq |x_i - y_i| + |y_i - z_i|$$

for $i = 1, \ldots, n$, so

$$d_1(x, z) = \sum_{i=1}^{n} |x_i - z_i| \leq \sum_{i=1}^{n} (|x_i - y_i| + |y_i - z_i|)$$

$$= \sum_{i=1}^{n} |x_i - y_i| + \sum_{i=1}^{n} |y_i - z_i| = d_1(x, y) + d_1(y, z).$$

Already in dimension $n = 2$ the Euclidean metric d and the taxicab metric differ, for

$$d((1, 1), (0, 0)) = \sqrt{2} \neq 2 = d_1((1, 1), (0, 0)).$$

In dimension $n = 1$, of course, they coincide:

$$d(x, y) = |x - y| = d_1(x, y) \qquad\qquad (x, y \in \mathsf{R}).$$

Hence both Euclidean distance and taxicab distance generalize to arbitrary dimension n, the usual distance in dimension 1. Next comes yet a third generalization to dimension n.

1.6 Definition. For points $x = (x_1, \ldots, x_n)$ and $y = (y_1, \ldots, y_n)$ in R^n, the *max distance from x to y* is defined to be the real number

$$d_\infty(x, y) = \max_{1 \le i \le n} |x_i - y_i|.$$

The function $d_\infty \colon \mathsf{R}^n \times \mathsf{R}^n \to \mathsf{R}$ so obtained is called the *max metric on* R^n.

The max distance in R^n also generalizes the usual distance in R, for when $n = 1$, each point in R^n has a single coordinate, and then $d_\infty(x, y) = |x - y|$.

The max metric has the same four properties that the Euclidean and taxicab metrics have:

(1) $d_\infty(x, y) \ge 0$.
(2) $d_\infty(x, y) = 0$ if and only if $x = y$.
(3) $d_\infty(x, y) = d_\infty(y, x)$.
(4) $d_\infty(x, z) \le d_\infty(x, y) + d_\infty(y, z)$.

This time let us write out the proofs of both (2) and (4). If $d_\infty(x, y) = 0$, then $0 \le |x_i - y_i| \le 0$, that is, $x_i = y_i$ for $i = 1, \ldots, n$, and hence $x = y$. Conversely, if $x = y$, then $|x_i - y_i| = 0$ for $i = 1, \ldots, n$ so $d_\infty(x, y) = 0$. This proves (2). To prove (4), from $|x_i - z_i| \le |x_i - y_i| + |y_i - z_i|$ for $i = 1, \ldots, n$ we deduce

$$
\begin{aligned}
d_\infty(x, z) &= \max_{1 \le i \le n} |x_i - z_i| \\
&= \max_{1 \le i \le n} (|x_i - y_i| + |y_i - z_i|) \\
&\le \max_{1 \le i \le n} |x_i - y_i| + \max_{1 \le i \le n} |y_i - z_i| \\
&= d_\infty(x, y) + d_\infty(y, z).
\end{aligned}
$$

In his short story *A Day's Wait*, Ernest Hemingway tells of the ailing boy Schatz waiting all day to die. The doctor has reported the boy's temperature to be 102°, and Schatz has remembered his school companions in France telling him one cannot live with a temperature over 44°. He is relieved only when his father explains that just as there are kilometers and miles, so there is more than one temperature scale. The doctor had used a Fahrenheit thermometer.

The reader who has grown up computing distance always by the "square root of the sum of the squares of the differences of the coordinates" formula may well believe that only the Euclidean metric gives the "true" distance. To be sure, there are contexts in which the Euclidean metric is the only appropriate measure of distance in R^n (see Exercise 4). Nonetheless, as we shall demonstrate later in this chapter, for purposes of defining continuity the taxicab and max metrics are just as good as the Euclidean metric.

Distances can also be measured in certain "spaces" whose points are functions.

1.7 Example. Let X be the set of all continuous real-valued functions on the closed unit interval $[0, 1]$ (the ordinary calculus sense of 'continuous' is meant here). Contrary to the usual practice in calculus, typical functions belonging to X will be denoted by the letters x, y, and z, whereas typical members of their domain $[0, 1]$ will be denoted by t and s.

We shall assume below, without proving it, the theorem that a continuous real-valued function on $[0, 1]$ must attain a maximum value.

Given functions

$$x: [0, 1] \rightarrow \mathsf{R}, \qquad y: [0, 1] \rightarrow \mathsf{R}$$

belonging to X, the function $t \mapsto |x(t) - y(t)|$ is continuous and consequently attains a maximum value on $[0, 1]$; we define $d_\infty(x, y)$ to be this maximum value, so that

$$d_\infty(x, y) = \max_{0 \leq t \leq 1} |x(t) - y(t)|.$$

Thus $d_\infty(x, y)$ is the greatest vertical distance between the graphs of x and y (see Fig. 1.3).

The *max metric* $d_\infty: X \times X \rightarrow \mathsf{R}$ has the properties:

(1) $d_\infty(x, y) \geq 0$.
(2) $d_\infty(x, y) = 0$ if and only if $x = y$.
(3) $d_\infty(x, y) = d_\infty(y, x)$.
(4) $d_\infty(x, z) \leq d_\infty(x, y) + d_\infty(y, z)$.

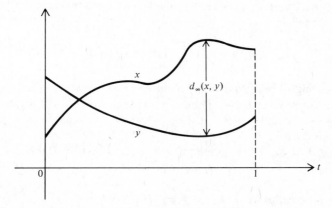

FIG. 1.3

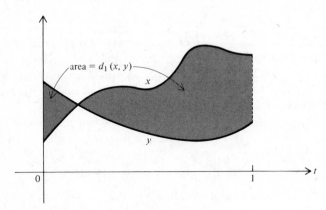

area = $d_1(x, y)$

x

y

0   1 t

FIG. 1.4

We prove only (4). For each $t \in [0, 1]$,

$$|x(t) - z(t)| \leq |x(t) - y(t)| + |y(t) - z(t)|$$

$$\leq \max_{0 \leq s \leq 1} |x(s) - y(s)| + \max_{0 \leq s \leq 1} |y(s) - z(s)|$$

$$= d_\infty(x, y) + d_\infty(y, z),$$

so

$$d_\infty(x, z) = \max_{0 \leq t \leq 1} |x(t) - z(t)| \leq d_\infty(x, y) + d_\infty(y, z).$$

1.8 Example. Again let X be the set of all continuous functions $x: [0, 1] \to \mathsf{R}$. For $x, y \in X$ define

$$d_1(x, y) = \int_0^1 |x(t) - y(t)| \, dt$$

(this is meaningful because the function $t \mapsto |x(t) - y(t)|$, being continuous on $[0, 1]$, is integrable there). Thus $d_1(x, y)$ is the area of the region enclosed by the graphs of x and y (see Fig. 1.4), which is a reasonable measure of how far apart x and y are.

Note that $d_1 \neq d_\infty$. For example, if $x(t) = t$ and $y(t) = 0$ for all $t \in [0, 1]$, then

$$d_1(x, y) = \int_0^1 t \, dt = \tfrac{1}{2} \neq 1 = \max_{0 \leq t \leq 1} t = d_\infty(x, y).$$

By now it should come as no surprise that for all x, y, $z \in X$:

(1) $d_1(x, y) \geq 0$.
(2) $d_1(x, y) = 0$ if and only if $x = y$.
(3) $d_1(x, y) = d_1(y, x)$.
(4) $d_1(x, z) \leq d_1(x, y) + d_1(y, z)$.

This time we give a detailed proof of the 'only if' part of (2). Suppose $d_1(x, y) = 0$. Then

$$\int_0^1 f(t) \, dt = 0,$$

where f is the continuous nonnegative function on $[0, 1]$ given by

$$f(t) = |x(t) - y(t)|.$$

To show $x = y$ we must show $f(t) = 0$ for all $t \in [0, 1]$. Now on $[0, 1]$ the function f has an antiderivative g, and the fundamental theorem of calculus tells us

$$\int_0^t f(s) \, ds = g(t) - g(0) \qquad (0 \leq t \leq 1).$$

Since f is nonnegative, for each $0 \leq t \leq 1$

$$0 \leq \int_0^t f(s) \, ds \leq \int_0^t f(s) \, ds + \int_t^1 f(s) \, ds = \int_0^1 f(s) \, ds = 0,$$

so that $g(t) - g(0) = 0$. Thus g is constant on $[0, 1]$. But then f, the derivative of g, is identically zero on $[0, 1]$.

The next example will be an "infinite-dimensional" analog of the finite-dimensional Euclidean spaces $\mathbb{R}^n$. It will be a space consisting of certain sequences of real numbers in which, by analogy with Euclidean distance, the distance from a sequence $x = (x_i \mid i = 1, 2, \ldots)$ to a sequence $y = (y_i \mid i = 1, 2, \ldots)$ is defined by

$$d_2(x, y) = \left[\sum_{i=1}^{\infty} (x_i - y_i)^2 \right]^{1/2}.$$

Of course, this formula does not have meaning for arbitrary x and y, for the infinite series on the right need not converge. Our space will include the special sequence 0, all of whose values are the number zero. Then we must impose the restriction that any other point $x = (x_i \mid i = 1, 2, \ldots)$ of our space be one for which

$$d_2(x, 0) = \left[\sum_{i=1}^{\infty} x_i^2 \right]^{1/2}$$

has meaning. It will turn out that no further restrictions are needed; the formula for $d_2(x, y)$ will be meaningful for any two sequences x and y of this kind.

1.9 Example. Define H to be the set of all sequences $x = (x_i \mid i = 1, 2, \dots)$ of real numbers that are "square summable" in the sense that the series $\sum_{i=1}^{\infty} x_i^2$ converges. For example, $(1/i \mid i = 1, 2, \dots) \in \mathsf{H}$, but the alternating sequence $((-1)^i/\sqrt{i} \mid i = 1, 2, \dots) \notin \mathsf{H}$. The set H is called the *Hilbert sequence space* (the more standard notation for H is l^2: read "little ell two").

By analogy with the Euclidean norm, the *Hilbert norm* of an element $x = (x_i \mid i = 1, 2, \dots) \in \mathsf{H}$ is defined to be the real number

$$||x||_2 = \left[\sum_{i=1}^{\infty} x_i^2 \right]^{1/2}.$$

Clearly

$$||x||_2 \geq 0, \qquad ||x||_2 = 0 \iff x = 0$$

for each $x \in \mathsf{H}$. Moreover, if $x = (x_i \mid i = 1, 2, \dots) \in \mathsf{H}$ and $\lambda \in \mathsf{R}$, then the sequence

$$\lambda x = (\lambda x_i \mid i = 1, 2, \dots)$$

also belongs to H and satisfies

$$||\lambda x||_2 = |\lambda| \cdot ||x||_2.$$

Let $x = (x_i \mid i = 1, 2, \dots) \in \mathsf{H}$ and $y = (y_i \mid i = 1, 2, \dots) \in \mathsf{H}$. For each $n \geq 1$ the inequality 1.2 (4) says

$$\left[\sum_{i=1}^{n} (x_i + y_i)^2 \right]^{1/2} \leq \left[\sum_{i=1}^{n} x_i^2 \right]^{1/2} + \left[\sum_{i=1}^{n} y_i^2 \right]^{1/2}$$

$$\leq \left[\sum_{i=1}^{\infty} x_i^2 \right]^{1/2} + \left[\sum_{i=1}^{\infty} y_i^2 \right]^{1/2}$$

$$= ||x||_2 + ||y||_2.$$

This proves two things. First, since all the partial sums of the series $\sum_{i=1}^{\infty} (x_i + y_i)^2$ are bounded above by the number $(||x||_2 + ||y||_2)^2$, this series converges; hence the sequence

$$x + y = (x_i + y_i \mid i = 1, 2, \dots)$$

belongs to H. Second

$$||x + y||_2 \leq ||x||_2 + ||y||_2.$$

Given $x = (x_i \mid i = 1, 2, \dots)$ and $y = (y_i \mid i = 1, 2, \dots)$ in H, the sequence

$$x - y = x + (-1)y = (x_i - y_i \mid i = 1, 2, \dots)$$

belongs to H, so we may define the *Hilbert distance* from x to y to be the real number

$$d_2(x, y) = \|x - y\|_2 = \left[\sum_{i=1}^{\infty} (x_i - y_i)^2\right]^{1/2}.$$

We know that the four properties of the Euclidean norm on R^n listed in 1.2 hold as well for the Hilbert norm on H. Hence the same proof used to deduce Proposition 1.4 establishes that for all $x, y, z \in H$:

(1) $d_2(x, y) \geq 0$.
(2) $d_2(x, y) = 0$ if and only if $x = y$.
(3) $d_2(x, y) = d_2(y, x)$.
(4) $d_2(x, z) \leq d_2(x, y) + d_2(y, z)$.

It is time to let the secret out. All the preceding examples are instances of an abstract notion first defined by Maurice Fréchet in 1906.

1.10 Definition. A *metric* on a set X is a function

$$d: X \times X \to R$$

such that for all $x, y, z \in X$:

(M1) $d(x, y) \geq 0$.
(M2) $d(x, y) = 0$ if and only if $x = y$.
(M3) $d(x, y) = d(y, x)$ [*symmetry*].
(M4) $d(x, z) \leq d(x, y) + d(y, z)$ [*the triangle inequality*].

If $x, y \in X$, then $d(x, y)$ is called the *d-distance from x to y*.

A set X together with a metric d on X comprise a pair (X, d), called a *metric space*.

Property (M2) asserts two things: first,

$$d(x, x) = 0$$

and second,

$$x \neq y \implies d(x, y) \neq 0.$$

When we are showing that a particular d is a metric, it is ordinarily the second of these and the triangle inequality (M4) that cause any trouble; usually we do not even bother to mention that the remaining properties hold.

Different metrics d and d' on the same set X give rise to different metric spaces (X, d) and (X, d'). For example, when $n > 1$, the Euclidean metric d, the taxicab metric d_1, and the max metric d_∞ on R^n yield the three distinct metric spaces (R^n, d), (R^n, d_1), and (R^n, d_∞).

We close this section with additional examples of metric spaces.

1.11 Example. Let X be any set. Then the function $\delta\colon X \times X \to \mathsf{R}$ defined by

$$\delta(x, y) = \begin{cases} 0 & \text{if } x = y, \\ 1 & \text{if } x \neq y \end{cases}$$

is a metric on X, called the *discrete metric*. The triangle inequality

$$\delta(x, z) \leq \delta(x, y) + \delta(y, z)$$

certainly holds in case $x = y$ and $y = z$, for then both sides are 0; it holds as well in case $x \neq y$ or $y \neq z$, for then $\delta(x, y) + \delta(y, z) \geq 1$ whereas $\delta(x, z) \leq 1$.

The adjective 'discrete' is applied to δ to suggest that this metric isolates distinct points. Indeed, for $0 < \epsilon \leq 1$ the only point of X whose δ-distance from an $x \in X$ is less than ϵ is x itself.

This example shows that any set, no matter what the nature of its elements, can be made into a metric space by endowing it with its discrete metric. Of course, a particular set—R^n, for example—may well have metrics much more interesting than the discrete one.

1.12 Example. Given a metric space (X, d) and a subset Y of X, we may use d to measure distances just between points of Y by forming the restriction

$$d' = d\,|_{Y \times Y}\colon Y \times Y \to \mathsf{R}$$

of $d\colon X \times X \to \mathsf{R}$ to $Y \times Y$. Since $d'(x, y) = d(x, y)$ for all $x, y \in Y$, the function d' is clearly a metric on Y, which is said to be *induced by d*.

When $X = \mathsf{R}^n$ and d is the Euclidean metric on R^n, then the induced metric d' on $Y \subset X$ is called the *Euclidean metric on Y*, and Y becomes a metric space in its own right when provided with d'.

1.13 Example. Let $(X_1, d_1), (X_2, d_2), \ldots, (X_n, d_n)$ be finitely many metric spaces. On the product set

$$X = X_1 \times X_2 \times \cdots \times X_n$$

the *max metric d_∞* induced by $(d_1, d_2, \ldots, d_n)$ is obtained by setting

$$d_\infty(x, y) = \max_{1 \leq i \leq n} d_i(x_i, y_i)$$

for any two points $x = (x_1, x_2, \ldots, x_n)$ and $y = (y_1, y_2, \ldots, y_n)$ of X.

When $X_1 = X_2 = \cdots = X_n = \mathsf{R}$ and $d_1 = d_2 = \cdots = d_n =$ the Euclidean metric on R, then d_∞ is the same max metric on R^n defined earlier (1.6). Of course, the Euclidean and taxicab metrics on R^n likewise suggest constructing metrics on a product of metric spaces (Exercise 10), but these metrics are less convenient to work with than the one above.

1.14 Example. Let X be the set of all sequences $x = (x_i \mid i \in \mathsf{N})$ in a given set Y. We want to regard two points $x = (x_i \mid i \in \mathsf{N})$ and $y = (y_i \mid i \in \mathsf{N})$

of X to be close together if for some large n their first n coordinates x_0, x_1, ..., x_{n-1} and y_0, y_1, ..., y_{n-1} agree; the larger the value of n, the closer together the points are to be. To accomplish this we define a distance

$$d(x, y) = \begin{cases} 0 & \text{if } x = y, \\ [1 + \min \{i \mid x_i \neq y_i\}]^{-1} & \text{if } x \neq y. \end{cases}$$

The triangle inequality

$$d(x, z) \leq d(x, y) + d(y, z)$$

certainly holds if $x = z$ or $x = y$ or $y = z$. To verify it when $x \neq z$, $x \neq y$, and $y \neq z$, set

$$j = \min \{i \mid x_i \neq y_i\}, \qquad k = \min \{i \mid y_i \neq z_i\}, \qquad m = \min \{i \mid x_i \neq z_i\}.$$

For $i \in \mathbf{N}$,

$$x_i = y_i \quad \text{and} \quad y_i = z_i \implies x_i = z_i.$$

Then $m \geq \min \{j, k\}$, so

$$\frac{1}{1 + m} \leq \frac{1}{1 + j} + \frac{1}{1 + k}.$$

EXERCISES

1. Verify for all $x, y \in \mathbf{R}^n$ the identity

$$||x||^2 \cdot ||y||^2 - \left[\sum_{i=1}^{n} x_i y_i \right]^2 = \frac{1}{2} \sum_{i=1}^{n} \sum_{j=1}^{n} (x_i y_j - x_j y_i)^2$$

and from it derive anew the Cauchy-Schwarz inequality (1.3).

2. (**a**) For which pairs of points x and y of $\mathbf{R}^n$ does $||x + y|| = ||x|| + ||y||$? [*Hint:* Examine the proofs of 1.2 and 1.3.]
(**b**) Use (a) to determine when equality actually holds in the triangle inequality for the Euclidean metric on $\mathbf{R}^n$.

3. Verify the inequalities

$$d(x, y) \leq d_1(x, y) \leq \sqrt{n}\, d(x, y)$$

relating the Euclidean metric d and the taxicab metric d_1 on $\mathbf{R}^n$.

4. A *translation* of $\mathbf{R}^2$ is a map $T: \mathbf{R}^2 \to \mathbf{R}^2$ of the form

$$T(x) = x + z$$

for some fixed $z \in \mathbf{R}^2$, and a *rotation* of $\mathbf{R}^2$ is a map $R: \mathbf{R}^2 \to \mathbf{R}^2$ of the form

$$R(x_1, x_2) = (x_1 \cos \theta - x_2 \sin \theta, x_1 \sin \theta + x_2 \cos \theta)$$

for some fixed real number θ (the angle of rotation).
(**a**) Show that the Euclidean metric d on $\mathbf{R}^2$ is "translation–invariant" in the sense that

$$d(Tx, Ty) = d(x, y) \qquad\qquad (x, y \in \mathbf{R}^2)$$

for every translation T of R^2 and is "rotation–invariant" in the sense that

$$d(Rx, Ry) = d(x, y) \qquad\qquad (x, y \in \mathsf{R}^2)$$

for every rotation R of R^2.

(**b**) Are the taxicab and max metrics on R^2 also translation–invariant? rotation–invariant?

5. Construct a denumerable subset B of the Hilbert sequence space (1.9) such that $d_2(x, y) = 1$ whenever x and y are distinct points of B.

6. (**a**) Show that property (M1) in the definition of a metric is actually a consequence of the remaining properties (M2)–(M4).

(**b**) Prove that a function $d: X \times X \to \mathsf{R}$ is a metric on a set X if it satisfies (M2) together with the property

(M4*) $\qquad\qquad d(x, z) \le d(x, y) + d(z, y) \qquad$ for all $x, y, z \in X$.

7. Which of the following formulas define a metric on R?

(**a**) $d(x, y) = |x - y|^2$.
(**b**) $d(x, y) = |x^2 - y^2|$.
(**c**) $d(x, y) = |x - y|^3$.
(**d**) $d(x, y) = |x^3 - y^3|$.

8. For which real numbers p is it true that d^p is a metric whenever d is a metric? [Here d^p is the pth power $(x, y) \mapsto (d(x, y))^p$ of d.]

9. Given a metric d on a set X, prove that

$$|d(x, z) - d(y, z)| \le d(x, y)$$

for all $x, y, z \in X$.

10. Generalize the Euclidean metric on R^n by constructing a metric on the product of any n metric spaces $(X_1, d_1), \ldots, (X_n, d_n)$. Do the same for the taxicab metric. (Compare 1.13.)

11. For points x, y on the circle $X = \{x \in \mathsf{R}^2 \mid x_1^2 + x_2^2 = 1\}$ in the plane, define $d(x, y) = 0$ if $x = y$, and $d(x, y) =$ the length of the shorter arc of X joining x to y if $x \ne y$. Verify that d is a metric on X and compare it with the Euclidean metric on X.

12. Let X be a nonempty set. A function $f: X \to \mathsf{R}$ is said to be *bounded* if there is a constant c with $|f(x)| \le c$ for all $x \in X$. Show that the formula

$$d_\infty(f, g) = \sup\{|f(x) - g(x)| : x \in X\}$$

defines a metric on the collection $\mathcal{B}(X)$ of all bounded functions $f: X \to \mathsf{R}$.

13. A *pseudometric* on a set X is a function $d: X \times X \to \mathsf{R}$ satisfying properties (M1), (M3), and (M4) for a metric together with the following weakened form of (M2):

(M2*) $\qquad\qquad d(x, x) = 0 \qquad$ for all $x \in X$.

In each case below verify that d is a pseudometric on X and determine whether it is actually a metric.

(**a**) X is the set of all continuous functions $x: [0, 1] \to \mathsf{R}$, and $d(x, y) = |x(0) - y(0)|$.

(**b**) X is any set, and for some function $f: X \to \mathsf{R}$, $d(x, y) = |f(x) - f(y)|$.

(**c**) X is the set of all convergent sequences $x = (x_n \mid n \in \mathsf{N})$ of real numbers, and $d(x, y) = \lim_{n \to \infty} |x_n - y_n|$.

14. Let d be a pseudometric on a nonempty set X. Show that a metric space is obtained by identifying points of X zero distance apart, as follows.

(a) Show that the relation $\sim$ on X given by

$$x \sim y \quad \Leftrightarrow \quad d(x, y) = 0$$

is an equivalence relation on X.

(b) For each $x \in X$ let x^* denote the equivalence class $\{y \in X \mid x \sim y\}$ of x, and let X^* denote the quotient set $X/\!\sim = \{x^* \mid x \in X\}$ consisting of all these equivalence classes. Show there is a unique metric d^* on X^* such that

$$d^*(x^*, y^*) = d(x, y) \qquad\qquad (x, y \in X).$$

(c) Describe X^* and d^* in case d is already a metric.

15. Let p be a fixed prime number (that is, p is a positive integer, and the only positive integers dividing p are 1 and p itself). On the set $\mathbf{Q}$ of rational numbers the *p-adic metric* d_p is defined as follows. If $0 \neq x \in \mathbf{Q}$, write x in the form

$$x = p^k \frac{m}{n}$$

for integers k, m, n with p dividing neither m nor n, and set

$$v(x) = k.$$

Define $d_p \colon \mathbf{Q} \times \mathbf{Q} \to \mathbf{R}$ by

$$d_p(x, y) = \begin{cases} 0 & \text{if } x = y, \\ p^{-v(x-y)} & \text{if } x \neq y. \end{cases}$$

Clearly d_p has properties (M1)–(M3). Show that d_p satisfies the triangle inequality by proving the stronger inequality

$$d_p(x, z) \leq \max \{d_p(x, y), d_p(y, z)\}.$$

[*Hint*: Show that $v(a - b) \geq \min \{v(a), v(b)\}$ whenever a, $b \in \mathbf{Q}$ with $a \neq 0 \neq b$ and $a \neq b$.]

2. Open Sets and Closed Sets

The notion of continuity of a function $f \colon \mathbf{R}^n \to \mathbf{R}$ suggested in the preceding section will become a precise definition once we introduce metrics to measure distances. On $\mathbf{R}^n$ we have defined several different metrics, each well-behaved in its own way. But for any prejudice bred of familiarity with the Euclidean metric, there is no good reason to prefer one of these metrics over the others. Fortunately, it will turn out that the Euclidean, taxicab, and max metrics all yield the same continuous functions, because points that are close to one another for one of these metrics are close for the other two as well.

The extent to which continuity of a function is independent of the particular metrics used will be clarified in Section 4, where we begin to study continuity in a systematic way. As preparation, we first study sets of points that are close to a given point in a metric space.

1.15 Definition. Let (X, d) be a metric space. For a point $x \in X$ and a real number $\epsilon > 0$, the *d-ball of radius ϵ at x* is the set

$$B_\epsilon(x; d) = \{ y \in X \mid d(x, y) < \epsilon \},$$

the *d-disk of radius ϵ at x* is the set

$$D_\epsilon(x; d) = \{ y \in X \mid d(x, y) \leq \epsilon \},$$

and the *d-sphere of radius ϵ at x* is the set

$$S_\epsilon(x; d) = \{ y \in X \mid d(x, y) = \epsilon \}.$$

(This terminology is far from standard. Some authors use 'open ball' or 'open cell' or 'open sphere' for 'ball', and 'closed ball' or 'closed cell' or 'cell' or 'closed sphere', for 'disk'. Some even use 'sphere' for 'ball'.)

Observe that

$$x \in B_\epsilon(x; d), \qquad x \notin S_\epsilon(x; d),$$

$$B_\epsilon(x; d) \cup S_\epsilon(x; d) = D_\epsilon(x; d), \qquad B_\epsilon(x; d) \cap S_\epsilon(x; d) = \varnothing.$$

1.16 Examples

(1) Take d to be the Euclidean metric on $\mathbf{R}^n$ and let $x \in \mathbf{R}^n$ and $\epsilon > 0$. Then the d-sphere $S_\epsilon(x; d)$ is the set of all $y \in \mathbf{R}^n$ satisfying the quadratic equation

$$\sum_{i=1}^{n} (x_i - y_i)^2 = \epsilon^2,$$

and the d-ball $B_\epsilon(x; d)$ is the set of all $y \in \mathbf{R}^n$ satisfying the quadratic inequality

$$\sum_{i=1}^{n} (x_i - y_i)^2 < \epsilon^2,$$

that is, the set of all $y \in \mathbf{R}^n$ lying "inside" $S_\epsilon(x; d)$ (see Exercise 17).

In dimension $n = 1$, $B_\epsilon(x; d)$ is the interval $]x - \epsilon, x + \epsilon[$, and $S_\epsilon(x; d)$ is the set $\{x - \epsilon, x + \epsilon\}$ of the two endpoints of this interval. In dimension $n = 2$, $B_\epsilon(x; d)$ is the circular region of radius ϵ centered at x, and $S_\epsilon(x; d)$ is the circle surrounding this region (see Fig. 1.5). In dimension $n = 3$, $B_\epsilon(x; d)$ is the solid spherical region of radius ϵ centered at x, and $S_\epsilon(x; d)$ is the spherical surface enclosing this region (see Fig. 1.6).

The d-ball $B_\epsilon(x; d)$ is convex (0.14). In fact, let

$$y = (1 - t)z + tw, \qquad 0 \leq t \leq 1$$

be a point on the line segment joining two points $z, w \in B_\epsilon(x; d)$. From

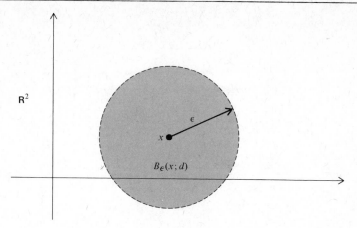

FIG. 1.5

properties 1.2 of the Euclidean norm,

$$d(x, y) = \|x - y\| = \|[(1 - t)x - (1 - t)z] + [tx - tw]\|$$
$$\leq (1 - t)\|x - z\| + t\|x - w\| = (1 - t)\,d(x, z) + t\,d(x, w)$$
$$< (1 - t)\epsilon + t\epsilon = \epsilon.$$

Hence $y \in B_\epsilon(x; d)$ also.

(2) Again let $x \in \mathsf{R}^n$ and $\epsilon > 0$, but now consider the max metric d_∞ on R^n (see 1.6). Since

$$d_\infty(x, y) = \max_{1 \leq i \leq n} |x_i - y_i| < \epsilon$$

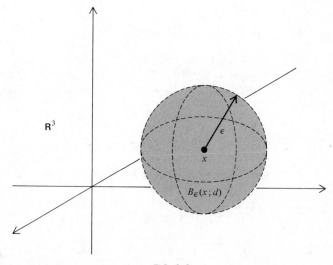

FIG. 1.6

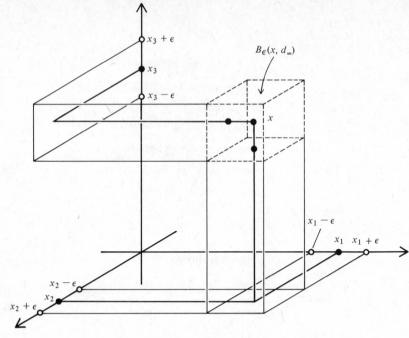

FIG. 1.7

if and only if

$$|x_i - y_i| < \epsilon \qquad (i = 1, \ldots, n),$$

that is,

$$x_i - \epsilon < y_i < x_i + \epsilon \qquad (i = 1, \ldots, n),$$

then $B_\epsilon(x; d_\infty)$ is the product

$$]x_1 - \epsilon, x_1 + \epsilon[\times]x_2 - \epsilon, x_2 + \epsilon[\times \cdots \times]x_n - \epsilon, x_n + \epsilon[$$

of n open intervals each of length 2ϵ, the ith interval being symmetric about x_i (see Fig. 1.7 for the case $n = 3$). Thus $B_\epsilon(x; d_\infty)$ is the ϵ-*cube centered at* x.

(3) Let δ be the discrete metric on a set X (1.11). For $x \in X$ and $\epsilon > 0$,

$$B_\epsilon(x; \delta) = \begin{cases} \{x\} & \text{if } \epsilon \leq 1, \\ X & \text{if } \epsilon > 1. \end{cases}$$

(4) Let (X, d) be any metric space, and let d' be the metric on a subset Y of X induced by d (1.12). If $x \in Y$, then

$$B_\epsilon(x; d') = B_\epsilon(x; d) \cap Y$$

for each $\epsilon > 0$.

(5) Let X be the set of all continuous real-valued functions on the closed interval $[0, 1]$, and let d_∞ be the metric on X defined in Example 1.7. Given

$x \colon [0, 1] \to \mathsf{R}$ belonging to X and $\epsilon > 0$, then

$$d_{\infty}(x, y) = \max_{0 \leq t \leq 1} |x(t) - y(t)| < \epsilon$$

if and only if

$$|x(t) - y(t)| < \epsilon \qquad\qquad (0 \leq t \leq 1).$$

Hence $B_{\epsilon}(x; d_{\infty})$ consists of all those continuous functions $y \colon [0, 1] \to \mathsf{R}$ whose graphs lie strictly between the graphs of the vertically translated functions

$$x_{-\epsilon} \colon [0, 1] \to \mathsf{R}, \qquad\qquad x_{\epsilon} \colon [0, 1] \to \mathsf{R}$$
$$t \mapsto x(t) - \epsilon \qquad\qquad t \mapsto x(t) + \epsilon$$

(see Fig. 1.8).

It is geometrically evident that if B is the region in the plane enclosed by (but not containing any of the points on) a circle, then each point of B is the center of another such circular region sufficiently small to be contained completely in B (see Fig. 1.9). A similar result is true of a ball in any metric space.

1.17 Proposition. Let x be a point of a metric space (X, d) and let $\epsilon > 0$. Then for each $y \in B_{\epsilon}(x; d)$ there is some $\eta > 0$ for which

$$B_{\eta}(y; d) \subset B_{\epsilon}(x; d)$$

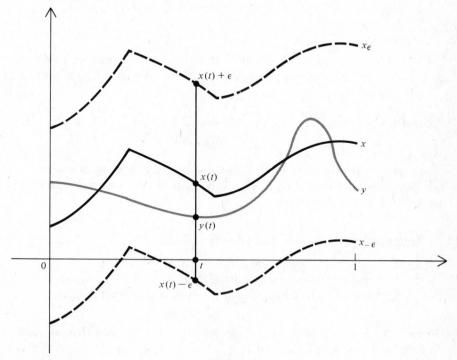

FIG. 1.8

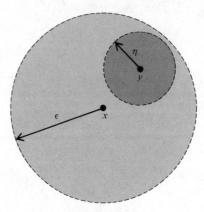

FIG. 1.9

[in fact, any $0 < \eta \le \epsilon - d(x, y)$ will do].

Proof. Let $y \in B_\epsilon(x; d)$. Then $\epsilon - d(x, y) > 0$. Choose any η with $0 < \eta \le \epsilon - d(x, y)$. If $z \in B_\eta(y; d)$, then $d(y, z) < \eta$, so by the triangle inequality

$$d(x, z) \le d(x, y) + d(y, z) < d(x, y) + [\epsilon - d(x, y)] = \epsilon,$$

and $z \in B_\epsilon(x; d)$.   ☐

This proposition says that a d-ball U at a point has the property that all points sufficiently close to any given point of U also belong to U. Let us look at other sets having this property.

1.18 Definition. Let (X, d) be a metric space. A subset U of X is said to be *d-open* if for each point $x \in U$ there is some $\epsilon > 0$ such that $B_\epsilon(x; d) \subset U$.

Thus *any d-ball in a metric space is d-open.* In order to construct examples of d-open sets other than d-balls, it will be convenient to establish the following properties of d-open sets.

1.19 Theorem. Let (X, d) be a metric space. Then:

    (1) The empty set $\varnothing$ and the entire set X are both d-open.
    (2) Any union of d-open sets is itself d-open.
    (3) Any intersection of *finitely many* d-open sets is itself d-open.

Proof. (1) Since there is no $x \in \varnothing$, it is vacuously true that at each $x \in \varnothing$ there is a d-ball contained in $\varnothing$. Hence $\varnothing$ is d-open. The set X is d-open since $B_\epsilon(x; d) \subset X$ for every $x \in X$ and every $\epsilon > 0$.

(2) Let $(U_i \mid i \in I)$ be any family of d-open sets, and let $U = \bigcup_{i \in I} U_i$. If $x \in U$, then $x \in U_j$ for some $j \in I$, $B_\epsilon(x; d) \subset U_j$ for some $\epsilon > 0$ since U_j is d-open, and hence $B_\epsilon(x; d) \subset U$. Thus U is d-open.

(3) Let $U_1, \ldots, U_n$ be d-open sets, and let $U = \bigcap_{i=1}^n U_i$. Let $x \in U$; we must show $B_\epsilon(x; d) \subset U$ for some $\epsilon > 0$. For each $1 \leq i \leq n$, $x \in U_i$, and since U_i is d-open,

$$B_{\epsilon_i}(x; d) \subset U_i$$

for some $\epsilon_i > 0$. Set

$$\epsilon = \min \{\epsilon_1, \ldots, \epsilon_n\} > 0.$$

Then

$$B_\epsilon(x; d) \subset B_{\epsilon_i}(x; d) \subset U_i \qquad (i = 1, \ldots, n),$$

so $B_\epsilon(x; d) \subset U$. $\quad\Box$

This theorem will be crucial for motivating the definition of a topological space in Chapter 2.

1.20 Examples

(1) Let δ be the discrete metric on a set X (1.11). Then each singleton $\{x\}$, $x \in X$, is δ-open, for $B_1(x; \delta) = \{x\}$. It follows from 1.19 (2) that *every* subset of X is δ-open, for if $A \subset X$, then

$$A = \bigcup_{x \in A} \{x\}.$$

(2) Let d be the Euclidean metric on R. We shall determine all d-open subsets of R. First, every open interval or open ray (0.36) in R is d-open. In fact, an open interval $]a, b[$ is d-open because it is the d-ball $B_\epsilon(x; d)$ of radius $\epsilon = \frac{1}{2}(b - a)$ at the midpoint $x = \frac{1}{2}(a + b)$ of $]a, b[$. An open ray $]a, \to [$ is d-open because if $x \in]a, \to [$, then for $\epsilon = x - a$ we have

$$B_\epsilon(x; d) =]x - \epsilon, x + \epsilon[\subset]a, \to [.$$

By 1.19 (2), the union of any countable collection of pairwise disjoint open intervals and open rays is d-open. We shall show that, conversely, *each proper d-open subset of R is the union of a countable collection of pairwise disjoint open intervals and open rays.*

Let U be a proper d-open subset of R. For each $x \in U$, define J_x to be the set of those $y \in \mathsf{R}$ such that there exist real numbers $a < b$ with

$$x \in]a, b[, \quad y \in]a, b[, \quad]a, b[\subset U.$$

We show

$$U = \bigcup_{x \in U} J_x.$$

If $x \in U$, then certainly $J_x \subset U$. Hence

$$\bigcup_{x \in U} J_x \subset U.$$

To prove the opposite inclusion, let $x \in U$. Since U is d-open, then $x \in$ $]x - \epsilon, x + \epsilon[$ for some $\epsilon > 0$. Hence $x \in J_x$.

The sets J_x need not be distinct for different values of x. For example, if U were itself an open interval $]a, b[$, then $J_x =]a, b[$ for all $x \in U$. However, if $x, z \in U$, then

$$z \in J_x \implies J_x = J_z.$$

In fact, let $z \in J_x$, so that

$$x, z \in]a, b[\subset U$$

for some $a < b$. If $y \in J_z$, also

$$y, z \in]c, d[\subset U$$

for some $c < d$, and then $]a, b[\cup]c, d[$ is an open interval $]u, v[$ with

$$x, y \in]u, v[\subset U,$$

so $y \in J_x$. Hence $J_z \subset J_x$. Similarly, $J_x \subset J_z$.

We can now show that the collection $\{J_x \mid x \in U\}$ is pairwise disjoint. Let $x, y \in U$ and suppose there is some $z \in J_x \cap J_y$. By what we just proved,

$$J_x = J_z, \qquad J_y = J_z.$$

Hence $J_x = J_y$.

Let $x \in U$. We show that J_x is an open interval or is an open ray. Let $u, v \in J_x$ with $u < v$. There are open intervals $]a, b[$, $]c, d[$ with

$$u, x \in]a, b[\subset U, \qquad v, x \in]c, d[\subset U.$$

Then $]a, b[\cup]c, d[$ is an open interval with

$$]u, v[\subset]a, b[\cup]c, d[\subset U,$$

so $]u, v[\subset J_x$. Hence J_x is an interval.

The interval $J_x \neq \mathbf{R}$ because $J_x \subset U \neq \mathbf{R}$. To see that J_x is an open interval or an open ray, we show it contains no endpoint. Just suppose it contained a left endpoint y. There exist a, b with

$$a < y < x < b, \qquad]a, b[\subset U.$$

Choose z with $a < z < y$. Since $z \notin J_x$ and $z \in U$, then $J_z \neq J_x$. However, $y \in J_z \cap J_x$, which is impossible. Similarly, J_x cannot contain a right endpoint.

Finally, we show that $\{J_x \mid x \in U\}$ is countable. In each open interval J_x we may choose some rational number $r(J_x)$. Since $\{J_x \mid x \in U\}$ is pairwise disjoint, $r(J_x) \neq r(J_y)$ if $J_x \neq J_y$. But there are only countably many rational numbers.

(3) Let (X, d) be a metric space, let $Y \subset X$, and let d' be the metric on Y induced by d (1.12). Suppose U is a d-open subset of X. Then

(*) $$V = U \cap Y$$

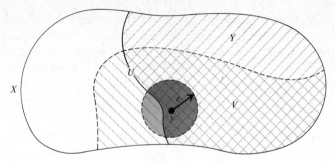

FIG. 1.10

is a d'-open subset of Y. To see this, let $y \in V$. Since $y \in U$ and U is d-open, $B_\epsilon(y; d) \subset U$ for some $\epsilon > 0$ (see Fig. 1.10). Then

$$B_\epsilon(y; d') = B_\epsilon(y; d) \cap Y \subset U \cap Y = V.$$

Hence V is d'-open.

Now let V be a d'-open subset of Y. We claim V has the form (*) for some d-open subset U of X. For each $y \in V$, there exists some $\epsilon(y) > 0$ with

$$B_{\epsilon(y)}(y; d') \subset V.$$

Define

$$U = \bigcup_{y \in V} B_{\epsilon(y)}(y; d).$$

From 1.19 (2) and the fact that any d-ball is d-open, it follows that U is d-open. An easy computation shows that $U \cap Y = V$.

(4) Let d' be the Euclidean metric on the closed interval $[0, 1]$, so that d' is induced by the Euclidean metric d on $\mathbf{R}$. Together (2) and (3) allow us to determine all d'-open subsets of $[0, 1]$. The intersection of an open interval or open ray in $\mathbf{R}$ with $[0, 1]$ is either $\varnothing$, $[0, 1]$, an interval of the form $[0, b[$ for some $0 < b < 1$, an interval of the form $]a, 1]$ for some $0 < a < 1$, or an interval of the form $]a, b[$ for some $0 < a < b < 1$. Hence a subset V of $[0, 1]$ is d'-open if and only if it has one of the forms

$$V = \varnothing,$$

$$V = [0, 1],$$

$$V = [0, b[\cup W,$$

$$V =]a, 1] \cup W,$$

$$V = [0, b[\cup]a, 1] \cup W,$$

$$V = W$$

where $0 < b < 1$, $0 < a < 1$, and W is a (possibly empty) union of a countable collection of disjoint open intervals contained in $]0, 1[$.

Notice that $[0, 1]$ is d'-open, but it is not d-open.

(5) Although in a metric space (X, d) the intersection of finitely many d-open sets will always be d-open, the intersection of even countably many d-open sets need not be d-open. For example, let d be the Euclidean metric on R. Then $]-1/n, 1/n[$ is d-open for every positive integer n, but

$$\bigcap_{n=1}^{\infty}]-1/n, 1/n[\, = \{0\}$$

is not d-open.

It is natural to single out, along with the d-open sets in a metric space, the complements of these sets.

1.21 Definition. Let (X, d) be a metric space. A subset E of X is said to be *d-closed* if its complement $X \setminus E$ in X is d-open.

Recalling what it means for $X \setminus E$ to be d-open, we see that a subset E of X is d-closed exactly when at each point of X not belonging to E there is some d-ball that is disjoint from E. Since $X \setminus (X \setminus A) = A$ for any subset of X, we also see that a subset A of X is d-open if and only if its complement $X \setminus A$ in X is d-closed.

The analog of 1.19 for d-closed sets is the following theorem.

1.22 Theorem. Let (X, d) be a metric space. Then:

(1) The empty set $\varnothing$ and the entire set X are both d-closed.
(2) Any intersection of d-closed sets is d-closed.
(3) Any union of *finitely many* d-closed sets is d-closed.

Proof. (1) By 1.19 (1), both $\varnothing$ and X are d-open. Hence $X = X \setminus \varnothing$ and $\varnothing = X \setminus X$ are d-closed.

(2) Let $(E_i \mid i \in I)$ be any family of d-closed sets, and let

$$E = \bigcap_{i \in I} E_i.$$

By one of the De Morgan laws (0.17),

$$X \setminus E = \bigcup_{i \in I} (X \setminus E_i).$$

Now $X \setminus E_i$ is d-open for each $i \in I$, so by 1.19 (2) the above union is d-open. Hence E is d-closed.

(3) Let $E_1, \ldots, E_n$ be d-closed sets, and let

$$E = \bigcup_{i=1}^{n} E_i.$$

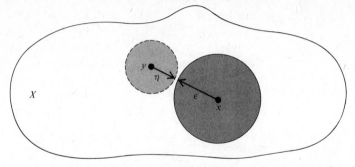

FIG. 1.11

By one of the De Morgan laws (0.17),

$$X \setminus E = \bigcap_{i=1}^{n} (X \setminus E_i).$$

Now use 1.19 (3).   ☐

1.23 Examples

(1) In any metric space (X, d), both the sets $\varnothing$ and X are simultaneously d-open and d-closed. (When these two are the only subsets of X having this property, X is said to be "connected." Connected spaces are studied in detail in Chapter 5.)

(2) Let δ be the discrete metric on a set X. Since every subset of X is δ-open, every subset of X is δ-closed as well.

(3) In the metric space (R, d), where d is the Euclidean metric, the set $]0, 1[$ is d-open but not d-closed.

(4) In the metric space (R, d) just considered, the set $[0, 1[$ is neither d-open nor d-closed.

This example should correct any mistaken idea that a set must be d-open if it is not d-closed, or d-closed if it is not d-open. Even a door can be neither open nor closed, but instead ajar (of course, a door cannot be both open and closed at the same time, but a set can be both d-open and d-closed).

(5) Any d-disk $D_\epsilon(x; d)$ at a point x in a metric space (X, d) is d-closed. In fact, let $y \in X$ with $y \notin D_\epsilon(x; d)$. Then $d(x, y) > \epsilon$. Define

$$\eta = d(x, y) - \epsilon > 0.$$

An easy use of the triangle inequality shows that $B_\eta(y; d)$ is disjoint from $D_\epsilon(x; d)$ (see Fig. 1.11).

A similar argument shows that any d-sphere $S_\epsilon(x; d)$ is d-closed.

(6) Let (X, d) be a metric space. Then any singleton $\{x\}$ is d-closed; in fact, if $y \neq x$, then $B_\epsilon(y; d)$ is disjoint from $\{x\}$ whenever $0 < \epsilon \leq d(x, y)$. From 1.22 (3) it follows that any finite subset of X is d-closed.

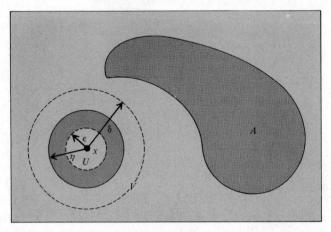

FIG. 1.12

It was noted above that a point x not belonging to a d-closed set A in a metric space (X, d) is contained in some d-open set (in fact, a d-ball) disjoint from A. Actually something more is true: x can be separated from A by d-open sets.

1.24 Proposition. Let (X, d) be a metric space. Suppose A is a d-closed subset of X and x is a point of X with $x \notin A$. Then there are disjoint d-open subsets U and V of X with

$$x \in U, \qquad A \subset V.$$

Proof. There exists $\delta > 0$ with $B_\delta(x; d)$ disjoint from A. Choose ϵ and η with

$$0 < \epsilon < \eta < \delta.$$

Define

$$U = B_\epsilon(x; d), \qquad V = X \setminus D_\eta(x; d)$$

(see Fig. 1.12). Then U is open by 1.17, and V is d-open by 1.23 (5). Since

$$B_\epsilon(x; d) \subset D_\eta(x; d) \subset B_\delta(x; d),$$

$A \subset V$ and $U \cap V = \varnothing$.   ☐

In view of 1.23 (6), we could take A to be a singleton $\{y\}$ in Proposition 1.24 to obtain the following corollary.

1.25 Corollary. If x and y are distinct points in a metric space (X, d), then there are disjoint d-open subsets U and V of X with $x \in U$ and $y \in V$.

Instead of relying on 1.24 to prove 1.25, the reader may show directly that if $x \neq y$, then the d-balls of radius $\epsilon = \frac{1}{2} d(x, y)$ at x and y are disjoint

An interesting criterion for sets to be d-closed can be formulated by introducing a concept of distance from a point to a set.

1.26 Definition. Let x be a point and A be a nonempty subset of a metric space (X, d). Then the *distance from x to A* is defined to be the nonnegative real number

$$d(x, A) = \inf \{d(x, y) \mid y \in A\}.$$

This definition makes sense since $\{d(x, y) \mid y \in A\}$ is nonempty, A being nonempty, and has 0 as a lower bound.

When A reduces to a singleton $\{y\}$, then $d(x, A) = d(x, y)$. If there happens to be a point $y \in A$ for which $d(x, y) = d(x, A)$ (so that the infimum in 1.26 is an actual minimum), then we may call y a *nearest point of A to x*. However, such a nearest point need not exist. Indeed, it can happen that $d(x, A) = 0$ while $x \notin A$. For example, take d to be the Euclidean metric on R and $A = \{1/n \mid n = 1, 2, \dots \}$. Then

$$0 \leq d(0, A) = \inf \{1/n \mid n = 1, 2, \dots \} \leq 1/n \quad (n = 1, 2, \dots),$$

so $d(0, A) = 0$, but $0 \notin A$. This example is explained by the following criterion alluded to above.

1.27 Proposition. Let (X, d) be a metric space and let A be a nonempty subset of X. Then a necessary and sufficient condition for A to be d-closed is that $d(x, A) > 0$ for each $x \in X$ not belonging to A.

Proof. Necessity. Assume A is d-closed. Let $x \in X \backslash A$. There exists $\epsilon > 0$ with

$$B_\epsilon(x; d) \subset X \backslash A.$$

If $y \in A$, then $y \notin B_\epsilon(x; d)$ so $d(x, y) \geq \epsilon$. Hence

$$d(x, A) = \inf \{d(x, y) \mid y \in A\} \geq \epsilon > 0.$$

Sufficiency. Assume the condition holds. Let $x \in X \backslash A$. Then $d(x, A) > 0$, and we may form the d-ball $B_\epsilon(x; d)$ of radius $\epsilon = d(x, A)$. Since $d(x, y) \geq d(x, A)$ for each $y \in A$, $B_\epsilon(x; d)$ is disjoint from A. Hence A is d-closed. $\square$

In generalizing from d-balls to d-open sets we neglected the "local" aspect of d-balls—the fact that a d-ball consists of points sufficiently close to a *given* point. Next we introduce a generalization of d-balls that does retain this aspect.

1.28 Definition. Let x be a point in a metric space (X, d). A subset V of X is called a *d-neighborhood of x* if there exists some d-open set U with $x \in U \subset V$.

Clearly V is a d-neighborhood of x if and only if there exists some d-ball $B_\epsilon(x; d)$ at x with $B_\epsilon(x; d) \subset V$. Any d-open set containing x, and in particular any d-ball at x, is a d-neighborhood of x. Any d-disk $D_\epsilon(x; d)$ at x is a d-neighborhood of x. Hence *a d-neighborhood of a point need not be d-open*. (*Caution*: Some authors include as part of the definition of a d-neighborhood the requirement that it be d-open.)

By their very definition, the d-neighborhoods of a point are determined by the d-open sets. Conversely, the d-open sets are determined by the d-neighborhoods of points.

1.29 Proposition. Let (X, d) be a metric space. Then a subset V of X is d-open if and only if V is a d-neighborhood of each of its points.

Proof. Clearly a d-open set is a d-neighborhood of each of its points. Conversely, let V be a subset of X that is a d-neighborhood of each of its points. Then for each $x \in V$ there is some d-open set U_x such that

$$x \in U_x \subset V.$$

Since then

$$V = \bigcup_{x \in V} U_x,$$

V is d-open by 1.19 (2). $\square$

1.30 Corollary. Let (X, d) be a metric space. Then a subset E of X is d-closed if and only if every point of X, each neighborhood of which intersects E, belongs to E.

EXERCISES

16. Given $x \in \mathbb{R}^2$ and $\epsilon > 0$, describe and draw a picture of $B_\epsilon(x; d_1)$, $D_\epsilon(x; d_1)$, and $S_\epsilon(x; d_1)$, where d_1 is the taxicab metric on $\mathbb{R}^2$.

17. Let d be the Euclidean metric on $\mathbb{R}^n$, let $x \in \mathbb{R}^n$, and let $\epsilon > 0$. Show that a point $y \in \mathbb{R}^n$ not belonging to $S_\epsilon(x; d)$ belongs to $B_\epsilon(x; d)$ if and only if it lies on some line segment joining x to a point of $S_\epsilon(x; d)$.

18. If d_1 is the taxicab metric on $\mathbb{R}^n$, is every d_1-ball convex? Answer the same question for the max metric d_∞.

19. Determine all d'-open subsets of $[0, 1[$, where d' is the Euclidean metric on $[0, 1[$. Do the same for $]0, 1]$.

20. Let X be the set of all continuous functions $x \colon [0, 1] \to \mathsf{R}$, and let d_∞ be the metric on X defined in Example 1.7. Suppose $x, y \in X$ with $x(t) < y(t)$ for all $0 \le t \le 1$. Is the subset

$$\{z \in X \mid x(t) < z(t) < y(t) \text{ for all } 0 \le t \le 1\}$$

of X a d_∞-ball?

21. Prove that a subset of a metric space (X, d) is d-open if and only if it is a union of d-balls.

22. Let d be the Euclidean metric on R^n, let $x \in \mathsf{R}^n$, and let $\epsilon > 0$. Suppose $B_\epsilon(x; d) \subset A \subset D_\epsilon(x; d)$. When is A d-open? d-closed?

23. Let (H, d_2) be the Hilbert sequence space of Example 1.9.
  (**a**) Show that each sequence $x = (x_i \mid i = 1, 2, \ldots)$ of real numbers with $|x_i| \le 1/i$ for $i = 1, 2, \ldots$ belongs to H.
  (**b**) Show that the *Hilbert cube*

$$\{x \in \mathsf{H} \colon |x_i| < 1/i \ (i = 1, 2, \ldots)\}$$

is d_2-closed in H.
  (**c**) Is the set

$$\{x \in \mathsf{H} \colon |x_i| < 1/i \ (i = 1, 2, \ldots)\}$$

d_2-open in H?

24. Let d' be the metric induced on a subset Y of a set X by a metric d on X.
  (**a**) Is every subset of Y of the form $E \cap Y$, E a d-closed subset of X, d'-closed? Is every d'-closed subset of Y of this form?
  (**b**) Let $y \in Y$. Is every subset of Y of the form $V \cap Y$, V a d-neighborhood of y, a d'-neighborhood of y? Is every d'-neighborhood of y of this form?

25. Let A be a nonempty subset of a metric space (X, d).
  (**a**) Show that

$$|d(x, A) - d(y, A)| \le d(x, y)$$

for all $x, y \in X$.
  (**b**) Prove that

$$\{x \in X \mid d(x, A) < \epsilon\}$$

is d-open for every $\epsilon > 0$.

26. A subset A of a metric space (X, d) is called a G_δ-*set* if it is the intersection of some sequence of d-open sets, and A is called an F_σ-*set* if it is the union of some sequence of d-closed sets.
  (**a**) Prove that every d-closed set A is a G_δ-set. [*Hint:* If $A \ne \varnothing$, consider the sets $\{x \in X \mid d(x, A) < 1/n\}$ for $n = 1, 2, \ldots$.]
  (**b**) Deduce from (a) that every d-open set is an F_σ-set.
  (**c**) For the Euclidean metric d on R, show that Q is an F_σ-set that is not d-open.

27. Let A and B be subsets of a metric space (X, d).
  (**a**) If A and B are nonempty, show that the set

$$\{x \in X \mid d(x, A) < d(x, B)\}$$

is d-open.
  (**b**) Using (a), prove that if A and B are disjoint d-closed sets, then there are disjoint d-open sets U and V with $A \subset U$ and $B \subset V$. (This generalizes 1.24.)

28. Given two nonempty sets A and B in a metric space (X, d), the *distance from A*

to B is defined to be the number

$$d(A, B) = \inf \{d(x, y) \mid x \in A, y \in B\}.$$

(a) Construct disjoint nonempty *d*-closed subsets A and B of $\mathbb{R}^2$, where d is the Euclidean metric, such that $d(A, B) = 0$ yet $d(x, y) > 0$ for all $x \in A, y \in B$.

(b) Show that, in general,

$$d(A, B) = \inf\{d(x, B) \mid x \in A\}.$$

29. If d is a pseudometric on a set X (Exercise 13), then the notions of '*d*-ball of radius ϵ at x', '*d*-open set', '*d*-closed set', and '*d*-neighborhood of x' may be defined exactly as for a metric d.

(a) Verify that all results of this section concerning arbitrary metrics, except 1.25, hold for pseudometrics as well.

(b) Given a pseudometric d on a set X, let (X^*, d^*) be the metric space constructed in Exercise 14 and let $p: X \to X^*$ be the quotient map defined by $p(x) = x^*$ for all $x \in X$. Show that a subset V of X^* is d^*-open if and only if the subset $p^{-1}(V)$ of X is *d*-open. If U is a *d*-open subset of X, must $p(U)$ be d^*-open? Must U be *d*-open if $p(U)$ is d^*-open?

30. Let d be a metric on a set X that satisfies the *ultrametric inequality*

(UM) $\qquad\qquad\qquad d(x, z) \leq \max \{d(x, y), d(y, z)\}$

for all $x, y, z \in X$. For example, d could be a *p*-adic metric on $\mathbb{Q}$ (Exercise 15) or the metric of Example 1.14. Prove:

(a) Equality holds in (UM) whenever $d(x, y) \neq d(y, z)$.

(b) Any *d*-ball is *d*-closed as well as *d*-open.

(c) Given $x \in X$ and $\epsilon > 0$, $B_\epsilon(y; d) = B_\epsilon(x; d)$ for every $y \in B_\epsilon(x; d)$. (Thus every point of a *d*-ball is a "center" of that *d*-ball.)

3. Equivalent Metrics

A definition of continuity of a function from one metric space to another will be stated in the next section. It will turn out that continuity depends not on the specific metrics themselves, but only on the open sets they define. For this reason we now investigate the question of when two metrics on the same set define the same open sets.

1.31 Definition. Metrics d and d' on the same set X are said to be *(topologically) equivalent* when each *d*-open subset of X is d'-open and each d'-open subset of X is *d*-open.

Recall that a subset U of a metric space (X, d) was defined to be *d*-open if at each point of U there is some *d*-ball contained in U. Hence we need only look at *d*-balls and d'-balls to tell whether two metrics on the same set are equivalent.

1.32 Proposition. Let d and d' be metrics on the same set X. Then a necessary and sufficient condition for d to be equivalent to d' is that, given any point $x \in X$, each d-ball at x contains some d'-ball at x, and each d'-ball at x contains some d-ball at x.

Proof. Necessity. Assume d is equivalent to d'. Let $x \in X$. Consider an arbitrary d-ball $B_\epsilon(x; d)$ at x. Since $B_\epsilon(x; d)$ is d-open, by our assumption it must be d'-open as well. Hence at the point $x \in B_\epsilon(x; d)$ there is some d'-ball $B_\eta(x; d')$ with $B_\eta(x; d') \subset B_\epsilon(x; d)$. Similarly, for each d'-ball $B_\epsilon(x; d')$ there is some d-ball $B_\eta(x; d)$ with $B_\eta(x; d) \subset B_\epsilon(x; d')$.

Sufficiency. Assume the condition holds. We show that each d-open set is d'-open; the proof that each d'-open set is d-open is similar. Let U be a d-open subset of X. Let $x \in U$. Since U is d-open, there is some $\epsilon > 0$ with

$$B_\epsilon(x; d) \subset U.$$

By assumption there is some $\eta > 0$ with

$$B_\eta(x; d') \subset B_\epsilon(x; d).$$

Then $B_\eta(x; d') \subset U$. Since x was an arbitrary point of U, we conclude that U is d'-open. $\quad\square$

It is geometrically evident that each square region centered at a point of R^2 contains some circular region centered at the same point and vice versa. The analogous statement for cubical regions and spherical regions in R^3 is also evident. Hence in dimensions $n = 2$ and $n = 3$ the Euclidean metric d and the max metric d_∞ are equivalent.

1.33 Proposition. The Euclidean metric d and the max metric d_∞ on R^n are equivalent.

Proof. In view of 1.32, it suffices to show that

$$B_\epsilon(x; d) \subset B_\epsilon(x; d_\infty)$$

and

$$B_{\epsilon/\sqrt{n}}(x; d_\infty) \subset B_\epsilon(x; d)$$

for any $x \in \mathsf{R}^n$ and any $\epsilon > 0$. Hence it is enough to show that

$$d_\infty(x, y) \leq d(x, y) \leq \sqrt{n}\, d_\infty(x, y)$$

for all $x, y \in \mathsf{R}^n$. To establish these inequalities, it is enough to prove that

$$\max_{1 \leq i \leq n} a_i \leq \left[\sum_{i=1}^n a_i^2 \right]^{1/2} \leq \sqrt{n} \max_{1 \leq i \leq n} a_i$$

for every n-tuple $(a_1, \ldots, a_n)$ of nonnegative real numbers. Given such an

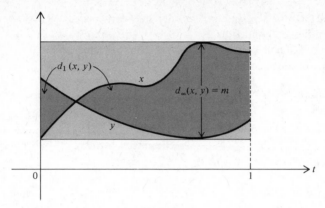

FIG. 1.13

n-tuple, let $1 \leq j \leq n$ with

$$a_j = \max_{1 \leq i \leq n} a_i.$$

Then

$$a_j = [a_j{}^2]^{1/2} \leq \left[\sum_{i=1}^{n} a_i{}^2 \right]^{1/2},$$

and since $a_i \leq a_j$ for each i,

$$\sum_{i=1}^{n} a_i{}^2 \leq n \cdot a_j{}^2. \quad \square$$

The taxicab metric is also equivalent to the Euclidean metric on R^n (Exercise 32). Of course, not every metric on R^n is equivalent to the Euclidean metric. For example, each singleton $\{x\}$ is δ-open for the discrete metric δ on R^n, but no singleton is d-open for the Euclidean metric d.

Here is a more substantial example of two inequivalent metrics on the same set.

1.34 Example. Let X be the set of all continuous functions $x \colon [0, 1] \to \mathsf{R}$. Consider the metrics d_∞ and d_1 of Examples 1.7 and 1.8 given by

$$d_\infty(x, y) = \max_{0 \leq t \leq 1} |x(t) - y(t)|, \qquad d_1(x, y) = \int_0^1 |x(t) - y(t)| \, dt.$$

Note that if $m = d_\infty(x, y)$, then (see Fig. 1.13)

$$d_1(x, y) \leq \int_0^1 m \, dt = m = d_\infty(x, y),$$

and hence

$$B_\epsilon(x; d_\infty) \subset B_\epsilon(x; d_1)$$

for any $x \in X$ and any positive ϵ. Nonetheless, d_1 is not equivalent to d_∞.

Define $x \colon [0, 1] \to \mathsf{R}$ to be the function that is constantly zero. We claim that

$$B_\eta(x; d_1) \not\subset B_1(x; d_\infty) \qquad\qquad (\eta > 0).$$

Let $\eta > 0$. Define $y \colon [0, 1] \to \mathsf{R}$ to be the function that takes the values 2 at $t = 0$ and 0 at $t = \eta/2$, is linear on $[0, \eta/2]$, and is constantly zero on $[\eta/2, 1]$ (see Fig. 1.14). Then $d_1(x, y)$ is just the area of a right triangle of height 2 and base $\eta/2$, so

$$d_1(x, y) = \frac{1}{2}\,(2)\left(\frac{\eta}{2}\right) = \frac{\eta}{2} < \eta.$$

Then $y \in B_\eta(x; d_1)$. However,

$$d_\infty(x, y) = 2 > 1,$$

so $y \notin B_1(x; d_\infty)$.

The next example of equivalent metrics will be a general one concerning metrics for which points cannot be arbitrarily far apart.

1.35 Definition. Let (X, d) be a metric space. A subset A of X is said to be *d-bounded* if there is some constant m for which $d(x, y) \leq m$ for all $x, y \in A$.

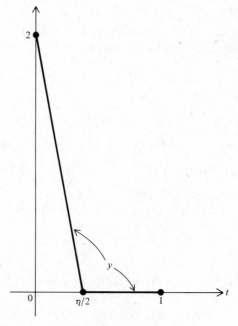

FIG. 1.14

If A is d-bounded and nonempty, then the real number

$$\text{diam } A = \sup \{d(x, y) \mid x \in A, y \in A\}$$

is called the d-*diameter of* A.

When X itself is d-bounded, then both the metric space (X, d) and the metric d on X are said to be *bounded*.

Any finite set in a metric space (X, d) is d-bounded. Any d-disk, and *a fortiori* any d-ball, in a metric space (X, d) is d-bounded, for $y, z \in D_\epsilon(x; d)$ implies

$$d(y, z) \leq d(y, x) + d(x, z) \leq \epsilon + \epsilon = 2\epsilon.$$

A subset of R is d-bounded for the Euclidean metric d if and only if it has both an upper and a lower bound in R.

The discrete metric on any set is bounded. However, the Euclidean metric d on R^n is not bounded, since for each $c > 0$ the d-distance from $(c, 0, \dots, 0)$ to $(0, 0, \dots, 0)$ is c.

We show now that from any metric d we can construct a bounded metric equivalent to d. The idea is simply to redefine to be 1 any distance that exceeds 1.

1.36 Proposition. Let (X, d) be any metric space. Then there is a bounded metric on X which is equivalent to d and for which diam $X \leq 1$. One such metric d^* is given by

$$d^*(x, y) = \min \{1, d(x, y)\} \qquad\qquad (x, y \in X).$$

Proof. Because the other properties of a metric are evidently satisfied by d^*, we verify only the triangle inequality

$$d^*(x, z) \leq d^*(x, y) + d^*(y, z).$$

If $d(x, y) \leq 1$ and $d(y, z) \leq 1$, then

$$d^*(x, z) \leq d(x, z) \leq d(x, y) + d(y, z) = d^*(x, y) + d^*(y, z).$$

Now suppose $d(x, y) > 1$ or $d(y, z) > 1$. If $d(x, y) > 1$, then

$$d^*(x, z) \leq 1 \leq 1 + d^*(y, z) = d^*(x, y) + d^*(y, z).$$

The case $d(y, z) > 1$ is treated similarly.

Since $d^*(x, y) \leq 1$ for all $x, y \in X$, d^* is bounded and diam $X \leq 1$ for d^*. Since $d^*(x, y) = d(x, y)$ whenever $d(x, y) < 1$,

$$B_\epsilon(x; d^*) = B_\epsilon(x; d)$$

for all $x \in X$ and all $0 < \epsilon \leq 1$. Hence d^* is equivalent to d. $\square$

Proposition 1.36 provides, in particular, a bounded metric equivalent to the Euclidean metric d on the real line R. In the example that follows an entirely different metric equivalent to d will be constructed. This same example will be important later in showing that infinite limits and limits at infinity, familiar from calculus, are just special cases of a more general concept of limit.

1.37 Example. The *extended real line* is the set

$$R^* = R \cup \{-\infty, +\infty\}$$

consisting of all real numbers together with two distinct objects $-\infty$ and $+\infty$ not belonging to R. (What the particular objects $-\infty$ and $+\infty$ really are does not matter, only that they are different from one another and are not real numbers.)

We extend the usual total ordering of R to a total ordering of R* by setting

$$-\infty < +\infty$$

$$-\infty < x < +\infty \qquad\qquad (x \in R).$$

Thus we have adjoined a largest element $+\infty$ and a smallest element $-\infty$ to R. Then

$$R^* = [-\infty, +\infty], \qquad R =]-\infty, +\infty[.$$

Since the intervals $]\leftarrow, a[$ and $]a, \rightarrow[$ in R are the intervals $]-\infty, a[$ and $]a, +\infty[$ in R* for each $a \in R$, each open ray in R becomes an interval in R* having both a left and a right endpoint, and similarly for closed rays.

We are going to make R* into a metric space that "looks like" $[-1, 1]$ with its Euclidean metric. First we construct a certain one-to-one correspondence between R* and $[-1, 1]$. Since

$$\left|\frac{x}{1 + |x|}\right| = \frac{|x|}{1 + |x|} < 1 \qquad\qquad (x \in R)$$

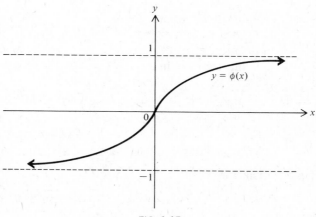

FIG. 1.15

we may define a map

$$\varphi: \mathsf{R} \to\ ]-1, 1[$$

by setting

$$\varphi(x) = \frac{x}{1 + |x|} \qquad\qquad (x \in \mathsf{R}).$$

The graph of φ, shown in Fig. 1.15, is easily sketched using the usual methods of elementary calculus. Direct computation shows that φ has the inverse

$$\psi:\ ]-1, 1[\to \mathsf{R}$$

given by

$$\psi(y) = \frac{y}{1 - |y|} \qquad\qquad (y \in\ ]-1, 1[).$$

Hence $\varphi: \mathsf{R} \to\ ]-1, 1[$ is a bijection.

Some manipulation of inequalities (or a calculation showing the derivatives of φ and ψ are positive) shows that both φ and its inverse ψ are strictly increasing functions. Then

$$\varphi(]a, b[) =]\varphi(a), \varphi(b)[$$

whenever $a < b$. In fact, $a < x < b$ implies $\varphi(a) < \varphi(x) < \varphi(b)$ since φ is strictly increasing. Conversely, $\varphi(a) < y < \varphi(b)$ implies $y = \varphi(x)$ for some $x \in \mathsf{R}$,

$$\psi(\varphi(a)) < \psi(\varphi(x)) < \psi(\varphi(b))$$

since ψ is strictly increasing, and hence $a < x < b$ since $\psi \circ \varphi$ is the identity map on R. In the same way one proves

$$\varphi(]a, +\infty[) =]\varphi(a), 1[, \qquad \varphi(]-\infty, a[) =]-1, \varphi(a)[$$

for each $a \in \mathsf{R}$.

The graph of φ has the lines $y = 1$ and $y = -1$ as horizontal asymptotes, for

$$\lim_{x \to +\infty} \varphi(x) = \lim_{x \to +\infty} \frac{x}{1 + x} = 1,$$

$$\lim_{x \to -\infty} \varphi(x) = \lim_{x \to -\infty} \frac{x}{1 - x} = -1.$$

Hence it is natural to extend $\varphi: \mathsf{R} \to\ ]-1, 1[$ to a map

$$\varphi^*: \mathsf{R}^* \to [-1, 1]$$

by setting

$$\varphi^*(-\infty) = -1, \qquad \varphi^*(+\infty) = 1.$$

Clearly φ^* is a bijection, and

$$\varphi^*(]a, +\infty]) =]\varphi(a), 1], \qquad \varphi^*([-\infty, a[) = [-1, \varphi(a)[$$

for each $a \in \mathsf{R}$.

We now define a metric D on the extended real line $\mathbf{R}^*$ by letting the D-distance between two points of $\mathbf{R}^*$ be the same as the Euclidean distance between the points in $[-1, 1]$ to which they correspond under φ. In other words, we let

$$D(x, y) = |\varphi^*(x) - \varphi^*(y)| \qquad (x, y \in \mathbf{R}^*).$$

The proof that D is actually a metric on $\mathbf{R}^*$ uses no property of φ^* except that it is injective. First, $D(x, y) \geq 0$ for all $x, y \in \mathbf{R}^*$. Since φ^* is an injective function,

$$D(x, y) = 0 \iff \varphi^*(x) = \varphi^*(y) \iff x = y.$$

The symmetry of D is obvious. Finally,

$$\begin{aligned}
D(x, z) &= |\varphi^*(x) - \varphi^*(z)| \\
&= |[\varphi^*(x) - \varphi^*(y)] + [\varphi^*(y) - \varphi^*(z)]| \\
&\leq |\varphi^*(x) - \varphi^*(y)| + |\varphi^*(y) - \varphi^*(z)| \\
&= D(x, y) + D(y, z),
\end{aligned}$$

which establishes the triangle inequality.

The metric space $(\mathbf{R}^*, D)$ is bounded, because

$$D(x, y) = |\varphi^*(x) - \varphi^*(y)| \leq |\varphi^*(x)| + |\varphi^*(y)| \leq 1 + 1 = 2$$

for all $x, y \in \mathbf{R}^*$. Since $D(-\infty, +\infty) = 2$,

$$\operatorname{diam} \mathbf{R}^* = 2.$$

1.38 Proposition. The metric d^* induced on $\mathbf{R}$ by D is a bounded metric equivalent to the Euclidean metric d on $\mathbf{R}$.

Proof. Let $x \in \mathbf{R}$. First we show that each d^*-ball at x of sufficiently small radius contains some d-ball at x. Since $B_\epsilon(x; d^*) = B_\epsilon(x; D) \cap \mathbf{R}$, we consider only d^*-balls at x of radius ϵ for which $-\infty, +\infty \notin B_\epsilon(x; D)$, that is, for which

$$B_\epsilon(x; d^*) = B_\epsilon(x; D).$$

Let

$$0 < \epsilon < \min\{\varphi(x) - (-1), 1 - \varphi(x)\}.$$

Then

$$-1 < \varphi(x) - \epsilon < \varphi(x) + \epsilon < 1,$$

and

$$B_\epsilon(x; D) =]\psi(\varphi(x) - \epsilon), \quad \psi(\varphi(x) + \epsilon)[$$

is an open interval in $\mathbf{R}$ which certainly contains a d-ball at x.

Next we show that each d-ball contains some d^*-ball at x. More generally, consider a d-open set containing x of the form $]a, b[$ with

$$-\infty < a < x < b < +\infty.$$

Choose
$$0 < \epsilon < \min \{\varphi(x) - \varphi(a), \varphi(b) - \varphi(x)\}.$$
Then
$$-1 < \varphi(a) < \varphi(x) - \epsilon < \varphi(x) + \epsilon < \varphi(b) < 1,$$
so
$$B_\epsilon(x; d^*) = B_\epsilon(x; D) =]\psi(\varphi(x) - \epsilon), \psi(\varphi(x) + \epsilon)[\subset]a, b[. \qquad \square$$

For later use we also establish a relationship between intervals in R^* and D-balls at the points $+\infty$ and $-\infty$.

1.39 Lemma. If $0 < \epsilon < 1$, then
$$B_\epsilon(+\infty; D) = \left]\frac{1}{\epsilon} - 1, +\infty\right],$$

and if $u > 0$, then
$$]u, +\infty] = B_{1/(1+u)}(+\infty; D).$$
If $0 < \epsilon < 1$, then
$$B_\epsilon(-\infty; D) = \left[-\infty, 1 - \frac{1}{\epsilon}\right[,$$

and if $u < 0$, then
$$[-\infty, u[= B_{1/(1-u)}(-\infty; D).$$

Proof. We prove only the assertions concerning $+\infty$, leaving proof of the others to the reader. Let $0 < \epsilon < 1$. Set
$$B = B_\epsilon(+\infty; D), \qquad J = \left]\frac{1}{\epsilon} - 1, +\infty\right].$$
We must show
$$(*) \qquad\qquad\qquad x \in B \iff x \in J.$$
Since $+\infty$ belongs to both B and J and the point $-\infty$ belongs to neither, we need only consider $x \in \mathsf{R}$. Now each $x \in J$ is positive since $(1/\epsilon) - 1 > 0$, and each $x \in B$ is positive since $\epsilon < 1$. Hence we need only prove $(*)$ for $x > 0$. But $x > 0$ implies
$$D(x, +\infty) = \left|\frac{x}{1+x} - 1\right| = \frac{1}{1+x},$$
and $(*)$ follows at once.
    Let $u > 0$. Set
$$\epsilon = \frac{1}{1+u}.$$
Then $0 < \epsilon < 1$, and by what we just proved,
$$]u, +\infty] = B_\epsilon(+\infty; D). \qquad \square$$

The metric D on R^* was so constructed that the D-distance between any two points x_1, x_2 of R^* is the same as the d-distance between the points $\varphi^*(x_1)$, $\varphi^*(x_2)$ of $[-1, 1]$ corresponding to them under the bijection $\varphi^*\colon \mathsf{R}^* \to [-1, 1]$. This relationship between (R^*, D) and $([-1, 1], d)$ suggests the following definition.

1.40 Definition. Let (X, d) and (Y, d') be metric spaces. A bijection $f\colon X \to Y$ is called an *isometry from* (X, d) *to* (Y, d') if

$$d(x_1, x_2) = d'(f(x_1), f(x_2)) \qquad\qquad (x_1, x_2 \in X).$$

The metric space (X, d) is said to be *isometric to* the metric space (Y, d') when there exists some isometry from (X, d) to (Y, d').

Thus an isometry from a metric space (X, d) to a metric space (Y, d') is a *distance-preserving* one-to-one correspondence between their points. Roughly speaking, one regards (X, d) and (Y, d') indistinguishable as metric spaces when (X, d) is isometric to (Y, d').

1.41 Examples

(1) Let d be the Euclidean metric on R^n. For each fixed $c \in \mathsf{R}^n$, the "translation"

$$T\colon \mathsf{R}^n \to \mathsf{R}^n$$

$$x \mapsto x + c$$

by c is an isometry from (R^n, d) to itself, because

$$d(T(x), T(y)) = d(x + c, y + c)$$
$$= \|(x + c) - (y + c)\|$$
$$= \|x - y\|$$
$$= d(x, y)$$

for all $x, y \in \mathsf{R}^n$. For each fixed $1 \le i \le n$, the "reflection"

$$\mathsf{R}^n \to \mathsf{R}^n$$

$$(x_1, \ldots, x_{i-1}, x_i, x_{i+1}, \ldots, x_n) \mapsto (x_1, \ldots, x_{i-1}, -x_i, x_{i+1}, \ldots, x_n)$$

in the ith coordinate is an isometry from (R^n, d) to (R^n, d).

There are many isometries from (R^n, d) to itself besides translations and reflections. With the aid of some linear algebra it is possible to determine all isometries from (R^n, d) to itself (see, for example, Jacob and Bailey [19, Theorems 8.1.13, 8.2.2] for dimensions $n = 2, 3$ and Fleming [11, pages 96–98] for arbitrary dimension n; see also Exercise 43).

(2) For $a, b \in \mathsf{R}^n$ with $a \ne b$, let d' be the Euclidean metric on the line segment

$$Y = \{(1 - t)a + tb \mid 0 \le t \le 1\}$$

joining a to b. Let d be the Euclidean metric on the closed interval

$$X = [0, c]$$

in R, where $c > 0$. If $c = d'(a, b)$, then the map

$$X \to Y$$

$$x \mapsto \left(1 - \frac{x}{c}\right) a + \frac{x}{c} b$$

is an isometry from (X, d) to (Y, d'). However, (X, d) is not isometric to (Y, d') if $c \neq d'(a, b)$, since in that case diam $X = c$ and diam $Y = d'(a, b)$.

(3) Let d be the Euclidean metric on R^n and let d' be the Euclidean metric on the subset

$$Y = \{ (x_1, \ldots, x_n, x_{n+1}) \in \mathsf{R}^{n+1} \mid x_{n+1} = 0 \}$$

of R^{n+1}. Then

$$\mathsf{R}^n \to Y$$

$$(x_1, \ldots, x_n) \mapsto (x_1, \ldots, x_n, 0)$$

is an isometry from (R^n, d) to (Y, d').

(4) For $n \neq m$, (R^n, d) is not isometric to (R^m, d'), where d and d' are the Euclidean metrics. The proof of this for $m = 1$ is easy: the d-sphere of radius 1 at the origin 0 of R^n contains infinitely many points, but the d'-sphere of radius 1 at any point of R^1 contains exactly two points. The proof for arbitrary m, although not difficult, uses some ideas from linear algebra and hence is omitted [the argument is a simple modification of the one used to determine all isometries from (R^n, d) to itself—see the reference in Example (1)].

(5) If (X, d) is any metric space, then the identity map $i \colon X \to X$ is an isometry from (X, d) to itself.

(6) Let f be an isometry from a metric space (X, d) to a metric space (Y, d'). Then $f^{-1} \colon Y \to X$ is an isometry from (Y, d') to (X, d). In fact, if $y_1, y_2 \in Y$, let $x_1 = f^{-1}(y_1)$, $x_2 = f^{-1}(y_2)$; then

$$d'(y_1, y_2) = d'(f(x_1), f(x_2)) = d(x_1, x_2) = d(f^{-1}(y_1), f^{-1}(y_2)).$$

(7) Let $f \colon X \to Y$ be an isometry from (X, d) to (Y, d'), and let $g \colon Y \to Z$ be an isometry from (Y, d') to (Z, d''). Then $g \circ f \colon X \to Z$ is an isometry from (X, d) to (Z, d'').

Examples (5)–(7) together say that the relation 'is isometric to' is an equivalence relation on the class of all metric spaces. Accordingly, metric spaces that are isometric to one another are often said to be *metrically equivalent*.

There is another sense in which metric spaces can be called equivalent to one another.

1.42 Definition. A metric space (X, d) is said to be *topologically equivalent to* a metric space (Y, d') if there is a bijection $f\colon X \to Y$ such that $f(U)$ is a d'-open subset of Y for each d-open subset U of X and $f^{-1}(V)$ is a d-open subset of X for each d'-open subset V of Y.

Thus (X, d) is topologically equivalent to (Y, d') if there is a one-to-one correspondence between X and Y under which the d-open subsets of X correspond to the d'-open subsets of Y.

1.43 Examples

(1) Metrically equivalent metric spaces are topologically equivalent. In fact, let $f\colon X \to Y$ be an isometry from (X, d) to (Y, d'). Then

$$f(B_\epsilon(x; d)) = B_\epsilon(f(x); d')$$

for all $x \in X$ and all $\epsilon > 0$, and

$$f^{-1}(B_\eta(y; d')) = B_\eta(f^{-1}(x); d)$$

for all $y \in Y$ and all $\eta > 0$.

Now let U be a d-open subset of X. We show that $f(U)$ is d'-open. Let $y \in f(U)$. Then $x = f^{-1}(y) \in U$, so

$$B_\epsilon(x; d) \subset U$$

for some $\epsilon > 0$. Hence

$$B_\epsilon(y; d') = f(B_\epsilon(x; d)) \subset f(U).$$

A similar argument shows that $f^{-1}(V)$ is d-open for each d'-open subset V of X.

(2) Consider two closed intervals $[0, c]$ and $[a, b]$ in R with their Euclidean metrics d and d', and suppose $c \neq a - b$. By 1.41 (2), the metric space $([0, c], d)$ is not metrically equivalent to the metric space $([a, b], d')$. However, $([0, c], d)$ is topologically equivalent to $([a, b], d')$. To see this, observe that the bijection

$$f\colon [0, c] \to [a, b]$$

$$x \mapsto a + \frac{x}{c}(b - a)$$

has the property

$$d'(f(x_1), f(x_2)) = \frac{b - a}{c} d(x_1, x_2),$$

and its inverse

$$f^{-1}\colon [a, b] \to [0, c]$$

$$y \mapsto \frac{c}{b - a}(y - a)$$

has the property

$$d(f^{-1}(y_1), f^{-1}(y_2)) = \frac{c}{b-a} d'(y_1, y_2).$$

(3) Let d and d' be metrics on the same set X that are equivalent in the sense of Definition 1.31. Then (X, d) is topologically equivalent to (X, d'); the identity map of X will do for the needed bijection $f \colon X \to X$.

We shall return to the topic of topological equivalence in a more general setting in Chapter 3.

EXERCISES

31. Given a metric d on a set X, show that αd is a metric on X which is equivalent to d for each real number $\alpha > 0$.

32. Prove that the taxicab metric d_1 is equivalent to the Euclidean metric d on R^n. (*Hint*: Use Exercise 3.)

33. Given a metric d on a set X, show that the formula

$$d'(x, y) = \frac{d(x, y)}{1 + d(x, y)}$$

defines a metric on X that is bounded and equivalent to d.

34. (**a**) Let (X_1, d_1), (X_2, d_2), ... be a sequence of bounded metric spaces, each of diameter at most 1. Show that

$$d'(x, y) = \sum_{i=1}^{\infty} \frac{d_i(x_i, y_i)}{2^i}$$

defines a bounded metric on the product set

$$X = \underset{i=1}{\overset{\infty}{\times}} X_i.$$

(**b**) If $(X_i, d_i) = (Y, \delta)$ for every i, where Y is some set and δ is its discrete metric, show that the metric d' in (a) is equivalent to the metric d of Example 1.14.

(**c**) Take $X_i = [-1/i, 1/i]$ and $d_i =$ the Euclidean metric on X_i for each i. Then X is the Hilbert cube (Exercise 23). Is d' equivalent to the metric induced on X by the metric d_2 on the Hilbert sequence space (Example 1.9)?

35. Let d and d' be metrics on a set X.

(**a**) Prove that a necessary condition for d and d' to be equivalent is that each d-closed set be d'-closed and each d'-closed set be d-closed. Is this condition also sufficient?

(**b**) Formulate a criterion for d and d' to be equivalent concerning d-neighborhoods and d'-neighborhoods of points of X.

36. (**a**) For the Euclidean metric d on R^n, show that

$$\operatorname{diam} B_\epsilon(x; d) = \operatorname{diam} D_\epsilon(x; d) = \operatorname{diam} S_\epsilon(x; d) = 2\epsilon$$

for each $x \in \mathbb{R}^n$ and each $\epsilon > 0$.

(**b**) Does (a) generalize to arbitrary metric spaces?

37. Let $\mathfrak{F}$ be the collection of all nonempty d-closed subsets of a bounded metric space (X, d). For $A, B \in \mathfrak{F}$, let

$$d^*(A, B) = \max \left\{ \sup_{x \in A} d(x, B), \sup_{y \in B} d(y, A) \right\}.$$

(**a**) Show that d^* is a metric on $\mathfrak{F}$; d^* is called the *Hausdorff metric* induced by d.

(**b**) Would d^* still be a metric if $\mathfrak{F}$ were the collection of all nonempty subsets of X?

(**c**) Compare $d^*(A, B)$ with $d(A, B)$ [Exercise 28] and with diam $(A \cup B)$.

(**d**) Take d to be the Euclidean metric on $[0, 1]$. Compute $d^*(A, B)$ if $A = [u, v]$ and $B = [t, s]$ are closed intervals in $[0, 1]$.

38. An *isometric embedding* of a metric space (X, d) into a metric space (Y, D) is an isometry from (X, d) to (Y', D'), where $Y' \subset Y$ and D' is the metric on Y' induced by D. Construct an isometric embedding of $(\mathbb{R}^n, d)$, where d is the Euclidean metric, into the Hilbert sequence space (H, d_2) of Example 1.9.

39. Let $X = \{x, y, z, w\}$ be a four point set.

(**a**) Show that there is a unique metric d on X for which

$$d(x, y) = d(y, z) = d(z, x) = 2,$$
$$d(x, w) = d(y, w) = d(z, w) = 1.$$

(**b**) Show that there is no isometric embedding of (X, d) into the Hilbert sequence space (H, d_2) of Example 1.9.

40. Must two metric spaces be isometric to one another if there are isometric embeddings of each into the other?

41. If d is the Euclidean metric and d_∞ is the max metric on $\mathbb{R}^2$, is $(\mathbb{R}^2, d)$ isometric to $(\mathbb{R}^2, d_\infty)$?

42. The *Euclidean inner product* on $\mathbb{R}^n$ is the map

$$\mathbb{R}^n \times \mathbb{R}^n \to \mathbb{R}$$
$$(x, y) \mapsto \langle x, y \rangle$$

defined by

$$\langle x, y \rangle = \sum_{i=1}^{n} x_i y_i.$$

Then the Euclidean inner product is related to the Euclidean norm (see the discussion preceding 1.2) by

$$\langle x, x \rangle = ||x||^2$$

so that $\langle x, x \rangle \geq 0$ for all $x \in \mathbb{R}^n$, and $\langle x, x \rangle = 0$ if and only if $x = 0$. The Cauchy-Schwarz inequality (1.3) says

$$|\langle x, y \rangle| \leq ||x|| \cdot ||y||.$$

Verify the following additional properties of the Euclidean inner product:

(**a**) $\langle x, y \rangle = \langle y, x \rangle$ for all $x, y \in \mathbb{R}^n$.

(**b**) $\langle x + y, z \rangle = \langle x, z \rangle + \langle y, z \rangle$ for all $x, y, z \in \mathbb{R}^n$.

(**c**) $\langle \lambda x, y \rangle = \lambda \langle x, y \rangle$ for all $\lambda \in \mathsf{R}$ and all $x, y \in \mathsf{R}^n$.

(**d**) $\langle x, y \rangle = \frac{1}{2} \{ ||x||^2 - ||x - y||^2 + ||y||^2 \}$ for all $x, y \in \mathsf{R}^n$.

43. (Continuation of Exercise 42.) A map $f : \mathsf{R}^n \to \mathsf{R}^n$ is said to "preserve" inner products if

$$\langle f(x), f(y) \rangle = \langle x, y \rangle \qquad\qquad (x, y \in \mathsf{R}^n).$$

(**a**) Show that each rotation of R^2 (Exercise 4) preserves inner products.

(**b**) If d is the Euclidean metric on R^n, prove that every isometry f from (R^n, d) to (R^n, d) with $f(0) = 0$ preserves inner products.

(**c**) Let f be an isometry from the Euclidean plane to itself. Show that the image under f of each line is a line. [*Hint:* If $f(0) = z$, then the translate g of f given by $g(x) = f(x) - z$ is also an isometry, but with $g(0) = 0$. The equation of a line can be put in the form $\langle (a, b), (x_1, x_2) \rangle = c$.]

44. Let B and B' be two balls in R^n (for the Euclidean metric on R^n), and let d and d' be their Euclidean metrics. Prove that (B, d) is topologically equivalent to (B', d'). When will (B, d) be metrically equivalent to (B', d')?

45. Let d be the Euclidean metric on R. Construct a metric d' on R such that (R, d) is topologically equivalent to (R, d') but d is *not* equivalent to d'. (*Moral:* The word 'equivalent' has several meanings.)

4. Continuity and Convergence

At long last we are ready to define continuity of a function from one metric space to another and to show that continuity depends not on the particular metrics involved but only on the open sets they determine. We also introduce the notion of sequential convergence and relate it to continuity.

The definition that follows is merely a generalization of the one suggested in Section 1 for continuity of a function from R^n to R.

1.44 Definition. Let (X, d) and (Y, d') be metric spaces and let $f : X \to Y$. Then f is said to be (d, d')-*continuous at* $x \in X$ when

for each $\epsilon > 0$ there exists some $\delta > 0$ such that for all $u \in X$,

$$d(x, u) < \delta \implies d'(f(x), f(u)) < \epsilon.$$

The map f is said to be (d, d')-*continuous* when it is (d, d')-continuous at every $x \in X$.

Loosely speaking, f is (d, d')-continuous at x if $f(u)$ can be made as d'-close to $f(x)$ as we wish by taking u sufficiently d-close to x.

1.45 Examples

(1) Let d be the Euclidean metric on R. Then a function $f : \mathsf{R} \to \mathsf{R}$ is (d, d)-continuous at a point $x \in \mathsf{R}$ precisely when for each $\epsilon > 0$ there exists $\delta > 0$ such that $|f(x) - f(u)| < \epsilon$ whenever $u \in \mathsf{R}$ with $|x - u| < \delta$.

Hence Definition 1.44 generalizes the familiar calculus definition of continuity of a real-valued function of a real variable.

We shall take as an established fact the continuity of the elementary functions encountered in calculus.

(2) For this example it is best we write out explicitly what it means for $f\colon X \to Y$ to be (d, d')-*discontinuous* at $x \in X$, that is, not (d, d')-continuous at x:

> There is some $\epsilon > 0$ such that for each $\delta > 0$, there is at least one $u \in X$ for which $d(x, u) < \delta$ but $d'(f(x), f(u)) \geq \epsilon$.

Now let d be the Euclidean metric on R, and define $f\colon \mathsf{R} \to \mathsf{R}$ by

$$f(u) = \begin{cases} u & \text{if } u \neq 0, \\ 1 & \text{if } u = 0. \end{cases}$$

Then f is (d, d)-discontinuous at 0. In fact, let $\epsilon = 1$. Take any $\delta > 0$. Then the point $u = -\delta/2$ is one for which $d(0, u) < \delta$ but $d(f(0), f(u)) = 1 + \delta/2 \geq \epsilon$.

(3) Let (X, d) and (Y, d') be any metric spaces and let

$$f\colon X \to Y$$
$$x \mapsto c$$

be a constant map. Then f is (d, d')-continuous since

$$d'(f(x), f(u)) = d'(c, c) = 0$$

for all $x, u \in X$.

(4) Let d' be the metric induced on a subset Y of a metric space (X, d). Then the inclusion map

$$j\colon Y \to X$$
$$x \mapsto x$$

is (d', d)-continuous, for

$$d'(x, u) < \epsilon \implies d(j(x), j(u)) = d(x, u) = d'(x, u) < \epsilon$$

for all $x, u \in Y$.

Specialize the above by taking $Y = X$, so that $d' = d$ and j is the identity map

$$i\colon X \to X$$
$$x \mapsto x$$

of X. Then i is (d, d)-continuous.

(5) Let (X, d), (Y, d'), and (Z, d'') be metric spaces, let

$$f\colon X \to Y, \qquad g\colon Y \to Z$$

be given maps, and form the composite map

$$g \circ f\colon X \to Z$$
$$x \mapsto g(f(x)).$$

Suppose f is (d, d')-continuous at a point $x \in X$ and g is (d', d'')-continuous at the point $f(x) \in Y$. Then the composite $g \circ f$ is (d, d'')-continuous at x. In fact, let $\epsilon > 0$. There exists $\eta > 0$ such that

$$d'(f(x), y) < \eta \;\; \Rightarrow \;\; d''(g(f(x)), g(y)) < \epsilon$$

for all $y \in Y$. Then there exists $\delta > 0$ such that

$$d(x, u) < \delta \;\; \Rightarrow \;\; d'(f(x), f(u)) < \eta.$$

Then

$$d(x, u) < \delta \;\; \Rightarrow \;\; d''(g \circ f(x), g \circ f(u)) < \epsilon.$$

By what we just proved, $g \circ f$ will be (d, d'')-continuous whenever f is (d, d')-continuous and g is (d', d'')-continuous.

(6) Let δ be the discrete metric on a set Z, and let (X, d) be any metric space. Then *every* map $f: Z \to X$ is (δ, d)-continuous. In fact, since $\delta(x, u) < 1$ precisely when $x = u$,

$$\delta(x, u) < 1 \;\; \Rightarrow \;\; d(f(x), f(u)) = 0 < \epsilon$$

for all $\epsilon > 0$ and all $x, u \in Z$.

(7) Let (X, d) be any metric space. The metric d is a map

$$d: X \times X \to \mathsf{R}$$

about whose continuity it makes sense to inquire. Of course, we provide R with its Euclidean metric d' and $X \times X$ with its max metric d_∞ given by

$$d_\infty((x_1, x_2), (u_1, u_2)) = \max \{d(x_1, u_1), d(x_2, u_2)\}$$

(see Example 1.13). We claim that the map d is (d_∞, d')-continuous.

Let $(x_1, x_2) \in X \times X$ and let $\epsilon > 0$ be given. We seek a $\delta > 0$ for which

$$d_\infty((x_1, x_2), (u_1, u_2)) < \delta \;\; \Rightarrow \;\; |d(x_1, x_2) - d(u_1, u_2)| < \epsilon.$$

Now

$$|d(x_1, x_2) - d(u_1, u_2)| \leq |d(x_1, x_2) - d(u_1, x_2)| + |d(u_1, x_2) - d(u_1, u_2)|$$

$$\leq d(x_1, u_1) + d(x_2, u_2)$$

$$\leq 2d_\infty((x_1, x_2), (u_1, u_2)).$$

Hence $\delta = \epsilon/2$ will do.

The implication

$$d(x, u) < \delta \;\; \Rightarrow \;\; d'(f(x), f(u)) < \epsilon$$

which appears in Definition 1.44 may be restated in the form

$$u \in B_\delta(x; d) \;\; \Rightarrow \;\; f(u) \in B_\epsilon(f(x); d')$$

or even more concisely in the form

$$f(B_\delta(x; d)) \subset B_\epsilon(f(x); d').$$

Hence $f: X \to Y$ is (d, d')-*continuous at* $x \in X$ *if and only if each* d'-*ball at* $f(x)$ *contains the image of some* d-*ball at* x.

The observation just made allows us to characterize continuity at a point in terms of arbitrary neighborhoods.

1.46 Theorem. Let (X, d) and (Y, d') be metric spaces, let $f: X \to Y$, and let $x \in X$. Then the following statements are equivalent:

(1) f is (d, d')-continuous at x.

(2) For each d'-neighborhood N of $f(x)$ in Y there exists some d-neighborhood M of x in X such that $f(M) \subset N$.

(3) The inverse image $f^{-1}(N)$ of each d-neighborhood N of $f(x)$ in Y is a d-neighborhood of x in X.

Proof. [We give a "circular proof" which consists in proving each of the implications (1) $\Rightarrow$ (2), (2) $\Rightarrow$ (3), (3) $\Rightarrow$ (1). It will then logically follow that all the equivalences (1) $\Leftrightarrow$ (2), (2) $\Leftrightarrow$ (3), (1) $\Leftrightarrow$ (3) are valid.]

(1) $\Rightarrow$ (2). Assume (1). Let N be a d'-neighborhood of $f(x)$. Choose $\epsilon > 0$ with

$$B_\epsilon(f(x); d') \subset N$$

(see Fig. 1.16). By (1) there exists $\delta > 0$ with

$$f(B_\delta(x; d)) \subset B_\epsilon(f(x); d').$$

Then $M = B_\delta(x; d)$ is a d-neighborhood of x for which $f(M) \subset N$.

(2) $\Rightarrow$ (3). Assume (2). Let N be a d'-neighborhood of $f(x)$. By (2) there exists a d-neighborhood M of x with

$$f(M) \subset N.$$

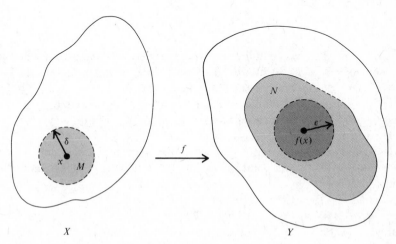

FIG. 1.16

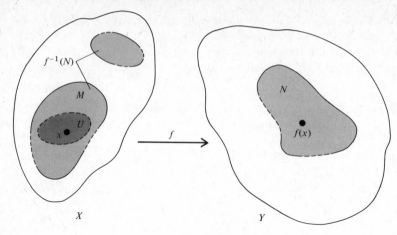

FIG. 1.17

By the definition of 'd-neighborhood at x', there is a d-open set U with

$$x \in U \subset M$$

(see Fig. 1.17). Since $M \subset f^{-1}(N)$,

$$x \in U \subset f^{-1}(N).$$

Hence $f^{-1}(N)$ is a d-neighborhood of x.

(3) $\Rightarrow$ (1). Assume (3). Let $\epsilon > 0$. Since $B_\epsilon(f(x); d')$ is a d'-neighborhood of $f(x)$, by (3) the set

$$M = f^{-1}(B_\epsilon(f(x); d'))$$

is a d-neighborhood of x. Then

$$B_\delta(x; d) \subset M$$

for some $\delta > 0$. Hence

$$f(B_\delta(x; d)) \subset B_\epsilon(f(x); d')). \qquad \square$$

Statement (3) above definitely does *not* say that the image of each d-neighborhood of x is a d'-neighborhood of $f(x)$. For example, let (X, d) be any metric space, let d' be the Euclidean metric on R, and let $f: X \to \mathsf{R}$ be the constant function with value 0. Then f is (d, d')-continuous at every $x \in \mathsf{R}$. However, $f(M) = \{0\}$ for each nonempty $M \subset X$, but $\{0\}$ is not a d'-neighborhood of 0. See also Exercise 49.

Because equivalent metrics determine the same open sets and hence the same neighborhoods of a point, we deduce the following corollary immediately from 1.46.

1.47 Corollary. In the notation of Theorem 1.46, let D be a metric on X equivalent to d and let D' be a metric on Y equivalent to d'. Then the map f is (d, d')-continuous at x if and only if it is (D, D')-continuous at x.

From Theorem 1.46 we can also deduce an especially simple criterion for continuity, a criterion involving only open sets.

1.48 Theorem. Let (X, d) and (Y, d') be metric spaces and let $f: X \to Y$. Then a necessary and sufficient condition for f to be (d, d')-continuous is that the inverse image $f^{-1}(V)$ of each d'-open subset V of Y be a d-open subset of X.

Proof. Necessity. Assume f is (d, d')-continuous. Let $V \subset Y$ be d'-open. To show that $f^{-1}(V)$ is d-open, it suffices by 1.29 to show that $f^{-1}(V)$ is a d-neighborhood of each of its points. Let $x \in f^{-1}(V)$. Then $f(x) \in V$, V is a d'-neighborhood of $f(x)$, and hence by 1.46 (3) the set $f^{-1}(V)$ is a d-neighborhood of x.

Sufficiency. Assume the condition holds. Let $x \in X$. To show that f is (d, d')-continuous at x we verify 1.46(2). Let N be a d'-neighborhood of $f(x)$. Choose a d'-open set V with

$$f(x) \in V \subset N.$$

By assumption, $M = f^{-1}(V)$ is d-open. Hence M is a d-neighborhood of x with

$$f(M) = V \subset N. \qquad \square$$

Suppose the map $f: X \to Y$ in Theorem 1.48 is a bijection, and let us apply 1.48 to the map $f^{-1}: Y \to X$. Then f^{-1} is (d', d)-continuous precisely when for each d-open subset U of X the subset

$$f(U) = (f^{-1})^{-1}(U)$$

is d'-open. Hence a metric space (X, d) is topologically equivalent to a metric space (Y, d') if and only if there is some bijection $f: X \to Y$ such that f is (d, d')-continuous and f^{-1} is (d', d)-continuous.

The study of continuity will be resumed in Chapter 3 in the more general setting of topological spaces. We turn now to the topic of sequential convergence.

Recall that a sequence $(x_n \mid n \in \mathsf{N})$ of real numbers is said to converge to the real number x if for each $\epsilon > 0$ there exists some index $m \in \mathsf{N}$ such that $|x - x_n| < \epsilon$ for all $n \geq m$. Now $|x - x_n|$ is just $d(x, x_n)$ for the Euclidean metric d on R, so it is obvious how to generalize the notion of sequential convergence to arbitrary metric spaces.

1.49 Definition. Let $(x_n \mid n \in \mathsf{N})$ be a sequence of points in a metric space (X, d). If $x \in X$, then we say that $(x_n \mid n \in \mathsf{N})$ *converges to* x *in* (X, d) provided that

for each $\epsilon > 0$ there exists some $m \in \mathsf{N}$ such that $d(x, x_n) < \epsilon$ for all $n \in \mathsf{N}$ with $n \geq m$.

When $(x_n \mid n \in \mathsf{N})$ converges to some point in (X, d), then we say that $(x_n \mid n \in \mathsf{N})$ *converges in* (X, d).

Thus $(x_n \mid n \in \mathsf{N})$ converges to x in (X, d) if we can make x_n as d-close to x as we wish by taking n sufficiently large.

1.50 Examples

(1) Let $(x_n \mid n \in \mathsf{N})$ be a convergent sequence in a metric space (X, d). Then the range $\{x_n \mid n \in \mathsf{N}\}$ of the sequence is d-bounded.

To see this, let $(x_n \mid n \in \mathsf{N})$ converge to x in (X, d). Corresponding to $\epsilon = 1$ there is an $m \in \mathsf{N}$ such that

$$(*) \qquad\qquad d(x_n, x) < 1 \qquad\qquad (n \geq m).$$

All we need do is find a constant c for which

$$d(x_n, x_m) \leq c \qquad\qquad (n \in \mathsf{N}),$$

since then

$$d(x_j, x_k) \leq d(x_j, x_m) + d(x_m, x_k) \leq 2c \qquad\qquad (j, k \in \mathsf{N}).$$

We have from $(*)$

$$d(x_n, x_m) \leq d(x_n, x) + d(x, x_m) < 2 \qquad\qquad (n \geq m).$$

Set

$$a = \max_{1 \leq n \leq m} d(x_n, x_m).$$

Then $c = \max \{2, a\}$ will do.

(2) Let $(x_n \mid n \in \mathsf{N})$ be a sequence in R that is "increasing", that is,

$$x_n \leq x_{n+1} \qquad\qquad (n \in \mathsf{N}).$$

Then $(x_n \mid n \in \mathsf{N})$ converges in (R, d), where d is the Euclidean metric, if and only if $\{x_n \mid n \in \mathsf{N}\}$ has an upper bound in R. Moreover, if $\{x_n \mid n \in \mathsf{N}\}$ has an upper bound in R, then $(x_n \mid n \in \mathsf{N})$ converges in (R, d) to

$$x = \sup \{x_n \mid n \in \mathsf{N}\}.$$

In view of Example (1), the range of $(x_n \mid n \in \mathsf{N})$ will certainly have an upper bound in R if this sequence converges in (R, d). Conversely, assume $\{x_n \mid n \in \mathsf{N}\}$ has an upper bound in R. We show that $(x_n \mid n \in \mathsf{N})$ converges to $x = \sup \{x_n \mid n \in \mathsf{N}\}$ in (R, d). Let $\epsilon > 0$. Since $x - \epsilon < x$, then $x - \epsilon$ is not an upper bound of $\{x_n \mid n \in \mathsf{N}\}$, that is,

$$x - \epsilon < x_m$$

for some $m \in \mathsf{N}$. Since our sequence is increasing and x is an upper bound of its range,

$$x - \epsilon < x_m \leq x_n \leq x < x + \epsilon \qquad\qquad (n \geq m).$$

Hence $d(x, x_n) < \epsilon$ for all $n \geq m$.

The result just established has an application to convergence of infinite series which was used in our discussion of the Hilbert sequence space (1.9). Suppose $\sum_{i=1}^{\infty} a_i$ is a series of *nonnegative* terms a_i. Then the sequence $(s_n \mid n = 1, 2, \dots)$ of its partial sums

$$s_n = \sum_{i=1}^{n} a_i$$

is increasing because

$$s_{n+1} = s_n + a_{n+1} \geq s_n \qquad\qquad (n = 1, 2, \dots).$$

Hence the series $\sum_{i=1}^{\infty} a_i$ converges if and only if the set $\{s_n \mid n = 1, 2, \dots\}$ has an upper bound in R, and in this event $s_n \leq \sum_{i=1}^{\infty} a_i$ for each n.

A corollary is the comparison test: Suppose $\sum_{i=1}^{\infty} b_i$ is another series that dominates $\sum_{i=1}^{\infty} a_i$ term by term, that is,

$$a_i \leq b_i \qquad\qquad (i = 1, 2, \dots),$$

and suppose $\sum_{i=1}^{\infty} b_i$ converges. Then the original series $\sum_{i=1}^{\infty} a_i$ also converges because for each n,

$$s_n = \sum_{i=1}^{n} a_i \leq \sum_{i=1}^{n} b_i \leq \sum_{i=1}^{\infty} b_i.$$

(3) Let $(y_n \mid n \in \mathsf{N})$ be a sequence in a subset Y of a metric space (X, d), and let d' be the metric on Y induced by d. If $y \in Y$, then clearly $(y_n \mid n \in \mathsf{N})$ converges to y in (Y, d') if and only if it converges to y in (X, d). Hence if $(y_n \mid n \in \mathsf{N})$ converges in (Y, d'), then it converges in (X, d). However, the converse of this last statement need not hold—see Example (4).

(4) Let d and d' be the Euclidean metrics on R and Q, respectively. We shall construct a sequence $(y_n \mid n \in \mathsf{N})$ in Q that is strictly increasing, has an upper bound in Q, does not converge in (Q, d'), and yet converges in (R, d).

Choose $y_1 \in \mathsf{Q}$ with

$$\sqrt{2} - 1 < y_1 < \sqrt{2},$$

and set $y_0 = y_1$. Next, choose $y_2 \in \mathsf{Q}$ with

$$\sqrt{2} - \tfrac{1}{2} < y_2 < \sqrt{2}, \qquad y_1 < y_2.$$

In general, once $y_1, \dots, y_{n-1}$ have been selected, choose $y_n \in \mathsf{Q}$ with

$$\sqrt{2} - \frac{1}{n} < y_n < \sqrt{2}, \qquad y_{n-1} < y_n.$$

At each stage the choice is possible because any open interval in R contains a rational number.

Clearly $(y_n \mid n \in \mathsf{N})$ is strictly increasing, and any rational number greater than $\sqrt{2}$ (for example, 2) is an upper bound of $\{y_n \mid n \in \mathsf{N}\}$ in Q. The sequence converges to $\sqrt{2}$ in (R, d) since

$$d\,(\sqrt{2},\, y_n) < \frac{1}{n} \qquad\qquad (n \geq 1).$$

Just suppose $(y_n \mid n \in \mathsf{N})$ converges in (Q, d') to some point $y \in \mathsf{Q}$. Since $\sqrt{2}$ is irrational, $y \neq \sqrt{2}$. Let

$$\epsilon = d\,(y, \sqrt{2}) > 0.$$

There is an $m \in \mathsf{N}$ for which

$$d\,(y, y_n) = d'\,(y, y_n) < \frac{\epsilon}{2} \qquad\qquad (n \geq m)$$

since $(y_n \mid \in \mathsf{N})$ converges to y in (Q, d'), and there is a $k \in \mathsf{N}$ for which

$$d\,(y_n, \sqrt{2}) < \frac{\epsilon}{2} \qquad\qquad (n \geq k)$$

since $(y_n \mid n \in \mathsf{N})$ converges to $\sqrt{2}$ in (R, d). Then for $n = \max \{m, k\}$,

$$d\,(y, \sqrt{2}) \leq d\,(y, y_n) + d\,(y_n, \sqrt{2}) < \frac{\epsilon}{2} + \frac{\epsilon}{2} = \epsilon,$$

which contradicts the definition of ϵ.

(5) Let δ be the discrete metric on a set X. For $x \in X$ and $\epsilon \leq 1$, the only point $y \in X$ satisfying $\delta(x, y) < \epsilon$ is x itself. Hence a sequence $(x_n \mid n \in \mathsf{N})$ converges in (X, δ) if and only if it is eventually constant, that is, for some $m \in \mathsf{N}$

$$x_n = x_m \qquad\qquad (n \geq m).$$

(6) Let d_∞ be the metric constructed in Example 1.7 on the set X of all continuous functions $x \colon [0, 1] \to \mathsf{R}$. Then to say that a sequence of functions $(x_n \mid n \in \mathsf{N})$ in X converges to a function x in (X, d_∞) means that for each $\epsilon > 0$, there is an index $m \in \mathsf{N}$ such that

$$|x\,(t) - x_n(t)| < \epsilon \qquad\qquad (n \geq m, 0 \leq t \leq 1).$$

Roughly speaking, this says that we can simultaneously make all the vertical distances $|x\,(t) - x_n(t)|$ between the graphs of x and x_n as small as we wish by taking n sufficiently large. For this reason convergence of a sequence in this metric space is said to be *uniform*.

(7) Let $(x_n \mid n \in \mathsf{N})$ be a sequence in an arbitrary metric space (X, d), and let d' be the Euclidean metric on R. Then $(x_n \mid n \in \mathsf{N})$ converges to a point x in (X, d) precisely when the sequence $(d\,(x, x_n) \mid n \in \mathsf{N})$ of real numbers converges to 0 in (R, d'). In fact, if

$$y_n = d\,(x, x_n) \qquad\qquad (n \in \mathsf{N}),$$

then

$$d'\,(0, y_n) = d\,(x, x_n) \qquad\qquad (n \in \mathsf{N}).$$

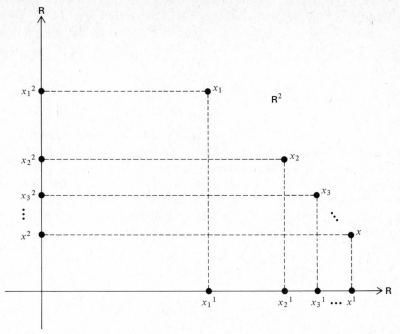

FIG. 1.18

The condition

$$d(x, x_n) < \epsilon,$$

which appears as part of Definition 1.49, means

$$x_n \in B_\epsilon(x; d).$$

Now each d-ball at a point x in a metric space (X, d) is a d-neighborhood of x, and each d-neighborhood of x contains some d-ball at x. Hence we have the following criterion.

1.51 Theorem. Let $(x_n \mid n \in \mathsf{N})$ be a sequence in a metric space (X, d), and let $x \in X$. Then a necessary and sufficient condition for $(x_n \mid n \in \mathsf{N})$ to converge to x in (X, d) is that for each d-neighborhood V of x there exist some $m \in \mathsf{N}$ with

$$x_n \in V \qquad\qquad\qquad (n \geq m).$$

1.52 Corollary. In the notation of Theorem 1.51, let D be a metric on X that is equivalent to d. Then $(x_n \mid n \in \mathsf{N})$ converges to x in (X, d) if and only if it converges to x in (X, D).

This independence of sequential convergence from the particular metric is exploited in the next example.

1.53 Example. Denote by d_k the Euclidean metric on R^k. Let $x \in \mathsf{R}^k$ and let $(x_n \mid n \in \mathsf{N})$ be a sequence of points in R^k. Denote the k coordinates of x by

$$x^1, \ldots, x^k$$

and for each $n \in \mathsf{N}$, the k coordinates of x_n by

$$x_n^1, \ldots, x_n^k$$

(our use of subscripts to denote the values of a sequence necessitates this departure from our usual practice of using subscripts for coordinates).

The sequence $(x_n \mid n \in \mathsf{N})$ converges to x in (R^k, d_k) if and only if for each coordinate $1 \le i \le k$, the sequence $(x_n^i \mid n \in \mathsf{N})$ of real numbers converges to x^i in (R, d), where d is the Euclidean metric. See Fig. 1.18 for the case of dimension $k = 2$, and examine the following diagram to keep track of what is going on in the general case.

$$\mathsf{R}^k = \mathsf{R} \times \mathsf{R} \times \cdots \times \mathsf{R} \times \cdots \times \mathsf{R}$$

$$x_1 = (x_1^1, \ x_1^2, \ldots, \qquad x_1^i, \ldots, \qquad x_1^k)$$

$$x_2 = (x_2^1, \ x_2^2, \ldots, \qquad x_2^i, \ldots, \qquad x_2^k)$$

$$\begin{matrix} \cdot & \cdot & \cdot & & \cdot & & \cdot \\ \cdot & \cdot & \cdot & & \cdot & & \cdot \\ \cdot & \cdot & \cdot & & \cdot & & \cdot \end{matrix}$$

$$x_n = (x_n^1, x_n^2, \ldots, \qquad x_n^i, \ldots, \qquad x_n^k)$$

$$\begin{matrix} \cdot & \cdot & \cdot & & \cdot & & \cdot \\ \cdot & \cdot & \cdot & & \cdot & & \cdot \\ \cdot & \cdot & \cdot & & \cdot & & \cdot \end{matrix}$$

$$\downarrow \quad \downarrow \quad \downarrow \qquad \downarrow \qquad \downarrow$$

$$x = (x^1, \ x^2, \ldots, . \qquad x^i, \ldots, \qquad x^k)$$

We know that the max metric d_∞ on R^k is equivalent to d_k (1.33), so to prove our assertion we consider convergence in (R^k, d_∞). Assume first that $(x_n \mid n \in \mathsf{N})$ converges to x in (R^k, d_∞). Fix $1 \le i \le k$. Let $\epsilon > 0$. There exists $m \in \mathsf{N}$ with

$$d_\infty(x, x_n) < \epsilon \qquad\qquad (n \ge m).$$

Since

$$d(x^i, x_n^i) \le d_\infty(x, x_n)$$

for all n,

$$d(x^i, x_n^i) < \epsilon \qquad\qquad (n \ge m).$$

Hence $(x_n^i \mid n \in \mathsf{N})$ converges to x^i in (R, d).

Conversely, assume that $(x_n^i \mid n \in \mathsf{N})$ converges to x^i in (R, d) for each $1 \le i \le k$. Let $\epsilon > 0$. For each $1 \le i \le k$, there exists $m(i) \in \mathsf{N}$ with

$$d(x^i, x_n^i) < \epsilon \qquad\qquad (n \ge m(i)).$$

Set
$$m = \max \{m(1), \ldots, m(k)\}.$$
Then
$$d(x^i, x_n{}^i) < \epsilon \qquad\qquad (1 \leq i \leq k, n \geq m),$$
so
$$d_\infty(x, x_n) < \epsilon \qquad\qquad (n \geq m).$$

Hence $(x_n \mid n \in \mathsf{N})$ converges to x in (R^k, d_∞).

Example 1.53 has an obvious generalization concerning the max metric on the product of any finite number of metric spaces (1.13).

The expression 'x is a limit of $(x_n \mid n \in \mathsf{N})$' is frequently used to mean '$(x_n \mid n \in \mathsf{N})$ converges to x'. Note that we did not say 'x is *the* limit of $(x_n \mid n \in \mathsf{N})$', for we have no right to do so until we know that a sequence cannot converge to two distinct points. Fortunately it is both true and easy to prove that limits of sequences in a metric space are unique.

1.54 Theorem. Let $(x_n \mid n \in \mathsf{N})$ be a sequence in a metric space (X, d) which converges in (X, d) both to a point x and to a point y. Then $x = y$.

Proof. Just suppose $x \neq y$. By 1.25, there are disjoint d-neighborhoods U of x and V of y. Since $(x_n \mid n \in \mathsf{N})$ converges to x, there is an $m \in \mathsf{N}$ with
$$x_n \in U \qquad\qquad (n \geq m).$$
Since the same sequence converges to y, there is a $k \in \mathsf{N}$ with
$$x_n \in V \qquad\qquad (n \geq k).$$
Then for $n = \max \{m, k\}$ we have
$$x_n \in U \cap V,$$
which contradicts the disjointness of U from V. [The alert reader will have noticed that essentially this argument was already used in the fourth paragraph of 1.50 (4).] $\square$

If we know which sets in a metric space (X, d) are d-open, then we know which sequences converge to which points in the space. Surprisingly enough, the converse is true as well; this is a consequence of the next theorem and the fact that the d-open sets are just the complements of the d-closed sets.

1.55 Theorem. Let A be a subset of a metric space (X, d). Then the following statements are equivalent:

(1) The set A is d-closed.

(2) Each point $x \in X$ to which some sequence of points of A converges in (X, d) belongs itself to A.

Proof. $(1) \Rightarrow (2)$. Assume (1). Let $x \in X$ and let $(x_n \mid n \in \mathsf{N})$ be a sequence in A such that $(x_n \mid n \in \mathsf{N})$ converges to x in (X, d). Just suppose $x \notin A$. Then $V = X \setminus A$ is a d-neighborhood of x such that $x_n \notin V$ for all n. This contradicts the convergence of $(x_n \mid n \in \mathsf{N})$ to x.

$(2) \Rightarrow (1)$. Assume (2). Just suppose A is not d-closed. Then $X \setminus A$ is not d-open. This means there is some point $x \in X \setminus A$ with the property that no d-ball at x is contained in $X \setminus A$. Hence for each $n \in \mathsf{N}$ the d-ball of radius $1/n$ at x meets A, and we may choose some

$$x_n \in A \cap B_{1/n}(x; d).$$

Then $(x_n \mid n \in \mathsf{N})$ is a sequence in A, $(x_n \mid n \in \mathsf{N})$ converges to x in (X, d) since

$$d(x, x_n) < \frac{1}{n} \qquad\qquad (n \in \mathsf{N}),$$

and $x \notin A$. This contradicts (2).   □

Thus sequential convergence in a metric space determines which sets are open there. The open sets in two metric spaces determine which functions from the one space to the other are continuous. Hence sequential convergence in two metric spaces should determine which functions from one to the other are continuous. This is indeed the case.

1.56 Theorem. Let (X, d) and (Y, d') be metric spaces, let $f : X \to Y$ be a map, and let $x \in X$. Then a necessary and sufficient condition for f to be (d, d')-continuous at x is that for each sequence $(x_n \mid n \in \mathsf{N})$ converging to x in (X, d), the sequence $(f(x_n) \mid n \in \mathsf{N})$ converge to $f(x)$ in (Y, d').

Proof. Necessity. Assume f is (d, d')-continuous at x. Let $(x_n \mid n \in \mathsf{N})$ be a sequence that converges to x in (X, d). To show that $(f(x_n) \mid n \in \mathsf{N})$ converges to $f(x)$ in (Y, d'), let V be any d'-neighborhood of $f(x)$. By 1.46, there is a d-neighborhood U of x with

$$f(U) \subset V.$$

Choose $m \in \mathsf{N}$ with

$$x_n \in U \qquad\qquad (n \geq m).$$

Then

$$f(x_n) \in V \qquad\qquad (n \geq m).$$

Sufficiency. Assume the condition holds. Let V be any d'-neighborhood of $f(x)$. According to 1.46 it suffices to show that V contains the image $f(U)$ of some d-neighborhood U of x. Just suppose this is not the case. In particular, then, for each $n \in \mathsf{N}$

$$f(B_{1/n}(x; d)) \not\subset V,$$

and we may choose a point x_n with

$$x_n \in B_{1/n}(x; d), \qquad f(x_n) \notin V.$$

Then $(x_n \mid n \in \mathsf{N})$ converges to x in (X, d), but certainly $(f(x_n) \mid n \in \mathsf{N})$ does not converge to $f(x)$ in (Y, d'). This contradicts our original assumption. ∎

Theorem 1.56 tells us that sequential convergence determines continuity. We close this section by showing that the reverse is true. More precisely, we shall demonstrate that any given sequence in a metric space converges precisely when a certain map taking values in that metric space is continuous. This fact is not at all essential to topology (and is therefore infrequently stated) but it does provide some additional insight into the meaning of sequential convergence.

For motivation, recall that a sequence in a set X is actually a map $n \mapsto x_n$ of $\mathsf{N} \to X$. If d is a metric on X, then to say that $(x_n \mid n \in \mathsf{N})$ converges in (X, d) is to say something about the behavior of this map for "large" numbers in its domain N. Now $\mathsf{N} \subset \mathsf{R} \cup \{+\infty\}$, and the "closer" an $n \in \mathsf{N}$ is to $+\infty$ in the extended real line R^*, the "larger" will n be. Hence it is natural to extend the map $n \mapsto x_n$ to all of $\mathsf{N} \cup \{+\infty\}$ by appropriately assigning a value to $+\infty$.

1.57 Remark. Denote by D' the metric induced on the subset

$$\mathsf{N}' = \mathsf{N} \cup \{+\infty\}$$

of the extended real line R^* by the metric D on R^* (see 1.37). Given a sequence $(x_n \mid n \in \mathsf{N})$ in a metric space (X, d) and a point $x \in X$, define a map

$$f : \mathsf{N}' \to X$$

by

$$f(n) = x_n \qquad (n \in \mathsf{N}), \qquad f(+\infty) = x.$$

Then $(x_n \mid n \in \mathsf{N})$ converges to x in (X, d) if and only if f is (D', d)-continuous at $+\infty$.

To prove this assertion observe first that each D'-neighborhood of $+\infty$ in N' contains a set of the form

$$U_m = \{n \in \mathsf{N} \mid n \geq m\} = \,]m - 1, +\infty] \cap \mathsf{N}'$$

for some integer $m \geq 1$, and each set of this form is a D'-neighborhood of $+\infty$ (see 1.39). For an arbitrary d-neighborhood V of x certainly $f(+\infty) = x \in V$, so

$$f(u) \in V \qquad (u \in U_m) \quad \Leftrightarrow \quad x_n \in V \qquad (n \geq m).$$

Now apply 1.46 and 1.51. ∎

EXERCISES

46. Suppose that in the condition

$$d(x, u) < \delta \quad \Rightarrow \quad d'(f(x), f(u)) < \epsilon$$

in Definition 1.44 one or the other or both of the strict inequalities $<$ were changed to the weak inequality $\leq$. Would this change which maps $f\colon X \to Y$ are (d, d')-continuous at x?

47. If d is the Euclidean metric on R, determine whether the given function $f\colon R \to R$ is (d, d)-continuous, and if it is not find all points at which it is (d, d)-discontinuous.

    (**a**) $f(x) = x \sin(1/x)$ if $x \neq 0$, and $f(0) = 0$.

    (**b**) $f(x) = x/|x|$ if $x \neq 0$, and $f(0) = 0$.

    (**c**) $f(x) = 0$ if x is irrational, and $f(x) = 1$ if x is rational.

    (**d**) $f(x) = 0$ if x is irrational, and $f(x) = 1/q$ if x is rational with q being that positive integer for which $x = p/q$, p being an integer that is relatively prime to q (that is, the only integers that divide both p and q are 1 and -1).

48. (**a**) If d is the Euclidean metric on R and d_∞ is the max metric on $R \times R$, show that the maps

$$R \times R \to R, \qquad R \times R \to R$$
$$(x, y) \mapsto x + y \qquad (x, y) \mapsto xy$$

are (d_∞, d)-continuous.

    (**b**) If d is the Euclidean metric on $R \backslash \{0\}$, show that the map

$$R \backslash \{0\} \to R \backslash \{0\}$$
$$x \mapsto \frac{1}{x}$$

is (d, d)-continuous.

49. Construct metric spaces (X, d) and (Y, d') and a map $f\colon X \to Y$ such that $f(U)$ is d'-open for each d-open subset U of X, yet f is (d, d')-continuous at no point of X.

50. (**a**) Given metric spaces $(X_1, d_1), \ldots, (X_n, d_n)$, let d_∞ be the max metric induced by $(d_1, \ldots, d_n)$ on $X = X_{i=1}^n X_i$ (see 1.13). Prove that for each $1 \leq j \leq n$ the jth projection

$$p_j\colon X \to X_j$$
$$x \mapsto x_j$$

is (d_∞, d_j)-continuous.

    (**b**) Give an analog of (a) for the product of a sequence of metric spaces (see Exercise 34).

51. If d' is the Euclidean metric on R, is it true that for all $n \in N$ the maps

$$X \to R$$
$$x \mapsto x_n$$

are (d, d')-continuous when:

    (**a**) X is the Hilbert sequence space and d is the metric d_2 on X constructed in Example 1.9?

    (**b**) X is the set of all sequences in R and d is the metric on X constructed in Example 1.14?

52. Let X be the set of all continuous functions $x: [0, 1] \to$ R, and let d' be the Euclidean metric on R.

(**a**) Is it true that for all $t \in [0, 1]$ the maps

$$X \to \text{R}$$

$$x \mapsto x(t)$$

are (d, d')-continuous if d is the max metric d_∞ on $X(1.7)$? If d is the metric d_1 constructed in Example 1.8?

(**b**) Is the map

$$X \to \text{R}$$

$$x \mapsto \int_0^1 x(t) \, dt$$

(d_∞, d')-continuous? Is it (d_1, d')-continuous?

53. Let (X, d) and (Y, d') be metric spaces, and let $f: X \to Y$. Prove that if f is (d, d')-continuous, then the inverse image $f^{-1}(E)$ of each d'-closed subset E of Y is a d-closed subset of X. Does the converse hold?

54. Let d be the Euclidean metric on R^n, and let d_∞ be the max metric that it induces on R$^n \times$ R^n. Prove that the Euclidean inner product on R^n (Exercise 42) is (d_∞, d)-continuous.

55. A *hyperplane in* R^n is a set of the form

$$\{x \in \text{R}^n \mid \langle x, a \rangle = c\}$$

for some real number c (here $\langle x, a \rangle$ is the Euclidean inner product as defined in Exercise 42); an *open-halfspace in* R^n is a set of the form

$$\{x \in \text{R}^n \mid \langle x, a \rangle < c\}$$

for some real number c; and a *closed-halfspace in* R^n is a set of the form

$$\{x \in \text{R}^n \mid \langle x, a \rangle \leq c\}$$

for some real number c. A *polyhedron in* R^n is a set that is the intersection of finitely many closed-halfspaces in R^n.

(**a**) Show that every hyperplane, open-halfspace, closed-halfspace, or polyhedron in R^n is convex.

(**b**) Prove that each open-halfspace in R^n is d-open, where d is the Euclidean metric on R^n. Prove also that each hyperplane, each closed-halfspace, and each polyhedron in R^n is d-closed. (*Hint*: Use Exercises 53 and 54.)

56. Let A be a subset of a metric space (X, d). Let d' be the Euclidean metric on R. Prove:

(**a**) The map

$$X \to \text{R}$$

$$x \mapsto d(x, A)$$

is (d, d')-continuous.

(**b**) Given $x \in X$, there is a sequence $(x_n \mid n \in \text{N})$ in A such that $(d(x, x_n) \mid n \in \text{N})$ converges to $d(x, A)$ in (R, d').

57. Let $(x_n \mid n \in \mathsf{N})$ and $(y_n \mid n \in \mathsf{N})$ be two sequences in a metric space (X, d), and suppose there is an index $k \in \mathsf{N}$ such that $x_n = y_n$ for all $n \geq k$. Show that one of these sequences will converge in (X, d) if and only if the other does, and that in this case they converge to the same point.

58. Establish an analog of 1.50 (2) for decreasing sequences of real numbers.

59. Let p be a prime number. Show that $(p^n \mid n \in \mathsf{N})$ converges in the metric space (Q, d_p), where d_p is the p-adic metric (Exercise 15), but not in (R, d), where d is the Euclidean metric.

60. (**a**) Prove an analog of 1.53 for the product of a sequence of metric spaces (see Exercise 34).

   (**b**) Does the analog of 1.53 hold for the Hilbert sequence space (H, d_2) [1.9]?

61. Let X be the set of all continuous functions $x \colon [0, 1] \to \mathsf{R}$. Denote the Euclidean metric on R by d.

   (**a**) Construct a sequence $(x_n \mid n \in \mathsf{N})$ in X with the property that for each $t \in [0, 1]$ the sequence $(x_n(t) \mid n \in \mathsf{N})$ converges in (R, d), yet $(x_n \mid n \in \mathsf{N})$ does not converge in (X, d_∞), where d_∞ is the max metric on X (1.7).

   (**b**) Does the sequence you constructed converge in (X, d_1), where d_1 is the metric defined in 1.8?

62. (**a**) Write down the calculus definition of

$$\lim_{n \to \infty} x_n = +\infty$$

for a sequence $(x_n \mid n \in \mathsf{N})$ of real numbers. Explain what this statement means in the context of sequential convergence in metric spaces.

   (**b**) Do the same for the statement

$$\lim_{n \to \infty} x_n = -\infty$$

63. The definitions of continuity and sequential convergence given in this section still make sense if we deal with pseudometrics (Exercise 13) instead of metrics. Check that all results of this section except 1.54 extend to the pseudometric case.

5. Completeness

A typical procedure for proving that a given sequence $(x_n \mid n \in \mathsf{N})$ converges in a metric space is first to make an educated guess at the point x to which the sequence ought to converge and then to check that $(x_n \mid n \in \mathsf{N})$ actually converges to x by using the definition of convergence to x. This procedure suffers the limitation that we must be able to guess the point x, and we often have no good reason to guess one point instead of another. Moreover, it is often of greater importance to know *that* $(x_n \mid n \in \mathsf{N})$ converges than to know to which particular point it converges.

What is needed to remedy this situation is a criterion for testing whether a sequence $(x_n \mid n \in \mathsf{N})$ converges by looking only at the values x_n of the sequence, not at some other point x as well. Now a sequence $(x_n \mid n \in \mathsf{N})$ converges when the values x_n get closer and closer to some point x for larger and larger values of n. Hence a convergent sequence has the property that its values x_n get closer and closer *to one another* for larger and larger values of n. It was precisely this property that was introduced in the early nineteenth century by Augustin Cauchy as a criterion for convergence of sequences of real numbers. In this section we study metric spaces in which this property always guarantees convergence.

1.58 Definition. A sequence $(x_n \mid n \in \mathsf{N})$ in a metric space (X, d) is said to be a *Cauchy sequence in* (X, d) when for each $\epsilon > 0$ there exists some $m \in \mathsf{N}$ such that

$$d(x_n, x_k) < \epsilon \qquad\qquad (n \geq m, k \geq m).$$

Many examples of Cauchy sequences are furnished by the first part of the following result, which also formalizes the idea that values of a sequence get close to one another if they get close to some fixed point.

1.59 Proposition. Let $(x_n \mid n \in \mathsf{N})$ be a sequence in a metric space (X, d).
(1) If $(x_n \mid n \in \mathsf{N})$ converges in (X, d), then it is a Cauchy sequence in (X, d).
(2) If $(x_n \mid n \in \mathsf{N})$ is a Cauchy sequence in (X, d), then its range $\{x_n \mid n \in \mathsf{N}\}$ is d-bounded.

Proof. (1) Assume $(x_n \mid n \in \mathsf{N})$ converges in (X, d) to a point x. Let $\epsilon > 0$. There exists $n \in \mathsf{N}$ with

$$d(x_n, x) < \frac{\epsilon}{2} \qquad\qquad (n \geq m).$$

Then $n \geq m$ and $k \geq m$ implies

$$d(x_n, x_k) \leq d(x_n, x) + d(x, x_k) < \frac{\epsilon}{2} + \frac{\epsilon}{2} = \epsilon.$$

(2) Assume $(x_n \mid n \in \mathsf{N})$ is a Cauchy sequence in (X, d). Corresponding to $\epsilon = 1$ there is an $m \in \mathsf{N}$ with

$$d(x_n, x_k) < 1 \qquad\qquad (n \geq m, k \geq m).$$

Set

$$a = \max \{ d(x_n, x_k) \mid 1 \leq n \leq m, 1 \leq k \leq m \}.$$

Then

$$d(x_n, x_k) \leq 1 + a$$

for all $n, k \in \mathsf{N}$.   ☐

Neither of the implications above is reversible, as the following examples demonstrate.

1.60 Examples

(1) Let (Y, d') be a metric space and let d be the metric induced by d' on a subset X of Y. Suppose $(x_n \mid n \in \mathsf{N})$ is a sequence in X that converges to a point $y \in Y \backslash X$ in (Y, d'). By uniqueness of limits (1.54), $(x_n \mid n \in \mathsf{N})$ does not converge in (X, d). However, $(x_n \mid n \in \mathsf{N})$ is a Cauchy sequence in (X, d) since by 1.59 (1) it is a Cauchy sequence in (Y, d').

As a specific example, take $Y = \mathsf{R}$ and $X = \mathsf{Q}$, let d' and d be the Euclidean metrics, and let $(x_n \mid n \in \mathsf{N})$ be the sequence constructed in 1.50 (4).

As another example, take $Y = \mathsf{R}$ and $X = \,]0, 1[$, let d' and d be the Euclidean metrics, and consider the sequence

$$\left(\frac{1}{n+1} \,\middle|\, n \in \mathsf{N} \right).$$

(2) Let d be the Euclidean metric on R. Then the sequence $(x_n \mid n \in \mathsf{N})$ given by

$$x_n = (-1)^n \qquad\qquad (n \in \mathsf{N})$$

is not a Cauchy sequence in (R, d) since $d(x_n, x_{n+1}) = 2$ for all n, but $\{x_n \mid n \in \mathsf{N}\} = \{-1, 1\}$ is certainly d-bounded.

Let us name those metric spaces which, unlike the ones in 1.60 (1), do contain all the points that ought to be there from the viewpoint of convergence.

1.61 Definition. Both a metric space (X, d) and its metric d are said to be *complete* if each Cauchy sequence in (X, d) converges in (X, d).

The simplest example of a complete metric is the discrete metric δ on any set X. Suppose $(x_n \mid n \in \mathsf{N})$ is a Cauchy sequence in (X, δ). Corresponding to $\epsilon = 1$ there is an $m \in \mathsf{N}$ with $\delta(x_n, x_m) < 1$ for all $n \geq m$, that is, $x_n = x_m$ for all $n \geq m$. According to 1.50 (5), this is precisely the condition needed for $(x_n \mid n \in \mathsf{N})$ to converge in (X, δ).

The single most important example of a complete metric space is the real line with its Euclidean metric.

1.62 Theorem. The Euclidean metric d on R is complete.

Proof. Let $(x_n \mid n \in \mathsf{N})$ be a Cauchy sequence in (R, d). Suppose first that the range

$$R = \{x_k \mid n \in \mathsf{N}\}$$

of the sequence is finite. Then there is an $x \in \mathsf{R}$ with $x_n = x$ for infinitely many values of n. We show $(x_n \mid n \in \mathsf{N})$ converges to x in (R, d). Let $\epsilon > 0$.

There is an $i \in \mathbb{N}$ with

$$|x_n - x_m| < \epsilon \qquad\qquad (n \geq i, m \geq i).$$

Since $\{m \in \mathbb{N} \mid x_m = x\}$ is infinite whereas $\{m \in \mathbb{N} \mid m < i\}$ is finite, there is an $m \geq i$ with $x_m = x$. Then $n \geq m$ implies $|x_n - x| < \epsilon$.

Now suppose R is infinite. Since R is d-bounded [1.59 (2)],

$$R \subset [a, b]$$

for some $a < b$. We apply a bisection argument to $[a_0, b_0] = [a, b]$. Since R is infinite, bisection of $[a_0, b_0]$ produces two intervals one of which contains infinitely many points of R; call such an interval $[a_1, b_1]$. Since $R \cap [a_1, b_1]$ is infinite, bisection of $[a_1, b_1]$ produces two intervals one of which contains infinitely many points of R; call such an interval $[a_2, b_2]$. Continuing in this way, we obtain a nested sequence

$$[a_0, b_0] \supset [a_1, b_1] \supset \cdots \supset [a_n, b_n] \supset [a_{n+1}, b_{n+1}] \supset \cdots$$

of intervals with the nth interval having length

$$b_n - a_n = \frac{b_0 - a_0}{2^n}$$

and containing infinitely many points of R. By 0.44 there exists a point

$$x \in \bigcap_{n=0}^{\infty} [a_n, b_n].$$

To finish the proof we show that $(x_n \mid n \in \mathbb{N})$ converges to x. Let $\epsilon > 0$. Because we have a Cauchy sequence there is an $i \in \mathbb{N}$ with

$$|x_n - x_k| < \frac{\epsilon}{2} \qquad\qquad (n \geq i, k \geq i).$$

By the Archimedean property (0.41), there is an $m \in \mathbb{N}$ with

$$m \geq i, \qquad b_m - a_m < \frac{\epsilon}{2}.$$

Since $R \cap [a_m, b_m]$ is infinite, there is a $k \in \mathbb{N}$ with

$$k \geq i, \qquad x_k \in [a_m, b_m].$$

Now $x \in [a_m, b_m]$ also, so

$$|x_k - x| < \frac{\epsilon}{2}.$$

Then $n \geq m$ implies

$$|x_n - x| = |x_n - x_k| + |x_k - x| < \frac{\epsilon}{2} + \frac{\epsilon}{2} = \epsilon$$

(see Fig. 1.19).   □

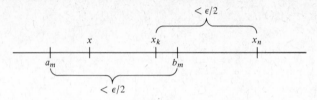

FIG. 1.19

The preceding proof used in a crucial way the nested interval and Archimedean properties of the real numbers. Hence the completeness of the Euclidean metric on R is a consequence of the order-completeness of R.

According to Example 1.60(1), the set Q of all rational numbers with its Euclidean metric d is not complete. We have just proved that the larger set R of all real numbers with its Euclidean metric d' is complete. Thus each Cauchy sequence in the smaller space (Q, d) converges in the larger space (R, d'). Moreover, it is clear from the order-density of Q in R that Q is "d'-dense" in R in the sense that for each point x of the larger set R there is some sequence in the smaller set Q which converges to x in (R, d'). Hence by passing from Q to R we supply all the points that are "missing" from Q, and we say that (R, d') *completes* (Q, d).

More generally, let (X, d) be any metric space. Then we can construct a *completion* of (X, d), that is, a complete metric space (X', d') such that $X \subset X'$, d' induces d on X, and X is d'-dense in the same sense as above. The construction of a completion is outlined in Exercise 90 (for technical reasons, the definition of a completion there is slightly different from the one here).

From 1.62 we shall deduce that the Euclidean metric on R^k is complete for all $k \geq 1$. First we establish a more general result.

1.63 Proposition. Let $(X_1, d_1), \ldots, (X_k, d_k)$ be complete metric spaces. Then the max metric d_∞ induced by $(d_1, \ldots, d_k)$ on

$$X = \underset{i=1}{\overset{k}{\times}} X_i$$

is also complete.

Proof. Let $(x_n \mid n \in N)$ be a Cauchy sequence in (X, d_∞). As in 1.53, denote the ith coordinate of x_n by $x_n{}^i$, so that

$$x_n = (x_n{}^1, x_n{}^2, \ldots, x_n{}^i, \ldots, x_n{}^k).$$

We show that for each $1 \leq i \leq k$ the sequence $(x_n{}^i \mid n \in N)$ is a Cauchy sequence in (X_i, d_i). Let $\epsilon > 0$. There exists $m \in N$ with

$$d_\infty(x_n, x_j) < \epsilon \qquad (n \geq m, j \geq m).$$

By definition of d_∞, for each $1 \leq i \leq k$ we have

$$d_i(x_n{}^i, x_j{}^i) < \epsilon \qquad (n \geq m, j \geq m).$$

For each $1 \leq i \leq k$ the metric space (X_i, d_i) is complete and so the Cauchy sequence $(x_n{}^i \mid n \in \mathsf{N})$ in this space converges to some point x^i in (X_i, d_i). Set

$$x = (x^1, x^2, \ldots, x^i, \ldots, x^k) \in X.$$

Then exactly as in 1.53 we see that $(x_n \mid x \in \mathsf{N})$ converges to x in (X, d_∞).   ☐

1.64 Corollary. The Euclidean metric d on R^k is complete.

Proof. Let $(x_n \mid n \in \mathsf{N})$ be a Cauchy sequence in (R^k, d). Using the inequality

$$d(x, y) \leq \sqrt{k} d_\infty(x, y) \qquad\qquad (x, y \in \mathsf{R}^k)$$

established in the course of proving 1.33, we see that $(x_n \mid n \in \mathsf{N})$ is also a Cauchy sequence in (R^k, d_∞), where d_∞ is the max metric. By 1.62 and 1.63, $(x_n \mid n \in \mathsf{N})$ converges in (R^k, d_∞). Now by 1.33 the metrics d and d_∞ are equivalent. From 1.52 we conclude that $(x_n \mid n \in \mathsf{N})$ converges in (R^k, d).   ☐

The reader must not be misled by the preceding proof into believing it is always true that if one of two equivalent metrics is complete, then the other is also.

1.65 Example. Let d^* be the bounded metric induced on R by the metric D on R^* constructed in 1.37. By 1.38, d^* is equivalent to the Euclidean metric d on R. We saw above that (R, d) is complete. We claim, however, that (R, d^*) is not complete.

To justify our claim, consider the bijection

$$\varphi: \mathsf{R} \to \,]-1, 1[$$

constructed in 1.37. Denote the Euclidean metric on $]-1, 1[$ by d'. The bijection φ is an isometry from (R, d^*) to $(]-1, 1[, d')$, for $x, y \in \mathsf{R}$ implies

$$d'(\varphi(x), \varphi(y)) = |\varphi(x) - \varphi(y)|$$

$$= D(x, y)$$

$$= d^*(x, y).$$

Then (R, d^*) will be complete if and only if $(]-1, 1[, d')$ is complete, but 1.60 (1) shows that the latter space is not complete.

This example demonstrates that completeness is a property which depends in an essential way on the particular metric d, not just on the d-open sets. Hence completeness of a metric space will have no generalization to the "topological spaces" we shall introduce in Chapter 2. (Completeness does generalize to "uniform spaces"—see [21].)

Despite Example 1.65 it is nonetheless true that any complete metric is equivalent to a bounded complete metric.

1.66 Proposition. Let (X, d) be a complete metric space. Then the equivalent bounded metric d^* of 1.36 is also complete.

Proof. By 1.52 the convergent sequences in (X, d) are the same as those in (X, d^*). Hence we need only show that the Cauchy sequences in (X, d) are the same as those in (X, d^*). But this follows from the fact that for all $0 < \epsilon < 1$,

$$d(x, y) < \epsilon \quad \Leftrightarrow \quad d^*(x, y) < \epsilon. \qquad \square$$

The infinite-dimensional analog of the finite-dimensional Euclidean spaces R^k is another complete metric space.

1.67 Theorem. The Hilbert sequence space (H, d_2) is complete.

Proof. Each point of H is itself a sequence of real numbers. Then in order to distinguish a point of H from a sequence of points in H, we use superscripts to denote the coordinates of a point of H. The strategy we use is similar to that used to prove 1.63. Denote the Euclidean metric on R by d.

Let $(x_n \mid n \in \mathsf{N})$ be a Cauchy sequence in (H, d_2). For each $n \in \mathsf{N}$ we write

$$x_n = (x_n{}^i \mid i = 1, 2, \ldots).$$

For all n, k, and i we have

$$|x_n{}^i - x_k{}^i| \leq \left[\sum_{j=1}^{\infty} (x_n{}^j - x_k{}^j)^2 \right]^{1/2} = d_2(x_n, x_k).$$

Then for each $i = 1, 2, \ldots$, the sequence $(x_n{}^i \mid n \in \mathsf{N})$ of real numbers is a Cauchy sequence in (R, d) and therefore converges to some real number x^i in (R, d). Set

$$x = (x^i \mid i = 1, 2, \ldots).$$

We shall complete the proof by showing, first, that $x \in \mathsf{H}$ and, second, that $(x_n \mid n \in \mathsf{N})$ converges to x in (H, d_2).

We show $x \in \mathsf{H}$. It suffices to find an m such that $x_m - x \in \mathsf{H}$, for then

$$x = x_m - (x_m - x)$$

will belong to H as well. Corresponding to $\epsilon = 1$ there is an $m \in \mathsf{N}$ with

(*) $$d_2(x_n, x_k) < 1 \qquad (n \geq m, k \geq m).$$

We claim that the sequence

$$x_m - x = (x_m{}^i - x^i \mid i = 1, 2, \ldots)$$

does indeed belong to H, that is, the series

$$\sum_{i=1}^{\infty} (x_m{}^i - x^i)^2$$

converges.

To show that this series of nonnegative numbers converges, it is enough to show that all its partial sums are bounded above by 1:

$$\sum_{i=1}^{p} (x_m{}^i - x^i)^2 \leq 1 \qquad (p = 1, 2, \dots).$$

Fix $p \geq 1$. From (*) we obtain

$$\sum_{i=1}^{p} (x_m{}^i - x_n{}^i)^2 < 1 \qquad (n \geq m).$$

Now for each $i = 1, 2, \dots, p$, the sequence $(x_n{}^i \mid n \in \mathbb{N})$ converges to x^i, and from some elementary properties of limits of sequences of real numbers we conclude

$$\sum_{i=1}^{p} (x_m{}^i - x^i)^2 \leq 1.$$

To complete the proof, we show that $(x_n \mid n \in \mathbb{N})$ converges to x in (H, d_2). Let $\epsilon > 0$. There exists $m \in \mathbb{N}$ with

$$d_2(x_n, x_k) < \epsilon \qquad (n \geq m, k \geq m).$$

Then for each $p = 1, 2, \dots$,

$$\sum_{i=1}^{p} (x_n{}^i - x_k{}^i)^2 < \epsilon^2 \qquad (n \geq m, k \geq m).$$

For each $i \geq 1$, the sequence $(x_k{}^i \mid k \in \mathbb{N})$ converges to x^i, so for each $p = 1, 2, \dots$, we have

$$\sum_{i=1}^{p} (x_n{}^i - x^i)^2 \leq \epsilon^2 \qquad (n \geq m).$$

Hence

$$d_2(x_n, x) = \left[\sum_{i=1}^{\infty} (x_n{}^i - x^i)^2 \right]^{1/2} \leq \epsilon \qquad (n \geq m). \qquad \square$$

From the complete spaces we have already considered, the first part of the next theorem supplies us with many more.

1.68 Theorem. Let (X, d) be a metric space, let $Y \subset X$, and let d' be the metric on Y induced by d.
  (1) If (X, d) is complete and Y is d-closed, then (Y, d') is complete.
  (2) If (Y, d') is complete, then Y is d-closed.

Proof. (1) Assume (X, d) is complete and Y is d-closed. Let $(y_n \mid n \in \mathbb{N})$ be a Cauchy sequence in (Y, d'). Then $(y_n \mid n \in \mathbb{N})$ is also a Cauchy sequence in (X, d), so by assumption it converges to some y in (X, d). By 1.55, $y \in Y$. Hence $(y_n \mid n \in \mathbb{N})$ converges to y in (Y, d').

(2) Assume (Y, d') is complete. Let $(y_n \mid n \in \mathsf{N})$ be a sequence in Y that converges to a point x in (X, d). By 1.55, it suffices to show $x \in Y$. Now $(y_n \mid n \in \mathsf{N})$, being a Cauchy sequence in (X, d), is a Cauchy sequence in (Y, d'). By assumption, $(y_n \mid n \in \mathsf{N})$ must converge to some y in (Y, d'), and hence in (X, d) as well. Then $x = y$ by uniqueness of limits (1.54), so $x \in Y$.   □

An important application of 1.68 (1) is the proof of completeness of the max metric d_∞ on the set of continuous functions from $[0, 1]$ to R (1.7). The larger complete space we need to apply 1.68 will consist of all bounded functions from $[0, 1]$ to R (we accept for the present without proof the facts that a continuous real-valued function on a closed interval is bounded and that it assumes a maximum value at some point of the interval—these facts will be proved in Chapter 4).

We shall actually consider, more generally, bounded real-valued functions on an arbitrary nonempty set X. Recall that a function $f \colon X \to \mathsf{R}$ is said to be *bounded* when there is a constant c for which $|f(x)| \leq c$ for all $x \in X$. Then for any two bounded functions $f \colon X \to \mathsf{R}$ and $g \colon X \to \mathsf{R}$, the set $\{|f(x) - g(x)| \colon x \in X\}$ has an upper bound in R, and it makes sense to let

$$d_\infty(f, g) = \sup_{x \in X} |f(x) - g(x)|.$$

This defines a metric on the set of all bounded functions from X to R (see Exercise 12) which we call the *sup metric*.

1.69 Theorem. Let $\mathcal{B}(X)$ be the set of all bounded real-valued functions on a nonempty set X. Then the sup metric d_∞ on $\mathcal{B}(X)$ is complete.

Proof. (The strategy is similar to that used to prove 1.67.) Let $(f_n \mid n \in \mathsf{N})$ be a Cauchy sequence in $(\mathcal{B}(X), d_\infty)$. Given $\epsilon > 0$, there is an $m \in \mathsf{N}$ with

$$d_\infty(f_n, f_k) < \epsilon \qquad\qquad (n \geq m, k \geq m)$$

and hence

$$|f_n(x) - f_k(x)| < \epsilon \qquad\qquad (n \geq m, k \geq m)$$

for each $x \in X$. Then for each $x \in X$ the sequence $(f_n(x) \mid n \in \mathsf{N})$ of real numbers is a Cauchy sequence in (R, d), d being the Euclidean metric, and therefore converges in (R, d) to a number we denote by $f(x)$. We obtain in this way a function $f \colon X \to \mathsf{R}$. It remains to show that $f \in \mathcal{B}(X)$ and $(f_n \mid n \in \mathsf{N})$ converges to f in $(\mathcal{B}(X), d_\infty)$.

We show that f is bounded. Corresponding to $\epsilon = 1$ there is an $m \in \mathsf{N}$ with

$$d_\infty(f_n, f_k) < 1 \qquad\qquad (n \geq m, k \geq m).$$

Let $c \in \mathsf{R}$ with $|f_m(x)| \leq c$ for all $x \in X$. Fix $x \in X$. Choose $n \geq m$ with

$$|f(x) - f_n(x)| < 1.$$

Then

$$|f(x)| \leq |f(x) - f_n(x)| + |f_n(x) - f_m(x)| + |f_m(x)| < 2 + c.$$

Since x was arbitrary, f is bounded.

Now that we know $f \in \mathcal{B}(X)$, we show, finally, that $(f_n \mid n \in \mathsf{N})$ converges to f in $(\mathcal{B}(X), d_\infty)$. Let $\epsilon > 0$. Choose $m \in \mathsf{N}$ with

$$(*) \qquad\qquad d_\infty(f_n, f_k) < \frac{\epsilon}{2} \qquad\qquad (n \geq m, k \geq m).$$

We claim

$$d_\infty(f, f_n) \leq \epsilon \qquad\qquad (n \geq m).$$

To prove this, we show

$$|f(x) - f_n(x)| < \epsilon \qquad\qquad (n \geq m)$$

for each $x \in X$. Fix $x \in X$. Since $(f_n(x) \mid n \in \mathsf{N})$ converges to $f(x)$, there exists $k \geq m$ with

$$|f(x) - f_k(x)| < \frac{\epsilon}{2}.$$

From this and $(*)$, $n \geq m$ implies

$$|f(x) - f_n(x)| \leq |f(x) - f_k(x)| + |f_k(x) - f_n(x)| < \frac{\epsilon}{2} + \frac{\epsilon}{2} = \epsilon. \qquad \square$$

1.70 Theorem. Let (X, d) be a metric space, let d' be the Euclidean metric on R, and let $\mathcal{C}^*(X)$ be the set of all bounded (d, d')-continuous functions from X to R. Then the sup metric on $\mathcal{C}^*(X)$ is complete.

Proof. By 1.68 (1), we need only show that $\mathcal{C}^*(X)$ is d_∞-closed in $\mathcal{B}(X)$, where $(\mathcal{B}(X), d_\infty)$ is the metric space of 1.69. We use the criterion 1.55 (2). Let $(f_n \mid n \in \mathsf{N})$ be a sequence in $\mathcal{C}^*(X)$ converging in $(\mathcal{B}(X), d_\infty)$ to a function $f \in \mathcal{B}(X)$. We must show $f \in \mathcal{C}^*(X)$, that is, f is (d, d')-continuous.

Let $x \in X$. We show that f is (d, d')-continuous at x. Let $\epsilon > 0$. There exists $n \in \mathsf{N}$ with $d_\infty(f, f_n) < \epsilon/2$, whence

$$|f(t) - f_n(t)| < \frac{\epsilon}{2} \qquad\qquad (t \in X).$$

By continuity of f_n at x, there is a d-neighborhood U of x such that

$$|f_n(x) - f_n(t)| < \frac{\epsilon}{2} \qquad\qquad (t \in U)$$

(see Fig. 1.20). Then $t \in U$ implies

$$|f(x) - f(t)| \leq |f(x) - f_n(x)| + |f_n(x) - f_n(t)| < \frac{\epsilon}{2} + \frac{\epsilon}{2} = \epsilon. \qquad \square$$

Succinctly put, the preceding proof consists of showing that *the limit of a uniformly convergent sequence of continuous functions is continuous.*

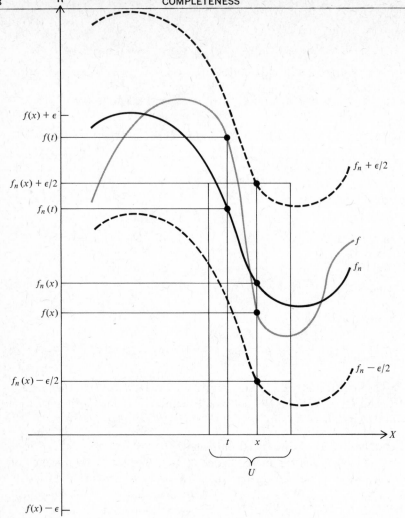

FIG. 1.20

Knowing now so many complete metric spaces, we deduce two of the most significant consequences of completeness. The first says in a new way that points which "ought to be there" really are there. It is a generalization of the nested interval property (0.44) and was first proved for arbitrary complete metric spaces by F. Hausdorff.

1.71 Theorem (nested set theorem). Let $(E_n \mid n \in \mathsf{N})$ be a decreasing sequence of nonempty, closed, d-bounded subsets of a complete metric space (X, d) such that

$$\lim_{n \to \infty} \operatorname{diam} E_n = 0.$$

Then there is a unique point

$$x \in \bigcap_{n=0}^{\infty} E_n.$$

Proof. Uniqueness. Suppose $x, y \in \bigcap_{n=0}^{\infty} E_n$ with $x \neq y$. Set

$$\epsilon = d(x, y) > 0.$$

Choose $n \in \mathbb{N}$ with diam $E_n < \epsilon$. Since $x \in E_n$ and $y \in E_n$, we have $d(x, y) < \epsilon$, a contradiction.

Existence. Arbitrarily choose a sequence $(x_n \mid n \in \mathbb{N})$ with

$$x_n \in E_n \qquad\qquad (n \in \mathbb{N}).$$

Then $(x_n \mid n \in \mathbb{N})$ is a Cauchy sequence in (X, d). In fact, given $\epsilon > 0$, we may choose $m \in \mathbb{N}$ with

$$\text{diam } E_m < \epsilon;$$

then $n \geq m$ and $k \geq m$ implies $x_n \in E_n \subset E_m$ and $x_k \in E_k \subset E_m$ so that

$$d(x_n, x_k) < \epsilon.$$

Since (X, d) is complete, $(x_n \mid n \in \mathbb{N})$ converges to some point x. We show that $x \in E_n$ for each $n \in \mathbb{N}$. Fix $n \in \mathbb{N}$. The sequence $(x_{n+i} \mid i \in \mathbb{N})$ also converges to x in (X, d), and each of its values x_{n+i} belongs to E_n. Because E_n is closed, it follows from 1.55 that $x \in E_n$, too.   ☐

For closed intervals in the real line and certain other complete metric spaces, the hypotheses on the sets E_n can be considerably relaxed. In a "compact" metric space (X, d) any sequence of d-closed sets will have a nonempty intersection provided each finite collection of these sets has nonempty intersection (see 4.6).

The second consequence of completeness concerns "dense" sets. A set D in a metric space (X, d) is said to be d-*dense in* X when it intersects every nonempty d-open set, or equivalently, when each d-neighborhood of each point of X intersects D. Roughly speaking, then, a set is d-dense in X when it contains points arbitrarily d-close to any given point of X. As examples, the set of rational numbers is d-dense in $\mathbb{R}$, where d is the Euclidean metric; the complement in $\mathbb{R}$ of any finite set is a d-open set that is d-dense in $\mathbb{R}$.

1.72 Theorem (Baire category theorem). In a complete metric space (X, d) the intersection of any sequence of d-open subsets of X that are d-dense in X is itself d-dense in X (and hence nonempty if X is nonempty).

Proof. Let

$$A = \bigcap_{n=0}^{\infty} A_n$$

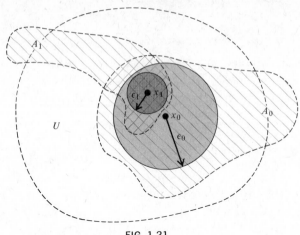

FIG. 1.21

with each A_n both d-open and d-dense in X. Let U be any nonempty d-open subset of X. To show that A intersects U we shall construct a sequence $(E_n \mid n \in \mathbf{N})$ of sets satisfying the hypotheses of 1.71 such that

$$E_n \subset A_n \cap U \qquad\qquad (n \in \mathbf{N}).$$

Then the point belonging to $\bigcap_{n=0}^{\infty} E_n$ will belong to $A \cap U$.

Since A_0 is d-dense in X, there is some point

$$x_0 \in A_0 \cap U,$$

and then since $A_0 \cap U$ is d-open, we may choose a number ϵ_0 with

$$0 < \epsilon_0 < \tfrac{1}{2}, \qquad D_{\epsilon_0}(x_0; d) \subset A_0 \cap U$$

(see Fig. 1.21). Since A_1 is d-dense in X, there is some point

$$x_1 \in A_1 \cap B_{\epsilon_0}(x_0; d),$$

and then we may choose ϵ_1 with

$$0 < \epsilon_1 < \tfrac{1}{4}, \qquad D_{\epsilon_1}(x_1; d) \subset A_1 \cap B_{\epsilon_0}(x_0; d).$$

Continuing in this way, we obtain a sequence $(x_n \mid n \in \mathbf{N})$ of points together with a sequence $(\epsilon_n \mid n \in \mathbf{N})$ of positive numbers such that

$$D_{\epsilon_{n+1}}(x_{n+1}; d) \subset A_{n+1} \cap B_{\epsilon_n}(x_n; d),$$

$$2\epsilon_n < \frac{1}{2^n}$$

for each n.

The decreasing sequence of sets to which we apply the nested set theorem (1.71) is given by

$$E_n = D_{\epsilon_n}(x_n; d) \qquad\qquad (n \in \mathbf{N}).$$

The requirement $\lim_{n \to \infty} \operatorname{diam} E_n = 0$ is satisfied because $\operatorname{diam} E_n \leq 2\epsilon_n < 2^{-n}$ for each n. $\square$

This theorem is named in honor of R. Baire, who proved it in 1899; the term "category" appears for reasons explained in 2.42. Observe that the hypothesis concerns completeness, a property of the particular metric d used, whereas the conclusion concerns only the d-open sets. (Essentially the same "topological" conclusion can be deduced from a completely different hypothesis not involving metrics—see Exercise 4.80.)

A d-dense set in a metric space contains "many" points of X. For this reason the Baire category theorem provides a method, and a powerful one, for proving that many points in a complete metric space have a certain property. Unfortunately most significant applications of this method are anything but elementary and would take us far afield of the subject of this book. One application, to topology, is made in Example 5.35 (5). Others may be found in Boas [5, Section 10] and Goldberg [13, Section 5.6].

We shall rest content with presenting here just one striking application of the Baire category theorem. The argument is complex and more analysis than topology, so the reader may wish to postpone studying it. It utilizes a result not proved until later (Example 4.44), namely, given any continuous f: $[0, 1] \to \mathsf{R}$ and any $\epsilon > 0$, there is a continuous piecewise linear p: $[0, 1] \to \mathsf{R}$ such that $|f(x) - p(x)| < \epsilon$ for all $x \in [0, 1]$. By a continuous function p: $[0, 1] \to \mathsf{R}$ that is *piecewise linear* we mean one whose graph consists of finitely many line segments.

1.73 Application. Amid all the computing of derivatives that goes on in calculus the reader has surely encountered continuous functions whose graphs have "corners" and which therefore fail to be differentiable at a few points (any continuous piecewise linear, but nonlinear function is of this type). We are going to establish the surprising fact that among all continuous functions those having a derivative at even a single point are rare! The argument is due to S. Banach.

Provide the set $\mathcal{C}([0, 1])$ of all continuous functions f: $[0, 1] \to \mathsf{R}$ with its max metric d_∞ (Example 1.7). Denote by $\mathcal{Q}$ the subset consisting of those functions having a derivative at no point in $[0, 1[$. (Conceivably the set $\mathcal{Q}$ could be empty, but our argument will show, in particular, that it is not.) We shall show that *the set $\mathcal{Q}$ is d_∞-dense in $\mathcal{C}([0, 1])$*, in other words, that *any continuous f: $[0, 1] \to \mathsf{R}$ can be approximated, uniformly over $[0, 1]$, to within any prescribed error ϵ by a continuous g: $[0, 1] \to \mathsf{R}$ which is nowhere differentiable on $[0, 1[$.*

If $f \in \mathcal{C}([0, 1])$ has a derivative at some $x \in [0, 1[$, then the right-hand difference quotients

$$\frac{f(x + h) - f(x)}{h} \qquad (0 < h < 1 - x)$$

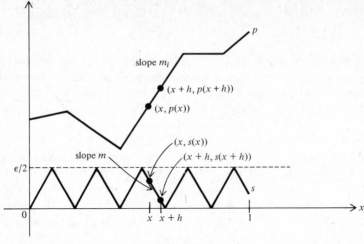

FIG. 1.22

are bounded; hence for a sufficiently large positive integer n the function f belongs to the complement of the set

$$\mathcal{a}_n = \{ f \in \mathcal{C}([0, 1]) : \text{for each } x \in [0, 1 - 1/n] \text{ there exists}$$
$$0 < h < 1 - x \quad \text{with } |f(x + h) - f(x)| > nh \}.$$

Thus

$$\mathcal{a} \supset \bigcap_{n=1}^{\infty} \mathcal{a}_n.$$

To show $\mathcal{a}$ is d_∞-dense in $\mathcal{C}([0, 1])$, it suffices to show the smaller set $\bigcap_{n=1}^{\infty} \mathcal{a}_n$ is d_∞-dense. Now d_∞ is a complete metric (1.70). According to the Baire category theorem, then, it suffices to show each $\mathcal{a}_n$ is both d_∞-open and d_∞-dense in $\mathcal{C}([0, 1])$. For the remainder of the argument the positive integer n is fixed.

To show that $\mathcal{a}_n$ is d_∞-open in $\mathcal{C}([0, 1])$ we show that its complement is d_∞-closed by using criterion 1.55. Let $(f_i \mid i \in \mathbb{N})$ be a sequence in $\mathcal{C}([0, 1]) \backslash \mathcal{a}_n$ and let $f \in \mathcal{C}([0, 1])$ with this sequence converging to f with respect to d_∞. We must show $f \notin \mathcal{a}_n$. For each $i \in \mathbb{N}$ there is a point x_i with

$$0 \leq x_i \leq 1 - \frac{1}{n}, \qquad |f_i(x_i + h) - f_i(x_i)| \leq nh \qquad (0 < h < 1 - x_i).$$

A bisection argument produces a point $x \in [0, 1 - 1/n]$ such that each d-neighborhood of x contains x_i for infinitely many values of i, where d is the Euclidean metric on $[0, 1]$.

We shall show that

$$|f(x + h) - f(x)| < nh + \epsilon$$

for each $\epsilon > 0$ and all $0 < h < 1 - x$. It will then follow that

$$|f(x+h) - f(x)| \leq nh$$

for all $0 < h < 1 - x$, so that $f \notin \mathcal{Q}_n$. Let $\epsilon > 0$ be given, and let $0 < h < 1 - x$. By continuity of f at x and $x + h$ there are d-neighborhoods U of x and V of $x + h$ such that

$$u \in U \;\Rightarrow\; |f(u) - f(x)| < \frac{\epsilon}{4}, \qquad v \in V \;\Rightarrow\; |f(x+h) - f(v)| < \frac{\epsilon}{4}.$$

Next, there is an $i \in \mathbb{N}$ such that

$$x_i \in U, \qquad x_i + h \in V, \qquad d_\infty(f, f_i) < \frac{\epsilon}{4}.$$

Then

$$|f(x+h) - f(x)| \leq |f(x+h) - f(x_i+h)| + |f(x_i+h) - f_i(x_i+h)|$$
$$+ |f_i(x_i+h) - f_i(x_i)| + |f_i(x_i) - f(x_i)|$$
$$+ |f(x_i) - f(x)|$$
$$< \frac{\epsilon}{4} + \frac{\epsilon}{4} + nh + \frac{\epsilon}{4} + \frac{\epsilon}{4} = nh + \epsilon.$$

To see that $\mathcal{Q}_n$ is d_∞-dense in $\mathcal{C}([0,1])$, let $f \in \mathcal{C}([0,1])$ and let $\epsilon > 0$. By a comment above, there is a piecewise linear $p \in \mathcal{C}([0,1])$ with $d_\infty(f, p) < \epsilon/2$. Hence we need only find some $g \in \mathcal{Q}_n$ with $d_\infty(p, g) \leq \epsilon/2$. The finitely many line segments comprising the graph of p have slopes $m_1, \ldots, m_k$. Choose a number m with

$$m > n + \max_{1 \leq i \leq k} |m_i|,$$

and let s be the "sawtooth" piecewise linear function whose line segments have slopes $\pm m$ and for which

$$0 \leq s(x) \leq \frac{\epsilon}{2} \qquad\qquad (0 \leq x \leq 1)$$

(see Fig. 1.22). Set

$$g = p + s.$$

Then $g \in \mathcal{C}([0,1])$ with

$$d_\infty(p, g) = \max_{0 \leq x \leq 1} s(x) = \frac{\epsilon}{2}.$$

We show $g \in \mathcal{Q}_n$. Let $0 \leq x \leq 1 - 1/n$. Choose $0 < h < 1 - x$ so small that the points $(x, p(x))$ and $(x+h, p(x+h))$ both lie on a line segment of the graph of p, of slope m_i, and at the same time the points $(x, s(x))$ and

$(x + h, s(x + h))$ both lie on a line segment of the graph of s. Then

$$|p(x + h) - p(x)| = |m_i|h,$$

$$|s(x + h) - s(x)| = mh > nh + |m_i|h.$$

Hence

$$|g(x + h) - g(x)| > nh.$$

Thus $g \in \mathcal{Q}_n$.

It must be noted that the Baire category theorem is not really needed to demonstrate the existence of everywhere continuous, nowhere differentiable functions. Such functions can be exhibited explicitly—see Goldberg [13, Section 9.7] or Boas [5, Section 21].

EXERCISES

64. What can be said about a Cauchy sequence in an arbitrary metric space if its range is finite?

65. A sequence $(x_n \mid n \in \mathsf{N})$ in a metric space (X, d) is said to *cluster at x in* (X, d), where $x \in X$, if for each $\epsilon > 0$ and each $m \in \mathsf{N}$ there is some $n \geq m$ with $d(x, x_m) < \epsilon$. Prove or disprove:

(**a**) If $(x_n \mid n \in \mathsf{N})$ clusters at both x and y in (X, d), then $x = y$.

(**b**) If $(x_n \mid n \in \mathsf{N})$ clusters at x and converges to y in (X, d), then $x = y$.

(**c**) If $(x_n \mid n \in \mathsf{N})$ converges to x in (X, d), then it clusters at x in (X, d).

(**d**) If $(x_n \mid n \in \mathsf{N})$ clusters at some point in (X, d) and if (X, d) is complete, then $(x_n \mid n \in \mathsf{N})$ converges in (X, d).

(**e**) If $(x_n \mid n \in \mathsf{N})$ is a Cauchy sequence clustering at some point in (X, d), then it converges in (X, d).

(**f**) If $(x_n \mid n \in \mathsf{N})$ is a sequence in a d-closed subset E of X which clusters at a point x in (X, d), then $x \in E$.

66. Given metric spaces (X, d) and (Y, d'), a map $f: X \to Y$ is said to be (d, d')-*uniformly continuous* when for each $\epsilon > 0$ there exists some $\delta > 0$ such that for all $x, t \in X$,

$$d(x, t) < \delta \implies d'(f(x), f(t)) < \epsilon.$$

For example, any isometry from (X, d) to (Y, d') is (d, d')-uniformly continuous.

(**a**) Prove: If f is (d, d')-uniformly continuous, then it is (d, d')-continuous.

(**b**) By considering the function $x \mapsto x^2$ from R to R, or the function $x \mapsto 1/x$ from $\{x \in \mathsf{R} \mid x > 0\}$ to R, show that the converse of (a) fails. [*Note:* The converse of (a) does hold for certain spaces among which are all closed bounded intervals in R—see 4.43.]

(**c**) In the notation of 1.45(7), show that $d: X \times X \to \mathsf{R}$ is (d_∞, d')-uniformly continuous.

(**d**) Give an analog of 1.45 (5) for uniform continuity.

(**e**) Which of the maps from X to R considered in Exercises 51 and 52 are (d, d')-uniformly continuous?

67. (Continuation of Exercise 66.) Suppose f is (d, d')-uniformly continuous.

(**a**) If $(x_n \mid n \in \mathsf{N})$ is a Cauchy sequence in (X, d), show that $(f(x_n) \mid n \in \mathsf{N})$ is a Cauchy sequence in (Y, d'). Is the same true if f is only (d, d')-continuous?

(**b**) If D and D' are metrics on X and Y equivalent to d and d', respectively, must f be (D, D')-uniformly continuous?

68. Let $(x_n \mid n \in \mathsf{N})$ and $(y_n \mid n \in \mathsf{N})$ be Cauchy sequences in a metric space (X, d). Prove that $(d(x_n, y_n) \mid n \in \mathsf{N})$ converges in (R, d'), where d' is the Euclidean metric.

69. If d is a complete metric on X, is the bounded metric d' on X given by $d'(x, y) = d(x, y)/[1 + d(x, y)]$ necessarily complete? (See Exercise 33.)

70. Is the metric D on the extended real line R^* (1.37) complete?

71. Is the p-adic metric d_p (Exercise 15) on Q complete?

72. Construct a metric space (X, d) in which every subset is d-open, yet (X, d) is not complete (and hence d is not the discrete metric).

73. In the notation of Exercise 34 (a), suppose the metrics $d_1, d_2, \ldots$ are all complete. Prove that the metric d' on the product set X is then complete.

74. Given a nonempty set X and a complete, bounded metric d on a set Y, show that the formula

$$d_\infty(f, g) = \sup \{d(f(x), g(x)) \mid x \in X\}$$

defines a complete metric on the set of all maps from X to Y.

75. Let V be a d-open set in a complete metric space (X, d) with $V \neq X$. Let d' be the metric induced on V by d. Prove that there is a complete metric d^* on V equivalent to d' as follows:
(**a**) The function $f: V \to \mathsf{R}$ defined by $f(x) = 1/d(x, X\backslash V)$ is (d', d_1) continuous, where d_1 is the Euclidean metric.
(**b**) The formula $d^*(x, y) = d(x, y) + |f(x) - f(y)|$ defines a metric on V.
(**c**) The metric d^* is equivalent to d.
(**d**) The metric space (V, d^*) is complete.

76. Show by example that the nested set theorem (1.71) no longer remains true if any of the hypotheses that (X, d) be complete, that each E_n be d-closed, or that $\lim_{n \to \infty} \operatorname{diam} E_n = 0$ is omitted.

77. Does the nested set theorem remain true if the hypothesis that $(E_n \mid n \in \mathsf{N})$ be decreasing is replaced by the weaker hypothesis that each finite collection of the sets E_n have nonempty intersection?

78. Generalize the nested interval property (0.44) to R^k. [*Note*: The analog of an interval $[a, b]$ in R is, of course, a cube (the product of k closed intervals) in R^k, ∿ you should consider a decreasing sequence of such cubes. No assumption is to be made, however, that the diameters of these cubes tend to zero, so the nested set theorem (1.71) cannot be applied directly to the cubes.]

79. Prove the following converse to the nested set theorem: A metric space (X, d) is complete if each decreasing sequence $(E_n \mid n \in \mathsf{N})$ of nonempty, d-closed, d-bounded sets with $\lim_{n \to \infty} \operatorname{diam} E_n = 0$ has nonempty intersection.

80. (**a**) Let $(E_n \mid n \in \mathsf{N})$ be a sequence of d-closed subsets of a complete metric space (X, d) such that no E_n contains any nonempty d-open set. Show that $\bigcup_{n=0}^{\infty} E_n$ contains no nonempty d-open set.
(**b**) Apply (a) to show that the plane is not the union of countably many lines.

81. The Euclidean metric d on $]0, 1[$ is not complete. Show that, nonetheless, the intersection of any sequence of d-open, d-dense sets in $]0, 1[$ is d-dense in $]0, 1[$.

82. Does there exist any complete metric on Q equivalent to the Euclidean metric?

83. The argument in 1.73 actually shows that the set of function $f \in \mathcal{C}([0, 1])$ for which the right-hand derivative $\lim_{h\to 0+} [f(x + h) - f(x)]/h$ exists at no $x \in [0, 1[$ is d_∞-dense in $\mathcal{C}([0, 1])$. Prove the analogous result concerning left-hand derivatives.

84. Given a metric space (X, d), a map $T: X \to X$ is called a *contraction* if there is a constant $c < 1$ with

$$d(Tx, Ty) \le c\, d(x, y) \qquad\qquad (x, y \in X).$$

Verify that T is a contraction when:

(**a**) $(X, d) = (\mathsf{H}, d_2)$, the Hilbert sequence space (1.9), and $Tx = (\frac{1}{2}x_i \mid i = 1, 2, \ldots)$ for each $x = (x_i \mid i = 1, 2, \ldots) \in \mathsf{H}$.

(**b**) $X = [0, 1]$ with its Euclidean metric, and $T: X \to X$ is a continuous function that is differentiable on $]0, 1[$ with $|T'(x)| \le c$ for some $c < 1$ and all $x \in]0, 1[$. (*Hint*: Use the mean-value theorem of calculus.)

85. (Continuation of Exercise 84.) Prove the *contraction mapping principle* of Banach: Let (X, d) be a complete metric space and let $T: X \to X$ be a contraction. Then T has a unique "fixed-point" $x \in X$ (that is, $Tx = x$) to which $(T^n y \mid n \in \mathsf{N})$ converges in (X, d) for each $y \in X$ (here $T^2 = T \circ T$, $T^3 = T \circ T^2$, etc.). [*Note*: An application of this theorem is given in the following exercise. Another application is the proof of the inverse-function theorem which says that a continuously differentiable function from R^n to R^n with a nonsingular Jacobian has at each point a local inverse—see Smith [30, Chapter 10, Section 7]. Other fixed-point theorems are considered in Chapter 5.]

86. (Continuation of Exercise 85.) Prove the *Picard existence theorem*: Let f be a real-valued continuous function on a rectangle

$$D = [x_0 - a, x_0 + a] \times [y_0 - b, y_0 + b]$$

in the plane satisfying a "Lipschitz condition"

$$|f(x, y) - f(x, z)| \le K|y - z| \qquad\qquad ((x, y), (x, z) \in D)$$

on D for some constant $K > 0$. Then on some interval $[x_0 - \delta, x_0 + \delta]$ containing x_0 there exists a solution g of the differential equation

(*) $g'(x) = f(x, g(x))$

satisfying the initial condition

(**) $g(x_0) = y_0.$

[*Method*: Show first that $g: [x_0 - \delta, x_0 + \delta] \to \mathsf{R}$ satisfies (*) and (**) if and only if

$$g(x) = y_0 + \int_{x_0}^{x} f(t, g(t))\, dt \qquad\qquad (|x - x_0| < \delta).$$

Choose a constant $M > 0$ with

$$|f(x, y)| \le M \qquad\qquad ((x, y) \in D)$$

and then a $\delta > 0$ such that

$$K\delta < 1, \qquad [x_0 - \delta, x_0 + \delta] \times [y_0 - M\delta, y_0 + M\delta] \subset D.$$

Define

$$\mathcal{E} = \{g \in \mathcal{C}([x_0 - \delta, x_0 + \delta]) : |g(x) - y_0| < M\delta \text{ for } |x - x_0| < \delta\}.$$

Show that $\mathcal{E}$ is d_∞-closed in $\mathcal{C}([x_0 - \delta, x_0 + \delta])$ and hence that the max metric on $\mathcal{E}$ is complete. Finally, show that

$$(Tg)(x) = y_0 + \int_{x_0}^{x} f(t, g(t)) \, dt$$

defines a map $T : \mathcal{E} \to \mathcal{E}$ which is a contraction. (For more about differential equations and the Picard theorem see Kreider, Kuller and Ostberg [22].)]

87. Let d be a pseudometric on a set X (Exercise 13). Cauchy sequences and completeness for (X, d) may be defined exactly as for the case that d is a metric. Show that (X, d) is complete if and only if the associated metric space (X^*, d^*) constructed in Exercise 14 is complete.

88. Let d be a metric (or pseudometric) on a set X. Suppose there is a d-dense set Z in X such that each Cauchy sequence in (X, d) consisting of points of Z converges in (X, d). Prove that (X, d) is complete.

89. Given a pseudometric d on X, let $(x_n \mid n \in \mathsf{N})$ be a Cauchy sequence in (X, d).
  (**a**) Show that the sequence $(d(y, x_k) \mid k \in \mathsf{N})$ of real numbers converges for each $y \in X$.
  (**b**) Show that $\lim_{n \to \infty} \lim_{k \to \infty} d(x_n, x_k) = 0$. [In view of (a), for each $n \in \mathsf{N}$ the limit $\lim_{k \to \infty} d(x_n, x_k)$ exists.]

90. Let (X, d) be any metric space. A *completion* of (X, d) is an isometric embedding f of (X, d) into a complete metric space (X', d') in which $f(X)$ is d'-dense. Carry out the following steps in Hausdorff's construction of a completion of (X, d):
  (**a**) Let Y be the set whose points are the Cauchy sequences in (X, d). Exercise 68 justifies forming

$$D(x, y) = \lim_{n \to \infty} d(x_n, y_n)$$

for any points $x = (x_n \mid n \in \mathsf{N})$ and $y = (y_n \mid n \in \mathsf{N})$ of Y. Verify that D is a pseudometric on Y.
  (**b**) Show that the map $g : X \to Y$ defined by the rule that $g(x) = (x_n \mid n \in \mathsf{N})$ with $x_n = x$ for all $n \in \mathsf{N}$ is an isometric embedding of (X, d) into (Y, D).
  (**c**) Show that $g(X)$ is D-dense in Y.
  (**d**) Prove that (Y, D) is complete. [*Hint*: Use Exercise 88. You may also want to use Exercise 89.]
  (**e**) Let (Y^*, D^*) be the metric space constructed from (Y, D) as in Exercise 14, so that (Y^*, D^*) is complete by Exercise 87. In the notation of Exercise 14, let $p : Y \to Y^*$ be the map sending each $y \in Y$ to its equivalence class y^* in Y. Show that $f = p \circ g : X \to Y^*$ is an isometric embedding of (X, d) into (Y^*, D^*) with $f(X)$ being D^*-dense in Y^*.
  (*Note*: This construction is based on Cantor's construction of the real numbers from the rational numbers.)

91. (Continuation of Exercise 90.) This exercise establishes the essential uniqueness— or as one says, "uniqueness up to isometry"—of a completion of an arbitrary metric space. Suppose f and f' are isometric embeddings of (X, d) into metric spaces (Y, D) and (Y', D'), respectively, which define completions of (X, d). Then there is a unique isometry h from (Y, D) to (Y', D') with $h \circ f = f'$. [*Hint:* To define $h(y)$ for a point $y \in Y$, use the fact that $f(X)$ is D-dense in Y to find a sequence $(x_n \mid n \in \mathsf{N})$ in X such that $(f(x_n) \mid n \in \mathsf{N})$ converges to y in (Y, D). Show that $(f'(x_n) \mid n \in \mathsf{N})$ converges to some point y' in (Y', D') and then set $h(y) = y'$. You will need to prove that the $y' \in Y'$ so obtained is independent of the choice of the sequence $(x_n \mid n \in \mathsf{N})$; in other words, if $(z_n \mid n \in \mathsf{N})$ is another sequence in X such that $(f(z_n) \mid n \in \mathsf{N})$ converges to y in (Y, D), then $(f'(z_n) \mid n \in \mathsf{N})$ also converges to y'.]

92. (Continuation of Exercise 90.) This exercise provides an alternative construction of a completioh of a metric space (X, d). Arbitrarily choose a point $z \in X$.

(**a**) For each $x \in X$ let $f_x \colon X \to \mathsf{R}$ be the function defined by

$$f_x(y) = d(x, y) - d(z, y) \qquad\qquad (y \in X).$$

Verify that each f_x is bounded and (d, d')-continuous, where d' is the Euclidean metric on R. Thus, in the notation of 1.70, $f_x \in \mathcal{C}^*(X)$ for each $x \in X$.

(**b**) Let $F \colon X \to \mathcal{C}^*(X)$ be the map $x \mapsto f_x$. Show that F is an isometric embedding of (X, d) into $(\mathcal{C}^*(X), d_\infty)$, where d_∞ is the sup metric. [*Note:* Although $(\mathcal{C}^*(X), d_\infty)$ is complete, we do not yet necessarily have a completion of (X, d), because $F(X)$ need not be d_∞-dense in $\mathcal{C}^*(X)$.]

(**c**) Define Y to be the set of those $g \in \mathcal{C}^*(X)$ to which some sequence $(g_n \mid n \in \mathsf{N})$ in $F(X)$ converges in $(\mathcal{C}^*(X), d_\infty)$. Let D be the metric on Y induced by d_∞. Prove that $F(X)$ is D-dense in Y and that Y is d_∞-closed in $\mathcal{C}^*(X)$. Conclude that (Y, D) is complete. [Thus F, regarded as a map from X into Y, defines a completion of (X, d).]

Topological Spaces

In the preceding chapter the notion of a metric space was abstracted from Euclidean spaces. In this chapter the process of abstraction is carried a step further to obtain the notion of a topological space. In a topological space there is no longer any quantitative measure of distance provided by a metric. Nonetheless, we can still talk about open and closed sets, neighborhoods, subspaces, and (as we shall see in the next chapter) limits and continuity.

1. Topologies

We have seen that a given set (even R^n) can have a number of different metrics yielding the same open sets. Now, our definition of continuity of a map from one metric space to another was phrased in terms of the particular metrics on the two spaces. According to 1.47, however, continuity of such a map depends not upon the metrics themselves, but solely upon the open sets they define. Similarly, for convergence of sequences it is the open sets and not the metrics that matter. All this suggests we ought to be able to discuss continuity and convergence without any recourse to metrics at all, just so long as we can still talk about open sets. Hence we shall discard what is not really essential for understanding continuity and convergence and shall study open sets in a general setting devoid of any metrics.

For a metric space the notion 'open set' was defined by means of the metric, but now we want to talk about open sets in a context where there are no distances and no ϵ-balls. Hence we no longer specify which sets are open by means of some prior notion. Rather, we suppose already given in a set certain subsets said to be "open"; all we presume to know about these subsets is specified by a few properties—taken as axioms—we *assume* they have. The properties we use as axioms are, of course, ones that are true of the open sets obtained from a metric.

2.1 Definition. Let X be a set. A *topology on* X is a collection $\mathfrak{I}$ of subsets of X such that:

>   (O1) $\varnothing \in \mathfrak{I}$ and $X \in \mathfrak{I}$.
>   (O2) Any union of sets belonging to $\mathfrak{I}$ also belongs to $\mathfrak{I}$.
>   (O3) Any intersection of *finitely many* sets belonging to $\mathfrak{I}$ also belongs to $\mathfrak{I}$.

The subsets of X belonging to a topology $\mathfrak{I}$ on X are variously said to be $\mathfrak{I}$-*open*, *open for* $\mathfrak{I}$, *open in* X, or simply *open*.

To verify that a collection $\mathfrak{I}$ is a topology it is enough to check instead of (O3) the simpler condition:

>   (O3′) The intersection of any *two* sets belonging to $\mathfrak{I}$ also belongs to $\mathfrak{I}$.

In fact, an easy application of mathematical induction shows that (O3′) implies (O3).

At last we come to the abstraction from metric spaces which is the principal object of study in general topology.

2.2 Definition. A *topological space* is an ordered pair $(X, \mathfrak{I})$ where X is a set, called the *underlying set of* $(X, \mathfrak{I})$, and $\mathfrak{I}$ is a topology on X, called the *topology of* $(X, \mathfrak{I})$.

Informally, then, a topological space consists of a set X provided with a specified collection of "open" subsets, subject only to the conditions that the empty set and the whole set X are open, any union of open sets is open, and any finite intersection of open sets is open.

2.3 Examples

(1) Let (X, d) be any metric space. Take $\mathfrak{I}$ to be the collection of all d-open subsets of X in the sense of 1.18; a subset U of X belongs to $\mathfrak{I}$ precisely when at each point of U there is a d-ball contained in U. In view of the motivation for Definition 2.1 it should come as no surprise that $\mathfrak{I}$ is a topology on X; Theorem 1.19 says just that.

The topology $\mathfrak{I}$ is said to be *induced by* d and is sometimes denoted by $\mathfrak{I}(d)$ to indicate its dependence on d. The topological space $(X, \mathfrak{I})$ is said to be *associated with* the metric space (X, d).

Unless otherwise indicated, any future reference to a topology on the set underlying a metric space shall be to the one induced by the metric.

Owing to the many examples of metric spaces presented in Chapter 1, numerous examples of topological spaces are now at hand.

(2) Specialize (1) by taking X to be a subset of some Euclidean space $\mathbf{R}^n$ and d to be the Euclidean metric on X. The topology $\mathfrak{I}(d)$ induced by d is called the *usual topology* on X. In particular, n-dimensional Euclidean space $\mathbf{R}^n$ becomes a topological space when provided with its usual topology.

Unless otherwise indicated, any future reference to a topology on a subset of a Euclidean space is to its usual topology.

(3) Let $X = \mathbf{R}^n$, let d be the Euclidean metric on X, and let d_∞ be the max metric on X. By 1.33, a subset of X is d-open if and only if it is d_∞-open. In other words the topology $\mathfrak{I}(d)$ on X induced by d is the same as the topology $\mathfrak{I}(d_\infty)$ induced by d_∞. However, $d \neq d_\infty$ if $n > 1$. Thus the metric spaces (X, d) and (X, d_∞) are different if $n > 1$, but their associated topological spaces $(X, \mathfrak{I}(d))$ and $(X, \mathfrak{I}(d_\infty))$ are the same.

(4) This example shows that any set can be made the underlying set of a topological space. Let X be a set. Then the collection $\mathcal{P}(X)$ of all subsets of X is obviously a topology on X, called the *discrete* topology. Every subset of X is open for this topology. Now the discrete metric δ on X (see 1.11) has the property that every subset of X is δ-open. Hence the discrete topology is induced by the discrete metric.

A topological space $(X, \mathfrak{I})$ is called a *discrete space* if its topology $\mathfrak{I}$ is the discrete topology on X.

(5) Let X be any set. Then $\{\varnothing, X\}$ is trivially a topology on X, called the *indiscrete topology* (the pun is intentional).

Of all topologies on a given set X the indiscrete topology is the "smallest" and the discrete topology is the "largest" in the sense that

$$\{\varnothing, X\} \subset \mathfrak{I} \subset \mathcal{P}(X)$$

for every topology $\mathfrak{I}$ on X. [See 3.8 (7) for further discussion of the comparison of topologies on the same underlying set.]

On a set of two points Examples (4) and (5) already give two different topologies. Here is a third one.

(6) Let $X = \{0, 1\}$. Then the collection $\mathfrak{I} = \{\varnothing, \{0\}, X\}$ is a topology on X. The topological space $(X, \mathfrak{I})$ is called the *Sierpinski space*.

(7) Let X be any set. Define $\mathfrak{I}$ to be the collection consisting of $\varnothing$ together with all subsets of X whose complements in X are finite. Then $\mathfrak{I}$ is a topology on X, called the *finite-complement topology*.

To see that $\mathfrak{I}$ is actually a topology on X we must, of course, verify that axioms (O1)–(O3) are satisfied. We verify, for example, axiom (O3). Let $(U_j \mid j \in J)$ be a nonempty finite family of sets belonging to $\mathfrak{I}$; we must show that $\bigcap_{j \in J} U_j \in \mathfrak{I}$. By definition of $\mathfrak{I}$, for each $j \in J$ either $U_j = \varnothing$ or $X \setminus U_j$ is finite. Then there are two cases to consider.

Case (i): $U_j = \varnothing$ for some $j \in J$. Then $\bigcap_{j \in J} U_j = \varnothing$ also, so this intersection belongs to $\mathfrak{I}$.

Case (ii): $U_j \neq \emptyset$ for all $j \in J$. Then $X \setminus U_j$ is finite for each j. Now

$$X \setminus \bigcap_{j \in J} U_j = \bigcup_{j \in J} (X \setminus U_j)$$

by one of De Morgan's laws (0.17), and the union on the right is finite. Hence $\bigcap_{j \in J} U_j \in \mathfrak{I}$ in this case too.

However peculiar or unnatural the finite-complement topology might seem, it satisfies the axioms for a topology and hence is just as genuine a topology as any other. In fact, it is interesting precisely because of its pathology.

Among all topological spaces those associated with metric spaces are important enough to deserve a special name.

2.4 Definition. A topological space $(X, \mathfrak{I})$ and its topology $\mathfrak{I}$ are both said to be *metrizable* if $\mathfrak{I}$ is the topology induced by some metric on X.

Note the distinction between 'metric space' and 'metrizable space'. A *metric* space (X, d) consists of a set X together with one particular metric d on X; it gives rise to the metrizable topological space $(X, \mathfrak{I}(d))$ but is not itself a topological space. By way of contrast, a *metrizable* space $(X, \mathfrak{I})$ is a topological space; its topology can be induced by some metric (and usually by many metrics), but no particular metric is specified.

Metrizable spaces are the "concrete" objects that topological spaces generalize. Hence they will be important for motivating definitions, suggesting theorems, and serving as test cases for conjectures about topological spaces in general.

There would be ample justification for studying topological spaces even were there not a single example of a nonmetrizable space (see the first paragraph of this section). The fact is that nonmetrizable spaces not only do exist but can behave in peculiar ways metrizable spaces cannot. Nonetheless, it is all too easy for the novice to fall into the psychological trap of believing that the only kind of topological space is a metrizable one. Machinery for dispelling this belief will very soon be at hand.

The definition of 'closed set' in a metric space made no direct reference to the metric, only to the open sets the metric determined. By using the very same definition, we can then talk about closed sets in an arbitrary topological space.

2.5 Definition. Given a topological space $(X, \mathfrak{I})$, a subset E of X is said to be *$\mathfrak{I}$-closed, closed for $\mathfrak{I}$, closed in X*, or simply *closed* when its complement $X \setminus E$ in X is $\mathfrak{I}$-open.

The basic properties (1.22) of closed sets in a metric space hold almost verbatim in any topological space.

2.6 Theorem. Let $(X, \Im)$ be a topological space. Then:
(1) The empty set $\varnothing$ and the entire set X are closed in X.
(2) Any intersection of closed subsets of X is closed in X.
(3) Any union of *finitely many* closed subsets of X is closed in X.

Aside from De Morgan's laws, the only thing used to prove 1.22 was Theorem 1.19. Since the properties of open sets listed in Theorem 1.19 hold by assumption in any topological space, the proof of 1.22 can be used for its generalization 2.6 as well.

We are now going to give a method for making each and every subset of a given topological space into a topological space in its own right. Let $(X, \Im)$ be a topological space and let Y be a subset of X. Of course there are topologies on Y—the discrete topology, for example—but we want to use the topology $\Im$ on X to construct a topology on Y.

For motivation, consider the special case in which $(X, \Im)$ is metrizable. Choose a metric d on X inducing $\Im$. Restrict $d: X \times X \to \mathsf{R}$ to $Y \times Y$ to obtain the metric d' on Y induced by d. The metric d' in turn induces a topology S on Y, yielding the new topological space (Y, S). By 1.20 (3), the d'-open subsets of Y are just the intersections with Y of the d-open subsets of X, that is,

$$\mathsf{S} = \{U \cap Y \mid U \in \Im\}.$$

Observe that metrics do not appear in this expression for S. Hence this expression could be used to define a topology S on Y even when $(X, \Im)$ is not metrizable, provided we could be assured that S is really a topology when there are no metrics.

2.7 Lemma. Let $(X, \Im)$ be an arbitrary topological space and let $Y \subset X$. Then the collection

$$\mathsf{S} = \{U \cap Y \mid U \in \Im\}$$

of subsets of Y is a topology on Y.

Proof. We verify axioms (O1)–(O3) for S.
(O1) Since $\varnothing = \varnothing \cap Y$ and $Y = X \cap Y$, both $\varnothing$ and Y belong to S.
(O2) Let $(V_i \mid i \in I)$ be any family of sets belonging to S. For each $i \in I$ we may write

$$V_i = U_i \cap Y$$

with $U_i \in \Im$. Since $\bigcup_{i \in I} U_i \in \Im$, the set

$$\bigcup_{i \in I} V_i = \bigcup_{i \in I} (U_i \cap Y) = \left(\bigcup_{i \in I} U_i \right) \cap Y$$

belongs to S.
(O3) In the notation of the preceding paragraph, suppose I is finite. Since

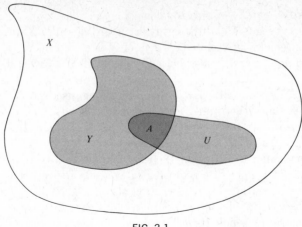

FIG. 2.1

$\bigcap_{i \in I} U_i \in \mathfrak{I}$, the set

$$\bigcap_{i \in I} V_i = \bigcap_{i \in I} (U_i \cap Y) = \left(\bigcap_{i \in I} U_i\right) \cap Y$$

belongs to $\mathcal{S}$. $\square$

This lemma justifies the following definition.

2.8 Definition. Let $(X, \mathfrak{I})$ be a topological space. If Y is a subset of X, then the topology $\mathcal{S} = \{U \cap Y \mid U \in \mathfrak{I}\}$ on Y is called the *relative topology on Y induced by* $\mathfrak{I}$, and the topological space $(Y, \mathcal{S})$ is said to be a *subspace of* $(X, \mathfrak{I})$.

Unless otherwise explicitly indicated, any future reference to a topology on a subset Y of a given topological space $(X, \mathfrak{I})$ is to the relative topology on Y induced by the given topology $\mathfrak{I}$ on X.

Thus for a set A in a subspace $(Y, \mathcal{S})$ of a topological space $(X, \mathfrak{I})$,

$\qquad A$ is open in $Y \quad \Leftrightarrow \quad A = U \cap Y$ for some set U open in X

(see Fig. 2.1).

2.9 Examples.

(1) To recapitulate the discussion preceding 2.7 in the language of Definition 2.8, let $(X, \mathfrak{I})$ be a metrizable topological space, let d be a metric on X inducing $\mathfrak{I}$, and let $Y \subset X$. If d' is the metric on Y induced by d, then the topology $\mathcal{S}$ on Y induced by d' is the same as the relative topology on Y induced by the topology $\mathfrak{I}$. This fact is indicated schematically in Fig. 2.2, where an arrow stands for 'induces'.

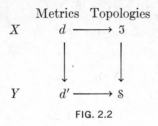

Metrics Topologies

FIG. 2.2

Metrizability is thus a "hereditary" property of topological spaces in the sense that *every subspace of a metrizable topological space is metrizable.*

(2) Earlier we made the convention that a subset Y of R^n is to be considered a topological space by using its "usual" topology, the topology induced by the Euclidean metric on Y. Hence the usual topology on Y is just the relative topology induced by the usual topology on R^n.

(3) Specialize (1) by taking $X = \mathsf{R}^*$ (the extended real line), $d = D$ (the metric on R^* defined in 1.37), and $Y = \mathsf{R}$. Let d' be the metric on R induced by D. By 1.38, both d' and the Euclidean metric on R induce the same topology on R. Hence the real line R (with its usual topology) is a subspace of the extended real line R^* (with its topology induced by D).

It is easy to tell which sets are closed in a subspace.

2.10 Theorem. Let $(Y, \mathcal{S})$ be a subspace of a topological space $(X, \mathcal{T})$, and let $A \subset Y$. Then A is closed in Y if and only if $A = E \cap Y$ for some closed subset E of X.

Proof. If A is closed in Y, then $Y \backslash A$ is open in Y, $Y \backslash A = U \cap Y$ for some open subset U of X, and hence

$$A = Y \backslash (U \cap Y) = Y \cap (X \backslash U)$$

with $X \backslash U$ closed in X. Conversely, if $A = E \cap Y$ with E closed in X, then $U = X \backslash E$ is open in X,

$$Y \backslash A = Y \backslash (E \cap Y) = Y \backslash [(X \backslash U) \cap Y] = U \cap Y$$

is open in Y, and hence A is closed in Y.   ⬜

According to 2.8 and 2.10, if $A \subset Y$ and if A is open or closed in the entire space X, then $A = A \cap Y$ is open or closed, respectively, in the subspace Y. The converse is not true in general; for example, $A = [0, 1[$ is both open and closed in $Y = [0, 1]$, but A is neither open nor closed in $X = \mathsf{R}$. The following partial converse is true, however.

2.11 Proposition. Let $(Y, \mathcal{S})$ be a subspace of a topological space $(X, \mathcal{T})$, and let $A \subset Y$. Then:

(1) If A is open in Y and Y is open in X, then A is open in X.

(2) If A is closed in Y and Y is closed in X, then A is closed in X.

Proof. (1) If A is open in Y, then $A = U \cap Y$ with U open in X. If in addition Y is open in X, then the intersection $A = U \cap Y$ of two open subsets of X is open in X.

(2) Replace 'open' by 'closed' above. $\square$

We have by now two good reasons for using at times such expressions as 'open for $\mathfrak{I}$' and 'closed in X' instead of the simpler 'open' and 'closed'. First, even so innocent a set as $X = \{0, 1\}$ has several different topologies, so a reference to an open or a closed subset of X will be ambiguous until a particular topology on X is singled out. Second, a set A in a topological space $(X, \mathfrak{I})$ may be a subset of a subspace $(Y, \mathfrak{S})$, and then being open or closed in Y (that is, for $\mathfrak{S}$) need not be the same thing as being open or closed in X (that is, for $\mathfrak{I}$).

Ordinarily, our discussion of open or closed sets will concern either an arbitrary topology or else some particular topology whose identity is clear from the context. In that case, when there is no danger of genuine ambiguity, we may suppress the $\mathfrak{I}$ and indulge in the following "abuse of language":

Given a set X underlying a topological space $(X, \mathfrak{I})$, we refer to X itself as a topological space.

As an example of this convention, we call a subset Y of a set X having a topology $\mathfrak{I}$ simply a subspace of X. Of course we must still distinguish being open in Y from being open in X.

EXERCISES

1. (**a**) Show that a set of one element has a unique topology. Does any set having more than one element have a unique topology?

   (**b**) Find a fourth topology on $\{0, 1\}$ different from the three obtained in 2.3. Are there more than four?

   (**c**) Find all topologies on the three-element set $\{0, 1, 2\}$. Are some of these like others of these in some sense?

   (**d**) Show that there are at most 2^{2^n} topologies on a finite set consisting of n elements. (*Note*: It is still an unsolved problem to find a nice formula telling how many topologies there are on a set of n elements, or even to calculate the actual number for any but small values of n.)

2. Is a discrete space the only kind of topological space having the property that each one-point set $\{x\}$ is open?

3. The Sierpinski space $[2.3(6)]$ can be generalized in two ways as follows. Let X be a set and let $x_0 \in X$.

   (**a**) Verify that $\{U \mid U \subset X, U = \varnothing \text{ or } x_0 \in U\}$ is a topology on X. This is called an *included point topology*.

(**b**) Verify that $\{U \mid U \subset X, x_0 \notin U\}$ is a topology on X. This is called an *excluded point topology*.

4. The condition $x_0 \in U$ used in Exercise 3 to define an included point topology may be written $\{x_0\} \subset U$. This suggests a generalization. Let A be a subset of a set X.
(**a**) Verify that $\{U \mid U = \varnothing \text{ or } A \subset U \subset X\}$ is a topology on X. This is called an "included set topology".
(**b**) Can excluded point topologies be similarly generalized to "excluded set topologies"?

5. Are there only countably many or uncountably many topologies on the set N of all natural numbers?

6. Let X be any set. Call a subset U of X "open" whenever $U = \varnothing$ or $X \setminus U$ is countable. Show that this collection of "open" subsets of X really is a topology on X. This topology is called the *countable-complement topology* on X.

7. Consider the usual topology, the finite-complement topology, and the countable-complement topology on R. Are any of these contained in others of these?

8. Is the intersection of any two topologies on a set necessarily a topology on the set? The union of two topologies?

9. Call a subset U of the plane $\mathbf{R}^2$ "linearly open" if its intersection with each line L in $\mathbf{R}^2$ is open in L (for the usual topology of L). Let S be the collection of all linearly open subsets of $\mathbf{R}^2$.
(**a**) Verify that S is a topology on $\mathbf{R}^2$.
(**b**) If K is a convex subset of $\mathbf{R}^2$, show that K is open for the topology S if and only if it is open for the usual topology, and closed for S if and only if it is closed for the usual topology.
(**c**) It is obvious from the definition that a subset of the plane will be linearly open if it is open for the usual topology. Is the converse true?

10. A topological space is called a T_1-*space* if $\{x\}$ is closed in X for each $x \in X$.
(**a**) Prove that every metrizable space is a T_1-space.
(**b**) Give an example of a topological space that is not a T_1-space.
(**c**) Show that a topological space X is a T_1-space if and only if the following condition holds: Given any two distinct points x and y, there is an open set U in X with $x \in U$ and $y \notin U$.

11. Which finite topological spaces, if any, are T_1-spaces?

12. Let d be a pseudometric on a set X; according to Exercise 1.29, the collection $\mathfrak{I}(d)$ of all d-open sets is a topology on X. We way that $\mathfrak{I}(d)$ is *induced by* d. A *pseudometrizable space* is a topological space whose topology is induced by some pseudometric.
(**a**) Give an example of a pseudometrizable space that is not metrizable.
(**b**) Is the Sierpinski space [2.3(6)] pseudometrizable?
(**c**) Prove: A pseudometrizable space is metrizable if and only if it is a T_1-space (Exercise 10).

13. (**a**) Are there topological spaces in which the intersection of *any* collection of open sets is always open?
(**b**) Answer the analog of (a) for unions of closed sets.

14. Let X be a set and let $\mathcal{E}$ be a given collection of subsets of X such that: (i) $\varnothing \in \mathcal{E}$

and $X \in \mathcal{E}$, (ii) any intersection of members of $\mathcal{E}$ is a member of $\mathcal{E}$, and (iii) any union of finitely many members of $\mathcal{E}$ is a member of $\mathcal{E}$. Prove that there is then a unique topology $\mathfrak{I}$ on X for which $\mathcal{E}$ is the collection of all $\mathfrak{I}$-closed sets.

15. Let Y be a subset of a topological space X.

(**a**) Show by example that the collection

$$\mathcal{S} = \{U \mid U \subset Y, U \text{ open in } X\}$$

need not be a topology on Y.

(**b**) Is $\mathcal{S}$ ever a topology on Y?

16. Let $\mathfrak{I}$ be the usual topology on R, and let $f: \text{R} \to \text{R}$ be a map.

(**a**) Is $\{f(U) \mid U \in \mathfrak{I}\}$ necessarily a topology on R? If not, is it ever a topology on R for special maps f?

(**b**) Repeat (a) for $\{f^{-1}(V) \mid V \in \mathfrak{I}\}$.

17. Prove: A subspace Z of a subspace Y of a topological space X is itself a subspace of X. In other words, if $(X, \mathfrak{I})$ is a topological space, if $Z \subset Y \subset X$, and if $\mathfrak{I}_Y$ is the relative topology on Y induced by $\mathfrak{I}$, then the relative topology on Z induced by $\mathfrak{I}_Y$ is the same as the relative topology on Z induced by $\mathfrak{I}$. (*Note*: This removes an apparent ambiguity in our convention that a subset of a space is to be provided with its relative topology.)

18. Can a subspace consisting of more than one point of a topological space X be discrete when X is not discrete? even when the subspace is infinite? when the subspace is uncountable?

19. Is the property of being a T_1-space (Exercise 10) hereditary? In other words, must a subspace of a T_1-space necessarily be a T_1-space?

20. Let $(X_i \mid i \in I)$ be a family of topological spaces with X_i disjoint from X_j whenever $i \neq j$, and let $X = \bigcup_{i \in I} X_i$.

(**a**) Show that there is a unique topology on X for which each X_i, with its given topology, is an open subspace. The resulting topological space is called the *sum of* $(X_i \mid i \in I)$.

(**b**) Prove that each X_i is closed in X.

(**c**) Suppose $I = \text{N}$ and $X_i = \text{R} \times \{i\}$ for each $i \in I$, so that $X \subset \text{R} \times \text{R}$. Verify that the sum topology on X is the same as the usual topology on X.

(**d**) Construct an example where each X_i is a subspace of $\text{R} \times \text{R}$ yet the sum X is not a subspace of $\text{R} \times \text{R}$.

(**e**) In general, is there a relationship between each of the spaces X_i being a T_1-space and the sum X being a T_1-space (Exercise 10)?

2. Neighborhoods

In a metric space, points ϵ-close to a given point x were those belonging to the d-ball of radius ϵ at x. More generally, points belonging to any neighborhood V of x could be thought of as being close to x, although V no longer had to provide a numerical measure of closeness to x. Now our definition of neighborhoods in a metric space referred only to open sets, not to the metric. Hence we can introduce into any topological space the idea of points close to a given point simply by extending that definition.

2.12 Definition. Let x be a point in a topological space X. Then a subset V of X is called a *neighborhood of x in X* when $x \in U \subset V$ for some open subset U of X. The collection of all neighborhoods of x in X is called the *neighborhood system at x.*

As usual, we may refer to a "$\mathfrak{I}$-neighborhood of x" or a "neighborhood of x for $\mathfrak{I}$" in order to indicate the particular topology $\mathfrak{I}$ on X under consideration.

2.13 Examples

(1) Let (X, d) be a metric space and let $(X, \mathfrak{I})$ be the associated topological space. Given $x \in X$, a subset V of X is a d-neighborhood of x in the metric space sense exactly when it is a $\mathfrak{I}$-neighborhood of x. Hence the neighborhoods of a point in a metrizable space are exactly the d-neighborhoods obtained from any metric d inducing the topology.

(2) In a discrete space X, every subset of X containing a point $x \in X$ is a neighborhood of x. At the opposite extreme, in an indiscrete space X the only neighborhood of a point is the entire space X.

(3) Let X be a set provided with its finite-complement topology [2.3 (7)] and let $x \in X$. Then a subset V of X containing x is a neighborhood of x if and only if $X \setminus V$ is finite. In fact, for $x \in U \subset V$ we have $X \setminus V \subset X \setminus U$, the set U is open if and only if $X \setminus U$ is finite, and in that event $X \setminus V$ is also finite. Note that then every neighborhood of a point in this space is open!

Recall that the open sets in a subspace Y of a space X are just the intersections with Y of the open sets in the entire space X. Hence a similar result holds for neighborhoods.

2.14 Proposition. Let y be a point in a subspace Y of a topological space X. Then the neighborhood system at y in Y is

$$\{ V \cap Y \mid V \text{ is a neighborhood of } y \text{ in } X \}$$

Proof. Let W be a neighborhood of y in the subspace Y. There is an open set U_1 in Y with

$$y \in U_1 \subset W.$$

Since the topology on Y is the relative one,

$$U_1 = U \cap Y$$

for some open set U in X. Because $y \in U$ and U is open, the set

$$V = W \cup U$$

is a neighborhood of y in X. Moreover,

$$V \cap Y = (W \cup U) \cap Y = (W \cap Y) \cup (U \cap Y) = W \cup U_1 = W.$$

Conversely, let V be a neighborhood of y in X. Then $y \in U \subset V$ for some open set U in X. Now $y \in U \cap Y$, $U \cap Y$ is open in Y, and $U \cap Y$ is contained in $V \cap Y$. Hence $V \cap Y$ is a neighborhood of y in Y. □

Among the neighborhoods of a point x in a topological space X are all the open subsets of X containing x. Of course, a neighborhood of x can be closed in X. Indeed, if V is a closed subset of X and if $x \in U \subset V$ for some open subset U of X, then V is a neighborhood of x.

The proof of 1.29 used only properties of open sets in a metric space that hold in any topological space. Hence the very same proof establishes the following proposition.

2.15 Proposition. A subset V of a topological space X is open in X if and only if V is a neighborhood of each of its points.

The fundamental properties of neighborhoods are listed in the next theorem.

2.16 Theorem. Let X be a topological space. For each $x \in X$ denote by $\mathfrak{N}_x$ the neighborhood system at x. Then for every $x \in X$:

> (N1) There exists at least one member of $\mathfrak{N}_x$.
> (N2) The point x belongs to each member of $\mathfrak{N}_x$.
> (N3) The intersection of any two members of $\mathfrak{N}_x$ also belongs to $\mathfrak{N}_x$.
> (N4) Each subset of X containing a member of $\mathfrak{N}_x$ also belongs to $\mathfrak{N}_x$.
> (N5) Each member V of $\mathfrak{N}_x$ contains some member U of $\mathfrak{N}_x$ such that U belongs to $\mathfrak{N}_y$ for each $y \in U$.

Proof. (N1) Let $x \in X$. Then $X \in \mathfrak{N}_x$. In fact, X is an open set with $x \in X \subset X$, so X is a neighborhood of x.

(N3) Let $x \in X$, let $V \in \mathfrak{N}_x$ and let $W \in \mathfrak{N}_x$. We must show that $V \cap W$ is a neighborhood of x. Since V and W are neighborhoods of x, there exist open sets U_1 and U_2 such that

$$x \in U_1 \subset V, \qquad x \in U_2 \subset W.$$

Then $x \in U_1 \cap U_2 \subset V \cap W$, and $U_1 \cap U_2$ is open in X. Hence $V \cap W$ is a neighborhood of x.

(N5) Let $x \in X$ and let $V \in \mathfrak{N}_x$. There exists an open set U with $x \in U \subset V$. By Proposition 2.15, for each $y \in U$ the set U is a neighborhood of y, that is, $U \in \mathfrak{N}_y$.

The verification of (N2) and (N4) is left to the reader. □

Theorem 2.16 describes the behavior of neighborhoods of points in a set on which a topology is already given. Suppose now we are given a set X, but without any topology. Suppose, further, that along with each point x of X we are given certain subsets of X which behave as they would were they really neighborhoods of x obtained from a topology on X. The next theorem says that then we can uniquely reconstruct a topology for which the neighborhoods of each $x \in X$ are precisely the given subsets associated with x.

2.17 Theorem. Let X be a set. For each $x \in X$ let $\mathfrak{N}_x$ be a given collection of subsets of X. Assume that these collections have properties (N1)–(N5). Then there exists a unique topology $\mathfrak{I}$ on X such that for each $x \in X$, $\mathfrak{N}_x$ is the collection of all $\mathfrak{I}$-neighborhoods of x. A subset V of X is $\mathfrak{I}$-open if and only if $V \in \mathfrak{N}_x$ for each $x \in V$.

Proof. (First we shall show that any such topology $\mathfrak{I}$ must have a specific form. This will establish uniqueness of $\mathfrak{I}$ and, at the same time, tell us actually how to construct $\mathfrak{I}$.)

Let $\mathfrak{I}$ be a topology on X such that for each $x \in X$, $\mathfrak{N}_x$ is the collection of all $\mathfrak{I}$-neighborhoods of x. Apply Proposition 2.15 to the topological space $(X, \mathfrak{I})$: A subset V of X is $\mathfrak{I}$-open if and only if V is a neighborhood of each $x \in V$. Hence V is $\mathfrak{I}$-open if and only if $V \in \mathfrak{N}_x$ for each $x \in V$. Thus

(*) $\mathfrak{I} = \{ V \mid V \subset X, V \in \mathfrak{N}_x \text{ for each } x \in V \}.$

This establishes uniqueness of $\mathfrak{I}$.

To prove existence of $\mathfrak{I}$, define $\mathfrak{I}$ by (*). We must show, first, that $\mathfrak{I}$ is a topology on X and, second, for each $x \in X$, $\mathfrak{N}_x$ is the collection of all $\mathfrak{I}$-neighborhoods of x.

[To show these two things, we must be careful to use only the definition (*) of $\mathfrak{I}$ and the assumed properties (N1)–(N5) of the collections $\mathfrak{N}_x$.]

We show that $\mathfrak{I}$ is a topology on X by verifying in turn each of the axioms (O1)–(O3) for a topology.

(O1) Since no $x \in \varnothing$, it is vacuously true that $\varnothing \in \mathfrak{N}_x$ for each $x \in \varnothing$. Hence $\varnothing \in \mathfrak{I}$. To see that $X \in \mathfrak{I}$, let $x \in X$. By (N1) there exists some $U \in \mathfrak{N}_x$. Since $U \subset X$, $X \in \mathfrak{N}_x$ by (N4).

(O2) Let $(V_i \mid i \in I)$ be a family of sets belonging to $\mathfrak{I}$ and let $V = \bigcup_{i \in I} V_i$. To see that $V \in \mathfrak{I}$, let $x \in V$; we must show $V \in \mathfrak{N}_x$. Choose $i \in I$ with $x \in V_i$. Since $V_i \in \mathfrak{I}$, $V_i \in \mathfrak{N}_x$. Since $V_i \subset V$, $V \in \mathfrak{N}_x$ by (N4).

(O3) is left to the reader to verify.

To complete the proof we show that for each $x \in X$, $\mathfrak{N}_x$ is the collection of all $\mathfrak{I}$-neighborhoods of x. Let $x \in X$. Let $V \in \mathfrak{N}_x$. By (N5) there is some $U \in \mathfrak{N}_x$ such that $U \subset V$ and $U \in \mathfrak{N}_y$ for each $y \in U$. By definition of $\mathfrak{I}$, $U \in \mathfrak{I}$, and by (N2), $x \in U$. Thus $x \in U \subset V$ and U is $\mathfrak{I}$-open. By definition of neighborhoods for a topology, V is a $\mathfrak{I}$-neighborhood of x.

Conversely, let V be a $\mathfrak{I}$-neighborhood of x. Then $x \in U \subset V$ for some $\mathfrak{I}$-open set U. By definition of $\mathfrak{I}$, $U \in \mathfrak{N}_x$. Hence, $V \in \mathfrak{N}_x$ by (N4).   □

We defined 'topological space' by using 'open set' as the primitive notion, subject to axioms (O1)–(O3). Then we defined 'neighborhood' in terms of 'open sets' and deduced from these axioms properties (N1)–(N5) of neigh-

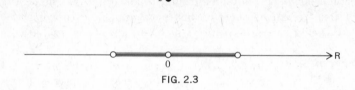

FIG. 2.3

borhoods. Theorem 2.17 indicates we would have arrived at the same notion of a topological space had we used 'neighborhood' as the primitive term and assumed (N1)–(N5) as axioms. Historically it was just this latter approach to topological spaces that was followed. In 1914 Felix Hausdorff abstracted from various "spaces" the concept of neighborhood, took 'neighborhood' as an undefined term, and set out as axioms for neighborhoods essentially properties (N1)–(N5) (plus a "separation axiom" discussed below).

In addition to this theoretical interest, Theorem 2.17 provides a method for constructing examples of topological spaces.

2.18 Examples

(1) Take $X = \mathsf{R}$. For each $x \in X$, define $\mathfrak{N}_x$ as follows: $V \in \mathfrak{N}_x$ if and only if V contains some left-closed, right-open interval $[x, z[$ with left endpoint x. Clearly properties (N1), (N2), and (N4) hold.

To verify (N3), let $V_1, V_2 \in \mathfrak{N}_x$. For some $z_1 > x$, $z_2 > x$, we have $[x, z_1[\subset V_1$ and $[x, z_2[\subset V_2$. Let z be the minimum of z_1 and z_2. Then $[x, z[\subset V_1 \cap V_2$, so $V_1 \cap V_2 \in \mathfrak{N}_x$.

To verify (N5), let $V \in \mathfrak{N}_x$. Choose $z > x$ with $[x, z[\subset V$. We show $[x, z[\in \mathfrak{N}_y$ for each $y \in [x, z[$. Let $y \in [x, z[$. Then $[y, z[\subset [x, z[$, so $[x, z[\in \mathfrak{N}_y$.

By 2.17 there is a unique topology on R for which the collection of all neighborhoods of an $x \in \mathsf{R}$ is exactly $\mathfrak{N}_x$. This topology is called the *right-interval topology* on R.

(2) More generally, let X be any totally ordered set (0.34) having no greatest element. For each $x \in X$, let $\mathfrak{N}_x$ consist of all sets containing intervals of the form $[x, z[$ with $z > x$. Just as in (1), the collections $\mathfrak{N}_x$ satisfy (N1)–(N5). The resulting topology given by 2.17 is called the *right-interval topology* on X.

(3) Let $X = \mathsf{R} \cup \{z\}$, where $z \notin \mathsf{R}$. For brevity call any neighborhood of an $x \in \mathsf{R}$ for the usual topology of R simply an "R-neighborhood of x." For each $x \in \mathsf{R}$, let $\mathfrak{N}_x$ consist of all subsets of X that contain some R-neighborhood of x. Let $\mathfrak{N}_z$ consist of all subsets of X that contain $V \cup \{z\} \setminus \{0\}$ for some R-neighborhood V of 0. If in the (x, y)-plane R is represented by the x-axis and z by the point $(0, 1)$, then a typical member of $\mathfrak{N}_z$ may be represented as in Fig. 2.3.

The topological space obtained by applying 2.17 is called the *line with two origins*. This space has the interesting property that each neighborhood U of

z intersects each neighborhood V of 0, for $U \cap V$ will contain a set of the form $]-\epsilon, 0[\cup]0, \epsilon[$ for some $\epsilon > 0$.

We now introduce the separation property that Hausdorff took as an additional axiom for neighborhoods.

2.19 Definition. A topological space X is called a *Hausdorff space* or a *T_2-space* and is said to be *separated* when each two distinct points of X have some disjoint neighborhoods.

Loosely speaking, X is a Hausdorff space when, given any $x, y \in X$ with $x \neq y$, it is impossible to find points arbitraily close both to x and to y.

2.20 Examples

(1) The line with two origins [2.18(3)] is not separated, for 0 and z fail to have any disjoint neighborhoods.

(2) Every metrizable space X is separated. In fact, let d be a metric inducing the topology of X and let $x, y \in X$ with $x \neq y$. Let $\epsilon = \frac{1}{2} d(x, y)$. Then $B_\epsilon(x; d)$ and $B_\epsilon(y; d)$ are disjoint neighborhoods of x and y.

(3) Let L be the x-axis and H be the open upper half-plane in the plane $\mathsf{R} \times \mathsf{R}$:

$$L = \mathsf{R} \times \{0\} = \{(x, 0) \mid x \in \mathsf{R}\}, \qquad H = \{(x, y) \in \mathsf{R} \times \mathsf{R} \mid y > 0\}.$$

Let

$$X = H \cup L = \{(x, y) \in \mathsf{R} \times \mathsf{R} \mid y \geq 0\},$$

the closed upper half-plane. We shall use 2.17 to construct a topology $\mathfrak{I}$ on X such that $(X, \mathfrak{I})$ is separated but not metrizable. The resulting topological space is called the *half-disk space*.

Denote by d the usual metric on $\mathsf{R} \times \mathsf{R}$.

Let $z \in H$. Let $\mathfrak{N}_z$ consist of all neighborhoods of z for the usual topology on X. Since z is at a positive d-distance above L, each member of $\mathfrak{N}_z$ contains a d-ball $B_\epsilon(z; d)$.

Now let $x \in L$. For each real number $r > 0$, define

$$H_r(x) = [B_r(x; d) \cap H] \cup \{x\}$$

(see Fig. 2.4). Let $\mathfrak{N}_x$ consist of all subsets of X that contain some $H_r(x)$.

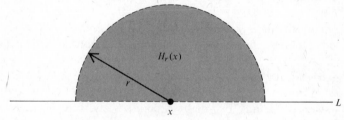

FIG. 2.4

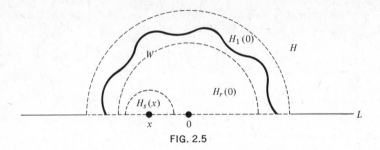

FIG. 2.5

The verification of (N1)–(N5) for these collections $\mathfrak{N}_z$ uses the corresponding properties of neighborhoods of points in the Euclidean plane. Let $\mathfrak{I}$ be the topology on X given by 2.17.

That $(X, \mathfrak{I})$ is separated follows from two facts. First, each d-ball at a point $z \in X$ contains some $\mathfrak{I}$-neighborhood of z. Second, any two distinct points of X have some disjoint d-balls about them.

Just suppose $(X,\mathfrak{I})$ is metrizable. Denote by 0 the origin $(0, 0) \in L$. By a remark following 1.28 the $\mathfrak{I}$-neighborhood $H_1(0)$ contains some $\mathfrak{I}$-neighborhood W of 0 which is $\mathfrak{I}$-closed (see Fig. 2.5). Since $W \in \mathfrak{N}_0$, then $H_r(0) \subset W$ for some $r > 0$. Take any $x \in L$ with $x \neq 0$ and $d(x, 0) < r$. Clearly $x \notin H_r(0)$. We shall obtain a contradiction by showing $x \in W$.

Just suppose $x \notin W$. Since W is closed in X, x has some $\mathfrak{I}$-neighborhood V that is disjoint from W. Choose $s > 0$ with $H_s(x) \subset V$. Now $H_s(x)$ intersects $H_r(0)$ at points of H. Hence V intersects W.

Being a Hausdorff space is a hereditary property.

2.21 Proposition. Every subspace of a Hausdorff space is a Hausdorff space.

Proof. Let X be a Hausdorff space, let $Y \subset X$, and let $x, y \in Y$ with $x \neq y$. There are disjoint neighborhoods U and V of x and y in X. Then $U \cap Y$ and $V \cap Y$ are disjoint, and by 2.14 these sets are neighborhoods of x and y in Y.   ☐

The terminology T_2 for a Hausdorff space comes from the German *Trennungsaxiom* meaning "separation axiom." The Hausdorff axiom is just one of the whole hierarchy of separation axioms we state here for reference.

A topological space X is:

(0) a T_0-*space* if, for any two distinct points x and y of X, either some neighborhood of x does not contain y or some neighborhood of y does not contain x;

(1) a T_1-*space* if, for any two distinct points x and y of X, there is a neighborhood of x not containing y and a neighborhood of y not containing x;

(2) a T_2-*space* if, for any two distinct points x and y of X, there are disjoint neighborhoods of x and y;

(3) *regular* (or a T_3-*space*) if, for each point $x \in X$ and each closed set $E \subset X$ with $x \notin E$, there are disjoint open sets containing x and E, respectively;

(4) *completely regular* if, for each point $x \in X$ and each closed set $E \subset X$ with $x \notin E$, there is a continuous function $f : X \to [0, 1]$ such that $f(x) = 0$ and $f(y) = 1$ for every $y \in E$;

(5) *normal* if, for any two disjoint closed sets E and F in X, there are disjoint open sets containing E and F, respectively.

T_0-spaces are the subject of Exercise 32, below. T_1-spaces were introduced in Exercise 10. Regular spaces are discussed in Exercises 60–63, below. Exercise 1.27 (b) said that any metrizable space is normal, and Theorem 4.16 will say that any "compact" Hausdorff space is normal. Aside from these items and a few other exercises, the separation axioms other than T_2 will not be pursued in this text. For a full treatment of all the separation axioms the reader should consult Kelley [21] or Willard [34].

EXERCISES

21. Describe explicitly the neighborhood system of each point in:
   (**a**) an indiscrete space;
   (**b**) the Sierpinski space [2.3(6)];
   (**c**) a set provided with its countable-complement topology (Exercise 6);
   (**d**) a set provided with an included point topology (Exercise 3).

22. Show that the following is a necessary and sufficient condition for a set E to be closed in a topological space X: Given any $x \in X$, if each neighborhood of x intersects E, then $x \in E$.

23. (**a**) Is the intersection of finitely many neighborhoods of a point necessarily a neighborhood of the point? the intersection of arbitrarily many neighborhoods?
   (**b**) Is the union of finitely many neighborhoods of a point necessarily a neighborhood of the point? the union of arbitrarily many neighborhoods?

24. Given a subset A of a topological space X, a point $x \in X$ is called a *limit point of* A if each neighborhood of x contains some point of A different from x.
   (**a**) Give an example of a subset A of R having exactly one limit point x, with $x \notin A$.
   (**b**) For $X = $ R^n, show that x is a limit point of A if and only if each neighborhood of x contains infinitely many points of A.
   (**c**) Prove that A is closed in X if and only if every limit point of A belongs to A.
   (The first attempt at an axiomatic treatment of spaces more general than metric spaces, due to Riesz and Fréchet, was based on limit points. Today limit points are out of fashion.)

25. Let A be a subset of a topological space X. A *neighborhood of A in X* is a subset V of X such that $A \subset U \subset V$ for some open set U in X. For example, a neighborhood of $A = \{x\}$ is just a neighborhood of the point x. Do analogs of (N1)–(N5) hold for neighborhoods of sets?

26. Let A be a subset of a topological space X. Suppose U is a neighborhood of A in X and V is a neighborhood of A in U. Show that V is then a neighborhood of A in X.

27. Let L, H, X, and d be as in Example 2.20 (3), and for each $z \in H$ define $\mathfrak{N}_z$ as in that example. For each $x = (x_1, 0) \in L$, however, redefine $\mathfrak{N}_x$ to consist of all sub-sets of X containing a set of the form

$$T_r(x) = \{x\} \cup B_r((x_1, r); d)$$

for some $r > 0$ (draw a picture).

    (a) Verify that (N1)–(N5) still hold. The topological space obtained with the use of 2.17 is called the *tangent disk space*.

    (b) Is this space a Hausdorff space?

    (c) Is this space metrizable?

    (d) What is the topology on the subspace L of X?

28. **(a)** Describe explcitly all open sets for the right-interval topology on R[2.18(1)].

    (b) Compare the right-interval topology with the usual topology on R.

    (c) Are (N1)–(N5) still satisfied if for each real number x we define $\mathfrak{N}_x$ to be the collection of all subsets of R that contain a closed interval of the form $[x, z]$ with $z > x$?

29. On $X = \mathsf{R} \times \mathsf{R}$, define a relation $\leq$ as follows: For $x = (x_1, x_2)$, $y = (y_1, y_2) \in X$,

$$x \leq y \quad \Leftrightarrow \quad (x_1 = y_1 \quad \text{and} \quad x_2 \leq y_2) \quad \text{or} \quad (x_1 < y_1).$$

    (a) Verify that this relation is a total ordering of X.

    (b) Describe geometrically an interval in X of the form $[x, z[$, where $z > x$, for this total ordering.

    (c) Compare the usual topology on X with the right-interval topology on X [2.18(2)] obtained from this total ordering.

    (d) When X is provided with the right-interval topology, what is the relative topology induced on $\{(x, y) \in \mathsf{R} \times \mathsf{R} \mid x = y\}$?

30. Must a discrete subspace of a Hausdorff space X be closed in X?

31. **(a)** Show that the following condition is both necessary and sufficient for a topological space X to be a T_1-space (Exercise 10): For each $x \in X$ the set $\{x\}$ is the intersection of all neighborhoods of x.

    (b) Prove that every T_2-space is a T_1-space.

    (c) Find among the examples already given in this chapter a T_1-space which is not a T_2-space.

32. A T_0-*space* is a topological space X in which, for any two distinct points x and y, either some neighborhood of x does not contain y or some neighborhood of y does not contain x.

    (a) Is every topological space a T_0-space?

    (b) Show that the Sierpinski space [2.3(6)] is a T_0-space but not a T_1-space (Exercise 10).

    (c) Prove that every T_1-space is a T_0-space.

    (d) Is a subspace of a T_0-space necessarily a T_0-space?

33. A point x in a topological space X is said to be *isolated* when $\{x\}$ is open in X. Let Y be a topological space and let E be any set disjoint from Y. Construct a topology on $X = Y \cup E$ making Y a subspace of X and each $x \in E$ an isolated point in X. (Thus any number of isolated points can be "adjoined" to a topological space.)

3. Boundary, Interior, and Closure

With the aid of neighborhoods we are going to describe three operations which may be applied to the subsets of a topological space to yield new subsets of the space. The first two, the operations of forming the boundary and the interior of a set, are natural extensions of familiar geometric ideas to topological spaces. The third, the operation of forming the closure of a set, will be of interest in connection with limits.

Consider the annular subset

$$A = \{x \in \mathsf{R}^2 \mid 1 \le d(x,c) < 2\}$$

of the plane R^2, where c is some point of R^2 and d is the Euclidean metric (see Fig. 2.6). The set A includes all points on the circle of radius 1 centered at c but none on the circle of radius 2. It is perfectly natural to call the set

$$B = \{x \in \mathsf{R}^2 \mid d(x,c) = 1\} \cup \{x \in \mathsf{R}^2 \mid d(x,c) = 2\}$$

of points on these two concentric circles the "boundary" of A. Clearly B consists of just those points $x \in \mathsf{R}^2$ each of whose neighborhoods intersects both A and $\mathsf{R}^2 \backslash A$. This characterization of B makes no direct mention of the metric. Hence it suggests a purely topological definition of 'boundary' meaningful in any topological space.

2.22 Definition. Let A be a subset of a topological space X. Then the *boundary of A (in X)*, denoted by bdy A, is defined to be the set of all $x \in X$ such that each neighborhood of x in X intersects both A and $X \backslash A$.

In checking that a point x in a metric space (X, d) belongs to bdy A, it is clearly enough to check whether each d-ball at x intersects both A and $X \backslash A$.

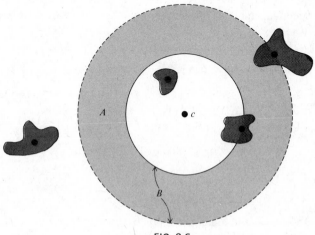

FIG. 2.6

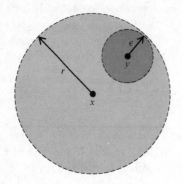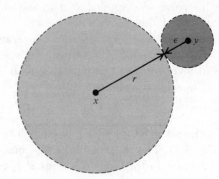

FIG. 2.7

2.23 Examples

(1) Let $X = \mathbf{R}$ and $A = \,]0, 1]$. Then bdy $A = \{0, 1\}$. In fact, for each $\epsilon > 0$ the neighborhood $]-\epsilon, \epsilon[$ of 0 intersects both A and $X \backslash A = \,]\leftarrow, 0] \cup \,]1, \rightarrow[$, as does the neighborhood $]1 - \epsilon, 1 + \epsilon[$ of 1, so $0 \in$ bdy A and $1 \in$ bdy A. However, if $x \neq 0, 1$, then for

$$\epsilon = \min \{|x|, |x - 1|\}$$

the neighborhood $]x - \epsilon, x + \epsilon[$ of x cannot intersect both A and $X \backslash A$.

This example shows that a point belonging to the boundary of a set A may, but need not, belong to A itself.

(2) In any topological space X, both $\varnothing$ and X have empty boundaries.

(3) Let $A = \mathbf{R} \times \{0\}$. Then A is the boundary of A in the plane $\mathbf{R} \times \mathbf{R}$. Now A is a topological space in its own right (under its usual topology induced by its Euclidean metric), and by (2) the boundary of A in the space A is empty. (This example will be elucidated by 2.43.)

(4) Let d be the Euclidean metric on $\mathbf{R}^n$, let $x \in \mathbf{R}^n$, and let $r > 0$. Then

$$\text{bdy } B_r(x; d) = S_r(x; d)$$

(recall that $S_r(x; d) = \{y \in \mathbf{R}^n \mid d(x, y) = r\}$, the d-sphere of radius r at x). To see this, let $y \in \mathbf{R}^n$.

Case (i): $d(x, y) < r$. Then

$$B_\epsilon(y; d) \subset B_r(x; d), \qquad \epsilon = r - d(x, y),$$

for $z \in B_\epsilon(y; d)$ implies

$$d(x, z) \leq d(x, y) + d(y, z) < d(x, y) + \epsilon = r$$

(see Fig. 2.7). Hence the neighborhood $B_\epsilon(y; d)$ of y does not intersect $\mathbf{R}^n \backslash B_r(x; d)$, so $y \notin$ bdy $B_r(x; d)$.

Case (ii): $d(x, y) > r$. Then

$$B_\epsilon(y; d) \subset \mathbf{R}^n \backslash B_r(x; d), \qquad \epsilon = d(x, y) - r$$

(see Fig. 2.7), so again $y \notin$ bdy $B_r(x; d)$.

Case (iii): $d(x, y) = r$. Let V be any neighborhood of y. Then $y \in V \cap [\mathbf{R}^n \backslash B_r(x; d)]$. To show $y \in$ bdy $B_r(x; d)$ it remains only to show that V intersects $B_r(x; d)$. Choose $0 < \epsilon < r$ with $B_\epsilon(y; d) \subset V$. We shall construct a point z on the line segment joining x to y with $z \in B_\epsilon(y; d) \cap B_r(x; d)$ (see Fig. 2.8). Each point z on this line segment has, in vector notation, the form

$$z = (1 - t)x + ty, \qquad 0 \le t \le 1.$$

For z of this form,

$$d(x, z) = tr, \qquad d(z, y) = (1 - t)r,$$

and we require a z with $d(x, z) < r$ and $d(z, y) < \epsilon$. Hence we take any t satisfying $1 - \epsilon/r < t < 1$.

Here are some elementary and geometrically reasonable properties of the boundary.

2.24 Proposition. Let A be any subset of a topological space X. Then:
  (1) Bdy $(X \backslash A)$ = bdy A.
  (2) Bdy A is closed in X.
  (3) A is open in X if and only if bdy $A \subset X \backslash A$.
  (4) A is closed in X if and only if bdy $A \subset A$.

Proof. (2) We show that $X \backslash$ bdy A is open in X. Let $x \in X \backslash$ bdy A. Then some open neighborhood V of x does not intersect both A and $X \backslash A$. If $y \in V$, then V is a neighborhood of y not intersecting both A and $X \backslash A$, so $y \notin$ bdy A. Hence $V \subset X \backslash$ bdy A.
  (3) Assume first A is open in X. We show that bdy $A \subset X \backslash A$ by showing

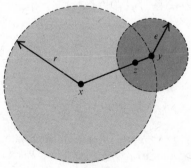

FIG. 2.8

$A \subset X \setminus \mathrm{bdy}\, A$. If $x \in A$, then some neighborhood V of x is contained in A, the set V does not intersect $X \setminus A$, and so $x \notin \mathrm{bdy}\, A$.

Conversely, assume $\mathrm{bdy}\, A \subset X \setminus A$. We show that A is open in X. Let $x \in A$. Then $x \notin X \setminus A$, by assumption $x \notin \mathrm{bdy}\, A$, and so x has a neighborhood V that does not intersect both A and $X \setminus A$. But $x \in A \cap V$, so V does not intersect $X \setminus A$, that is, $V \subset A$.

(4) Since statement (3) is true of any subset A of X, it is true of $X \setminus A$: The set $X \setminus A$ is open in X if and only if $\mathrm{bdy}\, (X \setminus A) \subset X \setminus (X \setminus A)$. But $\mathrm{bdy}\, (X \setminus A) = \mathrm{bdy}\, A$ and $X \setminus (X \setminus A) = A$. Hence $X \setminus A$ is open in X if and only if $\mathrm{bdy}\, A \subset A$. [Of course, (4) may also be proved directly without any use of (3).]   ⬚

We may regard points belonging to the boundary of a set A as lying on the periphery of A, arbitrarily close both to points in A and to points not in A. Then we should think of points of $A \setminus \mathrm{bdy}\, A$ as lying entirely inside A, away from all points not in A.

2.25 Definition. Let A be a subset of a topological space X. Then the *interior of A*, denoted by int A, is the subset $A \setminus \mathrm{bdy}\, A$ of A. An $x \in \mathrm{int}\, A$ is said to be *interior to A* and is called an *interior point of A*.

We have

(*) $x \in \mathrm{int}\, A \quad \Leftrightarrow \quad$ there exists a neighborhood V of x with $V \subset A$.

In fact, assume $x \in \mathrm{int}\, A$. Then $x \notin \mathrm{bdy}\, A$, so some neighborhood V of x does not intersect both A and $X \setminus A$. But V does intersect A since $x \in V \cap A$. Hence V does not intersect $X \setminus A$, that is, $V \subset A$. Conversely, assume there is a neighborhood V of x with $V \subset A$. Then $x \notin \mathrm{bdy}\, A$ since V does not intersect $X \setminus A$, and $x \in V \subset A$. Hence $x \in A \setminus \mathrm{bdy}\, A = \mathrm{int}\, A$.

2.26 Examples

(1) For the annular subset

$$A = \{x \in \mathbf{R}^2 \mid 1 \leq d(x, c) < 2\}$$

of $\mathbf{R}^2$ considered at the beginning of this section,

$$\mathrm{int}\, A = \{x \in \mathbf{R}^2 \mid 1 < d(x, c) < 2\}.$$

(2) In any topological space X,

$$\mathrm{int}\, \varnothing = \varnothing, \qquad \mathrm{int}\, X = X.$$

(3) The subset $\mathbf{Q}$ of $\mathbf{R}$ has empty interior in $\mathbf{R}$, because any neighborhood of a rational number contains irrational numbers. Likewise, the set $\mathbf{R} \setminus \mathbf{Q}$ has empty interior in $\mathbf{R}$.

(4) Since an open set is a neighborhood of each of its points, it follows from (*) that the interior of an open subset A of a topological space is A itself. In particular,

$$\text{int } B_\epsilon(x; d) = B_\epsilon(x; d)$$

for any $\epsilon > 0$ and any point x of a metric space (X, d).

Although the boundary of a set can be disjoint from the set, the interior of a set is always contained in the set. The next theorem makes the relationship of a set to its interior more precise.

2.27 Theorem. Let A be a subset of a topological space X. Then int A is the largest open subset of X which is contained in A, that is:
 (1) The set int A is open in X and int $A \subset A$.
 (2) If U is open in X and $U \subset A$, then $U \subset$ int A.

Proof. (2) Let U be an open subset of X with $U \subset A$. If $x \in U$, then U is a neighborhood of x contained in A, so $x \in$ int A by (*) above. Hence $U \subset$ int A.
 (1) Since int $A = A \backslash \text{bdy } A$, we have int $A \subset A$. To see that int A is open in X, let $x \in$ int A. By (*) there is an open neighborhood U of x with $U \subset A$. Then $U \subset$ int A by part (2).   ▯

2.28 Corollary. A subset A of a topological space X is open in X if and only if $A = $ int A.

Proof. If $A = $ int A, then A is open in X since int A is open in X. Conversely, if A is open in X, then A is itself the largest subset of A which is open in X, so $A = $ int A.   ▯

2.29 Corollary. For all subsets A and B of a topological space X:
 (1) If $A \subset B$, then int $A \subset$ int B.
 (2) Int $(A \cap B) = $ int $A \cap$ int B.

Proof. (1) Assume $A \subset B$. Then int A is open in X and int $A \subset A \subset B$. Since int B is the largest open subset of X contained in B, it follows that int $A \subset$ int B.
 (2) Since $A \subset A \cap B$, part (1) says that int $(A \cap B) \subset$ int A; similarly, int $(A \cap B) \subset$ int B. Hence int $(A \cap B) \subset$ int $A \cap$ int B. The opposite inclusion is true because int $A \cap$ int B is an open set contained in $A \cap B$.   ▯

The analog of 2.29 (2) for unions is not true in general. For example if $X = \mathsf{R}$, $A = \mathsf{Q}$, and $B = \mathsf{R} \backslash \mathsf{Q}$, then

$$\text{int } (A \cup B) = \text{int } \mathsf{R} = \mathsf{R} \neq \varnothing = \text{int } A \cup \text{int } B$$

by example 2.26 (3). What is true in general is that

$$\text{int } A \cup \text{int } B \subset \text{int } (A \cup B)$$

because $A \subset A \cup B$ and $B \subset A \cup B$.

We now combine the interior and boundary of a set.

2.30 Definition. Let A be a subset of a topological space X. Then the subset int $A \cup$ bdy A of X is called the *closure of* A and is denoted by cls A (another common notation is A^-).

Examples 2.23 and 2.26 allow us to compute at once the closures of some sets.

2.31 Examples

(1) For the annular subset

$$A = \{x \in \mathsf{R}^2 \mid 1 \leq d(x, c) < 2\}$$

of R^2 we have

$$\text{cls } A = \{x \in \mathsf{R}^2 \mid 1 \leq d(x, c) \leq 2\}.$$

(2) In any topological space X,

$$\text{cls } \varnothing = \varnothing, \qquad \text{cls } X = X.$$

(3) In the real line R,

$$\text{cls } \mathsf{Q} = \mathsf{R} = \text{cls } (\mathsf{R} \backslash \mathsf{Q}).$$

(4) Let d be the Euclidean metric on R^n, let $x \in \mathsf{R}^n$, and let $r > 0$. Then

$$\text{cls } B_r(x; d) = D_r(x; d)$$

(recall that $D_r(x; d) = \{y \in \mathsf{R}^n \mid d(x, y) \leq r\}$, the d-disk of radius r at x).

Intuitively, the boundary of a set A consists of all points arbitrarily close both to A and to its complement, and the interior of A consists of all points in A not arbitrarily close to the complement of A. Hence it is intuitively reasonable that the closure of A consists of all points arbitrarily close to A. This intuition is correct.

2.32 Proposition Let A be a subset of a topological space X and let x be a point of X. Then $x \in$ cls A if and only if each neighborhood of x intersects A.

Proof. Assume first $x \in$ cls A. If $x \in$ int A, then $x \in A$, so each neighborhood of x intersects A at least at the point x. If $x \in$ bdy A, then each neighborhood of x intersects not only A, but $X \backslash A$ as well.

Conversely, assume each neighborhood of x intersects A. If some neighborhood of x is contained in A, then $x \in$ int $A \subset$ cls A. Otherwise, each neighborhood of x intersects $X \backslash A$ as well as A, so $x \in$ bdy $A \subset$ cls A.   □

2.33 Example. Let $A \subset \mathsf{R}$. Denote by cls A the closure of A not in R, but in the extended real line $\mathsf{R}^* = \mathsf{R} \cup \{-\infty, +\infty\}$ (Example 1.37). It readily follows from Proposition 2.32 and Lemma 1.39 that

$$+\infty \in \operatorname{cls} A \qquad \Leftrightarrow \qquad A \text{ is not bounded above in } \mathsf{R},$$

$$-\infty \in \operatorname{cls} A \qquad \Leftrightarrow \qquad A \text{ is not bounded below in } \mathsf{R}.$$

These facts will be used in Chapter 3 in the discussion of "limits at infinity".

For a metric space, the characterization 2.32 of the closure of a set A can be stated more concretely in terms of the distance of points from A (1.26).

2.34 Corollary. Let A be a subset of a metric space (X, d). Then

$$\operatorname{cls} A = \{x \in X \mid d(x, A) = 0\}.$$

Proof. Let $x \in X$ and assume $d(x, A) = \delta > 0$. Then $d(x, a) \geq \delta$ for all $a \in A$, the d-ball $B_{\delta/2}(x; d)$ does not intersect A, and so $x \notin \operatorname{cls} A$. Conversely, assume $x \in X$ with $x \notin \operatorname{cls} A$. Then some d-ball $B_{\epsilon}(x; d)$ at x does not intersect A, $d(x, a) \geq \epsilon$ for all $a \in A$, and so $d(x, A) \geq \epsilon > 0$. $\quad\square$

For an application of 2.32 let us generalize the notion of a d-dense set in a metric space (as defined in the paragraph preceding 1.72).

2.35 Definition. A subset D of a topological space X is said to be *dense in X* when D intersects each nonempty open subset of X.

Evidently D is dense in X if and only if each neighborhood of each point of X contains a point of D, or in looser language, there are points of D arbitrarily close to each point of X. According to 2.32, then,

$$D \text{ is dense in } X \qquad \Leftrightarrow \qquad \operatorname{cls} D = X.$$

2.36 Examples

(1) Examples 2.31 (3) says that the set Q of all rational numbers and the set $\mathsf{R} \backslash \mathsf{Q}$ of all irrational numbers are both dense in the real line R.

(2) The only subset of a discrete space X that is dense in X is the entire space X.

(3) In the space $[0, 1]$ each of the following sets is dense: $]0, 1[$, $[0, 1[$, $]0, \frac{1}{2}[\cup]\frac{1}{2}, 1[$, $]0, 1[\backslash \{1/n \mid n = 1, 2, \ldots\}$.

(4) By Example 2.33, the real line R is dense in the extended real line R^*.

Proposition 2.32 allows us to express the boundary of a set A in a space X in terms of closures:

$$\operatorname{bdy} A = \operatorname{cls} A \cap \operatorname{cls}(X \backslash A).$$

More significantly, it allows us to express closures in terms of interiors and vice versa.

2.37 Theorem. For every subset A of a topological space X:

(1) $\qquad\qquad\qquad X \setminus \text{cls } A = \text{int } (X \setminus A)$
(2) $\qquad\qquad\qquad \text{int } A = X \setminus \text{cls } (X \setminus A)$
(3) $\qquad\qquad\qquad X \setminus \text{int } A = \text{cls } (X \setminus A)$
(4) $\qquad\qquad\qquad \text{cls } A = X \setminus \text{int } (X \setminus A).$

Proof. (1) Let $x \in X$. Then $x \in X \setminus \text{cls } A$ if and only if some neighborhood V of x does not intersect A, that is, $V \subset X \setminus A$. But $x \in \text{int } (X \setminus A)$ if and only if some neighborhood of x is contained in $X \setminus A$.

(2) Replace A by $X \setminus A$ in (1)—this is legitimate because (1) holds for every subset of X—to get

$$X \setminus \text{cls } (X \setminus A) = \text{int } (X \setminus [X \setminus A]) = \text{int } (A).$$

(3) Take the complements in X of both sides of equation (2).
(4) Replace A by $(X \setminus A)$ in (3).　　$\square$

Since $\text{bdy } A = \text{cls } A \cap \text{cls } (X \setminus A)$, equation (3) yields

$$\text{bdy } A = \text{cls } A \setminus \text{int } A.$$

Theorem 2.37 can be put into an easily remembered form through the use of superscripts $^-$, $^\circ$, and $'$ to denote closure, interior, and complement in X, respectively:

$$A^{-\prime} = A'^{\circ}, \qquad A^{\circ} = A'^{-\prime}, \qquad A^{\circ\prime} = A'^{-}, \qquad A^{-} = A'^{\circ\prime}.$$

As a working rule of calculation, then, one simply interchanges $^-$ and $'$ and then replaces $^-$ by $^\circ$; one interchanges $^\circ$ and $'$ and then replaces $^\circ$ by $^-$.

Theorem 2.37 allows us to derive from each property of interiors a corresponding "dual" property of closures (and vice versa). For example, if A is a subset of a space X, then Theorem 2.27 (1) says that $\text{int } (X \setminus A)$ is open in X and is contained in $X \setminus A$; by 2.37 (4), $\text{cls } A = \text{int } (X \setminus A)$ is closed in X and contains A. This establishes part of the following dual of 2.27.

2.38 Theorem. Let A be a subset of a topological space X. Then $\text{cls } A$ is the smallest closed subset of X containing A, that is:

(1) The set $\text{cls } A$ is closed in X and $A \subset \text{cls } A$.
(2) If E is closed in X and $A \subset E$, then $\text{cls } A \subset E$.

Proof. (2) Let E be a closed subset of X with $A \subset E$. Then $X \setminus E$ is an open subset of X with $X \setminus E \subset X \setminus A$. By 2.27 (2),

$$X \setminus E \subset \text{int } (X \setminus A).$$

By 2.37 (4),

$$\operatorname{cls} A = X \backslash \operatorname{int} (X \backslash A) \subset X \backslash (X \backslash E) = E. \qquad \square$$

2.39 Corollary. A subset A of topological space X is closed in X if and only if $A = \operatorname{cls} A$.

Proof. If $A = \operatorname{cls} A$, then A is closed in X since $\operatorname{cls} A$ is. Conversely, if A is closed in X, then A is itself the smallest closed subset of X containing A, so $A = \operatorname{cls} A$. $\square$

Of course 2.39 can also be deduced from 2.28 by using 2.37.
Hopefully the reader has anticipated the following dual of 2.29.

2.40 Proposition. For all subsets A and B of a topological space X:

(1) If $B \subset A$, then $\operatorname{cls} B \subset \operatorname{cls} A$.
(2) Cls $(A \cup B) = \operatorname{cls} A \cup \operatorname{cls} B$.

Proof. (2) From 2.37 (4), De Morgan's laws, and 2.29 (2) we obtain

$$\begin{aligned}
\operatorname{cls} (A \cup B) &= X \backslash \operatorname{int} [X \backslash (A \cup B)] \\
&= X \backslash \operatorname{int} [(X \backslash A) \cap (X \backslash B)] \\
&= X \backslash [\operatorname{int} (X \backslash A) \cap \operatorname{int} (X \backslash B)] \\
&= [X \backslash \operatorname{int} (X \backslash A)] \cup [X \backslash \operatorname{int} (X \backslash B)] \\
&= \operatorname{cls} A \cup \operatorname{cls} B. \qquad \square
\end{aligned}$$

The analog of (2) for intersection is not true in general, for if $A = \mathsf{R} \backslash \mathsf{Q}$ and $B = \mathsf{Q}$ in the space R, then

$$\operatorname{cls} (A \cap B) = \operatorname{cls} \varnothing = \varnothing \neq \mathsf{R} = \operatorname{cls} A \cap \operatorname{cls} B.$$

What is true in general is that

$$\operatorname{cls} (A \cap B) \subset \operatorname{cls} A \cap \operatorname{cls} B$$

for all subsets A and B of a topological space.

A set A is dense in a space X when its closure is the largest open subset of X—namely, X itself. At the opposite extreme, the closure of A may contain no open subset of X except the empty one.

2.41 Definition. A subset A of a topological space X is said to be *nowhere dense in X* when $\operatorname{cls} A$ contains no nonempty open subset of X, that is, when $\operatorname{int} (\operatorname{cls} A) = \varnothing$.

For example, each finite subset of R^n is nowhere dense in R^n. The infinite set $\{1/n : n = 1, 2, \ldots\}$ is nowhere dense in R.

By theorem 2.37 (3), a set A is nowhere dense in X if and only if

$$X = X \setminus \varnothing = X \setminus \text{int} \, (\text{cls} \, A) = \text{cls} \, (X \setminus \text{cls} \, A),$$

in other words, $X \setminus \text{cls} \, A$ is dense in X. In particular, A is a nowhere dense closed set in X if and only if its complement $X \setminus A$ is a dense open set in X. Dually, A is a dense open set in X if and only if $X \setminus A$ is a nowhere dense closed set in X.

The preceding observations suggest equivalent statements of the Baire category theorem (1.72), whose topological conclusion concerned dense open sets. For these we introduce Baire's classification of sets into two types: A subset B of a topological space X is said to be *first-category in* X if B is the union of some sequence of nowhere dense sets in X, and *second-category in* X if it is not first-category in X. We should think of a first-category set in X as having its points sparsely scattered through the space and hence as being "small" or "thin". Then we should think of a second-category set in X as being "big" or "thick".

2.42 Proposition. For a topological space X the following conditions are equivalent:

   (1) The intersection of each sequence of dense open sets in X is itself dense in X.
   (2) Each first-category set in X has empty interior.
   (3) Each nonempty open set in X is second-category in X.

Proof. Assume (1). We show (2). Let B be first-category in X, so that

$$B = \bigcup_{n=0}^{\infty} A_n$$

for some sequence $(A_n \mid n \in \mathbf{N})$ of nowhere dense sets in X. For each $n \in \mathbf{N}$ the set

$$D_n = X \setminus \text{cls} \, A_n$$

is dense and open in X. By (1),

$$\text{cls} \left(\bigcap_{n=0}^{\infty} D_n \right) = X.$$

Now

$$D_n \subset X \setminus A_n \qquad\qquad (n \in \mathbf{N}),$$

so also

$$\text{cls} \left[\bigcap_{n=0}^{\infty} (X \setminus A_n) \right] = X.$$

Using 2.37 we conclude

$$\text{int } B = \text{int} \left(\bigcup_{n=0}^{\infty} A_n \right) = \text{int} \left(X \setminus \bigcap_{n=0}^{\infty} [X \setminus A_n] \right)$$

$$= X \setminus \text{cls} \bigcap_{n=0}^{\infty} [X \setminus A_n] = \emptyset.$$

A similar argument shows that (2) implies (1). The proof that (2) is equivalent to (3) is easy.　　☐

Already in Example 2.23 (3) we observed that a set can have different boundaries in a space and in a subspace of that space; the same is true of interiors and closures. We can make this observation more precise.

2.43 Proposition. Let A be a subset of a subspace Y of a topological space X. Use subscripts X and Y to refer to the spaces in which the closure, interior, and boundary are computed. Then:

(1)　　　　　　　　　　$\text{cls}_Y A = Y \cap \text{cls}_X A$
(2)　　　　　　　　　　$\text{int}_Y A \supset Y \cap \text{int}_X A$
(3)　　　　　　　　　　$\text{bdy}_Y A \subset Y \cap \text{bdy}_X A$

Proof. (1) In view of Theorem 2.38 we need only show that $Y \cap \text{cls}_X A$ is the smallest subset of Y which contains A and is closed in Y. Since $A \subset \text{cls}_X A$ and $\text{cls}_X A$ is closed in X, then $A \subset Y \cap \text{cls}_X A$ and $Y \cap \text{cls}_X A$ is closed in Y. Now let E be any subset of Y with $A \subset E$ and E closed in Y. Write $E = Y \cap F$ with F closed in X. Then $\text{cls}_X A \subset F$ because $A \subset F$. Hence $Y \cap \text{cls}_X A \subset Y \cap F = E$.

The proof of (2) is similar to the above but employs Theorem 2.27. Statement (3) follows from (1) and (2):

$$\begin{aligned}
\text{bdy}_Y A &= \text{cls}_Y A \setminus \text{int}_Y A \\
&\subset \text{cls}_Y A \setminus (Y \cap \text{int}_X A) \\
&= (Y \cap \text{cls}_X A) \setminus (Y \cap \text{int}_X A) \\
&= Y \cap (\text{cls}_X A \setminus \text{int}_X A) \\
&= Y \cap \text{bdy}_X A. \qquad ☐
\end{aligned}$$

Inclusions (2) and (3) cannot be strengthened to equalities. For example, take

$$X = \mathbb{R}^2, \qquad A = Y = \mathbb{R} \times \{0\}.$$

Then

$$\text{int}_X A = \emptyset, \qquad \text{bdy}_X A = A,$$

but

$$\text{int}_Y A = A, \qquad \text{bdy}_Y A = \emptyset.$$

EXERCISES

34. Find the boundary, interior, and closure of the given set A in the given space X:

(**a**) $A = \{1/n : n = 1, 2, \ldots\}$, $X = \mathsf{R}$.

(**b**) $A = \{0\} \cup \{1/n : n = 1, 2, \ldots\}$, $X = \mathsf{R}$.

(**c**) $A = \mathsf{N}$, $X = \mathsf{R}$.

(**d**) $A = \mathsf{R} \backslash \mathsf{N}$, $X = \mathsf{R}$.

(**e**) $A = \{(x, y) \in \mathsf{R}^2 \mid (x^2/a^2) + (y^2/b^2) \leq 1\}$, where a and b are positive real numbers, $X = \mathsf{R}^2$.

(**f**) $A = \{(x, y) \in \mathsf{R}^2 \mid x > 0, 0 < y \leq 1/x\}$, $X = \mathsf{R}^2$.

(**g**) $A = \{(x, y) \in \mathsf{R}^2 \mid 0 < x^2 + y^2 \leq 1\}$, $X = \mathsf{R}^2$.

(**h**) $A = \{(x, y) \in \mathsf{R}^2 \mid 0 < x \leq 1, y = \sin (1/x)\}$, $X = \mathsf{R}^2$.

(**i**) $A = \{(x, y) \in \mathsf{R}^2 \mid 0 < x \leq 1, y = x \sin (1/x)\}$, $X = \mathsf{R}^2$.

(**j**) $A = \{0\}$, $X =$ the Sierpinski space [2.3(6)].

(**k**) $A =]0, 1]$, $X = \mathsf{R}$ with its right-interval topology [2.18(1)].

(**l**) $A =]0, 1]$, $X =$ the line with two origins [2.18(3)].

(**m**) $A = H$, $X =$ the half-disk space [2.20(3)].

(**n**) $A = H$, $X =$ the tangent disk space (Exercise 27).

35. What is the relationship between bdy(bdy A) and bdy A?

36. For the Euclidean metric d on $X = \mathsf{R}^n$ we know

$$\text{bdy } B_r(x; d) = S_r(x; d), \qquad \text{cls } B_r(x; d) = D_r(x; d)$$

for each $x \in X$ and each $r > 0$.

(**a**) Show that these equations do not hold in general for an arbitrary metric space (X, d).

(**b**) Do they hold for $X = \mathsf{R}^n$ and d an arbitrary metric inducing the usual topology on X?

37. Prove that A is both open and closed in X if bdy $A = \varnothing$. Is the converse true?

38. Let A be a nonempty subset of R having both an upper bound and a lower bound in R.

(**a**) Show that sup $A \in$ bdy A and inf $A \in$ bdy A.

(**b**) Deduce that sup $A \in A$ and inf $A \in A$ if A is closed in R. Does the converse hold?

39. Prove that X is a Hausdorff space if and only if for each $x \in X$ and each $y \in X$ with $y \neq x$ there is a neighborhood V of x such that $y \notin$ cls V.

40. Let A be a d-bounded set in a metric space (X, d). Prove: The set cls A is also d-bounded, and diam (cls A) = diam A. Discuss the analogous statement for int A.

41. (**a**) If $A \subset B \subset$ cls A, what can be said about cls B?

(**b**) What can be said about a closed subset E of X when E contains a dense subset of X?

42. Prove or disprove: If Y is a dense subspace of a topological space X and if $D \subset Y$ is dense in Y, then D is dense in X.

43. Let D_1 and D_2 be two dense sets in a topological space X. Show that $D_1 \cap D_2$ need not be dense in X but will be in case D_1 or D_2 is open in X.

44. Let $K \subset \mathsf{R}^n$ with K convex (0.14)

(**a**) Show that cls K must also be convex. Are bdy K and int K necessarily convex?

(**b**) Suppose K is closed in $\mathbf{R}^n$. An *extreme point* of K is a point $x \in K$ that does *not* have the form

$$x = (1 - t)a + tb$$

for any $0 < t < 1$ and any $a,\ b \in K$. Prove that bdy K contains the set of all extreme points of K but need not equal this set.

45. Let A be a subset of a topological space X.
(**a**) How can 'x is a limit point of A' (Exercise 24) be phrased in terms of closures?
(**b**) What is the relationship between cls A and the set D of all limit points of A in X?

46. The *exterior* of a set A in a space X is the set ext $A = \operatorname{int}(X \backslash A)$. Show that the three sets int A, bdy A, ext A are always pairwise disjoint and have the entire space X as their union.

47. (**a**) Prove: For a subset A of a topological space X,

$$\operatorname{cls} A = \bigcap \{E \mid A \subset E \subset X,\ E \text{ closed in } X\}.$$

(**b**) Formulate and prove the dual of (a) for int A.

48. (**a**) Prove that

$$\operatorname{cls} \bigcup_{i \in I} A_i = \bigcup_{i \in I} \operatorname{cls} A_i$$

for any *finite* family $(A_i \mid i \in I)$ of subsets of a space X, thereby generalizing 2.40 (2), but that this equation does not remain true for an infinite (even a denumerable) family.
(**b**) Prove that

$$\operatorname{cls} \bigcap_{i \in I} A_i \subset \bigcap_{i \in I} \operatorname{cls} A_i$$

for an arbitrary family $(A_i \mid i \in I)$ of subsets of X.
(**c**) Discuss the analogs of (a) and (b) for interiors.

49. A family $(A_i \mid i \in I)$ of subsets of a topological space X is said to be *locally finite* when each $x \in X$ has some neighborhood that intersects A_i for only finitely many $i \in I$.
(**a**) Verify that if $A_i = [i, \rightarrow[$ for each $i \in \mathbf{N}$, then the family $(A_i \mid i \in \mathbf{N})$ of subsets of $\mathbf{R}$ is locally finite even though for each $j \in \mathbf{N}$ the set A_j intersects A_i for infinitely many i.
(**b**) Construct a family $(A_i \mid i \in I)$ of subsets of $\mathbf{R}$ which is not locally finite but such that each point $x \in \mathbf{R}$ belongs to A_i for only finitely many $i \in I$.
(**c**) If $(A_i \mid i \in I)$ is locally finite, show that the family $(\operatorname{cls} A_i \mid i \in I)$ is also locally finite.
(**d**) Generalize Exercise 48 (a) by proving that

$$\operatorname{cls} \bigcup_{i \in I} A_i = \bigcup_{i \in I} \operatorname{cls} A_i$$

whenever the family $(A_i \mid i \in I)$ is locally finite.

50. Let A be a subset of a topological space X. Prove:
(**a**) If A is nowhere dense in X, then $X \backslash A$ is dense in X. The converse need not hold unless A is closed in X.
(**b**) If A is closed in X, then A is nowhere dense in X if and only if $A = \operatorname{bdy} A$.

(**c**) The set A is nowhere dense in X exactly when cls A is nowhere dense in X.

(**d**) The boundary of A is nowhere dense in X in case A is either open or closed in X, but otherwise bdy A need not be nowhere dense in X.

(**e**) The union of finitely many nowhere dense subsets of X is itself nowhere dense in X.

51. Verify that for each $1 \leq m < n$ the subspace

$$\{(x_1, \ldots, x_n) \in \mathsf{R}^n \mid x_{m+1} = x_{m+2} = \cdots = x_n = 0\}$$

of R^n is nowhere dense in R^n.

52. The complement of a first-category set in a topological space X is said to be *residual in* X. If the topology of X is induced by some complete metric, show that each residual set in X is dense in X.

53. Can the inclusions 2.43 (2) and (3) be strengthened to equalities when Y is closed in X? when Y is open in X?

54. Let X be a set. Suppose $c: \mathcal{P}(X) \to \mathcal{P}(X)$ is a map, assigning to each subset A of X a subset $c(A)$ of X, which satisfies the following four *Kuratowski closure axioms*:

(K_1) $c(\varnothing) = \varnothing$.

(K_2) For all $A \subset X$, $A \subset c(A)$.

(K_3) For all $A \subset X$, $c(c(A)) = c(A)$.

(K_4) For all $A \subset X$ and $B \subset X$, $c(A \cup B) = c(A) \cup c(B)$.

Prove that there is a unique topology $\mathcal{I}$ on X such that in the topological space $(X, \mathcal{I})$, cls $A = c(A)$ for each subset A of X. (*Hint*: A subset of X will be open for a topology $\mathcal{I}$ when its complement is its own closure.)

55. (Kuratowski closure-complement problem.) For a fixed subset A of a topological space X let $\mathcal{K}$ be the collection of all subsets of X obtained from A in a finite number of steps each of which consists of taking the closure or the complement in X of a set obtained in the preceding step. Thus $\mathcal{K}$ has as members the sets $A, X \backslash A$, cls A, cls$(X \backslash A)$, $X \backslash$ cls A, $X \backslash$ cls $(X \backslash A)$, etc.

(**a**) Prove that $\mathcal{K}$ can never contain more than 14 distinct subsets of X.

(**b**) Show by example (say in R) that $\mathcal{K}$ can actually contain 14 distinct sets.

4. Bases and Local Bases

In a metric space (X, d) the collection of all d-balls at a point x determines all the neighborhoods of x, and the collection of all d-balls determines all the open subsets of X. Likewise, in an arbitrary topological space a "local base" at a point will determine all the neighborhoods of the point, and a "base" will determine all the open sets. By examining a local base or a base, it is often possible to reduce a question concerning all neighborhoods of a point or all open sets to a simpler one concerning just certain "nice" neighborhoods or open sets.

2.44 Definition. Let x be a point of a topological space X. A collection $\mathfrak{M}$ of neighborhoods of x such that each neighborhood of x contains a member of $\mathfrak{M}$ is called a *local base at* x.

Synonyms for 'local base at x' are 'neighborhood base at x' and 'fundamental system of neighborhoods of x'.

2.45 Examples

(1) The entire neighborhood system at a point x of a topological space is itself a local base at x.

(2) For a point x of a discrete space, the collection having the single member $\{x\}$ is a local base at x.

(3) Let X be a metrizable space and let $x \in X$. Take any metric d inducing the topology of X. Then each of the following collections is a local base at x:

$\{B_\epsilon(x; d) \mid \epsilon > 0\}$,
$\{D_\epsilon(x; d) \mid \epsilon > 0\}$,
$\{B_r(x; d) \mid r \in \mathsf{Q}, r > 0\}$,
$\{B_{1/n}(x; d) \mid n = 1, 2, \ldots\}$.

(4) Specialize (3) by taking X to be the real line R. Then the four local bases at x given by (3) are the set of all open intervals $]x - \epsilon, x + \epsilon[$ symmetric about x, the set of all closed intervals $[x - \epsilon, x + \epsilon]$ of positive length symmetric about x, the set of all open intervals $]x - r, x + r[$ of rational length symmetric about x, and the set of all open intervals of the form $]x - 1/n, x + 1/n[$ for $n = 1, 2, \ldots$. Two more local bases at x are the set

$$\{]a, b[\mid a < x < b\}$$

of all open intervals containing x and the set

$$\{[a, b] \mid a < x < b\}$$

of all closed intervals of positive length containing x.

(5) Specialize (3) by taking $X = \mathsf{R}^n$. Since the max metric d_∞ induces the topology on R^n [see 2.3 (3) and 1.33], the collection

$$\{B_\epsilon(x; d_\infty) \mid \epsilon > 0\}$$

of all cubes centered at x is a local base at $x \in \mathsf{R}^n$.

(6) Consider the half-disk space of Example 2.20 (3). For $x \in L$, the collection $\{H_r(x) \mid r > 0\}$ is a local base at x, as is the countable collection $\{H_r(x) \mid 0 < r \in \mathsf{Q}\}$. For $z = (z_1, z_2) \in H$, the countable collection

$$\{B_r(z; d) \mid r \in \mathsf{Q}, 0 < r < z_2\}$$

is a local base at z.

Using local bases we now define the first of several *countability properties* a topological space may possess (the countability properties 'second-countable' and 'separable' are introduced below).

2.46 Definition. A topological space X is said to be *first-countable* if at each $x \in X$ there is some countable local base.

2.47 Examples

(1) *Every metrizable space X is first-countable.* In fact, if d is a metric inducing the topology of X and if $x \in X$, then both $\{B_r(x; d) \mid 0 < r \in \mathbf{Q}\}$ and $\{B_{1/n}(x; d) \mid n = 1, 2, \ldots\}$ are countable local bases at x.

(2) In view of 2.45 (6), the half-disk space, which is not metrizable, is first-countable.

(3) This is an example of a Hausdorff space which is not first-countable. Let X be a set and let $x \in X$ be a specific point. Define $\mathfrak{I}$ to be the collection of all subsets U of X such that $X \setminus U$ is finite or $x \notin U$. Then $\mathfrak{I}$ is a topology on X, called the *Fort topology* (named after M. K. Fort, Jr.).

We show that $(X, \mathfrak{I})$ is a Hausdorff space. Let $y, z \in X$ with $y \neq z$. One of the two points y and z, say y, is distinct from x. Then $\{y\}$ and $X \setminus \{y\}$ are disjoint neighborhoods of y and z.

Assume now X is uncountable. We show there is no countable local base at x in $(X, \mathfrak{I})$. Suppose, to the contrary, that $\{V_n \mid n \in \mathbf{N}\}$ is a local base at x for some sequence $(V_n \mid n \in \mathbf{N})$. Since $X \setminus V_n$ is finite for each $n \in \mathbf{N}$, then

$$X \setminus \bigcap_{n=0}^{\infty} V_n = \bigcup_{n=0}^{\infty} (X \setminus V_n)$$

is countable. Since X is uncountable, there exists $y \in \bigcup_{n=0}^{\infty} V_n$ with $y \neq x$. Then $X \setminus \{y\}$ is a neighborhood of x, so $V_m \subset X \setminus \{y\}$ for some $m \in \mathbf{N}$. This is impossible since $y \in V_m$.

Note that $(X, \mathfrak{I})$ does have a countable local base at each $y \neq x$, namely, the collection consisting of the single neighborhood $\{y\}$ of y.

As an example of a Hausdorff space which fails to have a countable local base at each of its points, we shall construct below [2.52 (3)] a "function space". That example is geometrically more natural than the Fort example, but more difficult to describe.

The collection of all d-balls at a point of a metric space (X, d) has been generalized to the concept of a local base at a point of a topological space. Next we introduce a topological generalization of the collection of all d-balls in a metric space.

2.48 Definition. Let X be a topological space with topology $\mathfrak{I}$. A collection $\mathfrak{B}$ of subsets of X is called a *base of X* (and *of $\mathfrak{I}$*) if (i) each member of $\mathfrak{B}$ is open in X, and (ii) each open subset of X is the union of some collection of sets belonging to $\mathfrak{B}$.

To check condition (ii), it suffices to check that each *nonempty* open subset of X is the union of some collection of sets belonging to $\mathfrak{B}$.

2.49 Examples

(1) The entire topology of a topological space X is itself a base of X.

(2) The collection $\{\{x\} \mid x \in X\}$ of all singletons in a discrete space X is a base of X.

(3) Let X be a metrizable space. Take any metric d inducing the topology of X. By 1.17 and Exercise 1.21 the collection $\{B_\epsilon(x; d) \mid x \in X, \epsilon > 0\}$ is a base of X.

(4) Specialize (3) by taking X to be the real line R. Then the collection

$$\{]a, b[\mid a \in \mathsf{R}, b \in \mathsf{R}, a < b\}$$

of all open intervals is a base of R.

(5) Specialize (3) by taking $X = \mathsf{R}^n$. Since the max metric d_∞ induces the usual topology on R^n, the collection

$$\{B_\epsilon(x; d_\infty) \mid x \in \mathsf{R}^n, \epsilon > 0\}$$

of all cubes is a base of R^n.

Comparison of 2.49 (1)–(5) with 2.45 (1)–(5) suggests the following characterization of a base in terms of local bases.

2.50 Proposition. Let $\mathcal{B}$ be a collection of open sets in a topological space X. Then the following two conditions are equivalent:
  (1) $\mathcal{B}$ is a base of X.
  (2) For each $x \in X$, the collection

$$\mathcal{B}_x = \{B \mid x \in B \in \mathcal{B}\}$$

of all members of $\mathcal{B}$ containing x is a local base at x.

Proof. Assume (1). We show (2). Let $x \in X$. If $B \in \mathcal{B}_x$, then B is an open set containing x, so B is a neighborhood of x. Now let V be any neighborhood of x. Choose an open set U with $x \in U \subset V$. Since $\mathcal{B}$ is a base of X, $U = \bigcup_{i \in I} B_i$ for some family $(B_i \mid i \in I)$ of sets belonging to $\mathcal{B}$. Then $x \in B_j$ for some $j \in I$. Hence $B_j \in \mathcal{B}_x$ and $B_j \subset V$.

Assume (2). We show (1). Let U be any nonempty open set in X. For each $x \in U$ the set U is a neighborhood of x, so by assumption there exists some $B_x \in \mathcal{B}_x$ with $B_x \subset U$. Then $U = \bigcup_{x \in U} B_x$ with $B_x \in \mathcal{B}$ for each $x \in U$. $\quad\square$

Condition (2) above for a collection $\mathcal{B}$ of open sets to be a base of X has the equivalent formulation: For each open set U in X and each $x \in U$, there exists $B \in \mathcal{B}$ with $x \in B \subset U$.

Proposition 2.50 furnishes additional examples. Thus if d is a metric inducing the topology of a space X, then 2.50 together with 2.47 (1) says that $\{B_r(x; d) \mid x \in X, 0 < r \in \mathsf{Q}\}$ is a base of X.

The next theorem gives us a new way of constructing topological spaces.

2.51 Theorem. Let X be a set. Let $\mathcal{B}$ be a collection of subsets of X having the two properties:

(B1) Each $x \in X$ belongs to some $B \in \mathfrak{B}$.

(B2) If $B_1, B_2 \in \mathfrak{B}$ and $x \in B_1 \cap B_2$, then $x \in B \subset B_1 \cap B_2$ for some $B \in \mathfrak{B}$.

Then there exists a unique topology $\mathfrak{I}$ on X of which $\mathfrak{B}$ is a base.

Proof. In view of 2.50, we shall be interested in the collections

$$\mathfrak{B}_x = \{ B \mid x \in B \in \mathfrak{B} \}$$
$$\mathfrak{N}_x = \{ V \mid \text{for some } B \in \mathfrak{B}, \quad x \in B \subset V \}$$

for all $x \in X$. Observe that these collections are completely determined by $\mathfrak{B}$ alone.

Uniqueness. Suppose $\mathfrak{I}$ is a topology on X of which $\mathfrak{B}$ is a base. By 2.50, for each $x \in X$ the collection $\mathfrak{B}_x$ is a local base at x for $\mathfrak{I}$, so $\mathfrak{N}_x$ is the $\mathfrak{I}$-neighborhood system at x. According to Theorem 2.17 the topology $\mathfrak{I}$ is uniquely determined by the $\mathfrak{I}$-neighborhood systems at all $x \in X$.

Existence. We shall construct the desired topology $\mathfrak{I}$ by applying Theorem 2.17 to the collections $\mathfrak{N}_x$ defined above. Conditions (N2) and (N4) needed to apply 2.17 are immediate consequences of the definition of $\mathfrak{N}_x$; (N1) and (N3) follow from (B1) and (B2), respectively. To verify (N5), let $x \in X$ and $V \in \mathfrak{N}_x$. Choose $B \in \mathfrak{B}$ with $x \in B \subset V$. Then $B \in \mathfrak{N}_y$ for each $y \in B$.

By 2.17 there is a topology $\mathfrak{I}$ on X such that for each $x \in X$ the collection $\mathfrak{N}_x$ is the $\mathfrak{I}$-neighborhood system at x. It remains to show $\mathfrak{B}$ is a base of $\mathfrak{I}$. If $B \in \mathfrak{B}$, then $B \in \mathfrak{N}_x$ for each $x \in X$, so B is $\mathfrak{I}$-open. For each $x \in X$ the collection $\mathfrak{B}_x$ is a local base at x by definition of $\mathfrak{N}_x$. It follows from 2.50 that $\mathfrak{B}$ is a base of $\mathfrak{I}$.   □

Note that conditon (B2) above is automatically satisfied in case the intersection of any two members of $\mathfrak{B}$ belongs to $\mathfrak{B}$.

2.52 Examples

(1) Let X be a totally ordered set under a total ordering $\leq$, and suppose X has at least two elements. Define $\mathfrak{B}$ to be the collection of all open rays and all open intervals in X, that is,

$$\mathfrak{B} = \{\,]x, \rightarrow[\ \mid x \in X\} \cup \{\,]\leftarrow, y[\ \mid y \in X\} \cup \{\,]x, y[\ \mid x, y \in X\}.$$

The collection $\mathfrak{B}$ satisfies (B1). In fact, if $x \in X$, there is a $y \in X$ with $y \neq x$; then $x \in\ ]\leftarrow, y[$ in case $y > x$, and $x \in\ ]y, \rightarrow[$ in case $y < x$. We leave the verification of (B2) to the reader, noting only that

$$]x, \rightarrow[\ \cap\ ]\leftarrow, y[\ =\ ]x, y[$$

for all $x, y \in X$.

The topology $\mathfrak{I}$ on X of which $\mathfrak{B}$ is a base is called the *order topology*.

Suppose X has no smallest element and no greatest element. Then for each $x \in X$,

$$]x, \rightarrow [\; = \bigcup_{y>x} \;]x, y[, \qquad]\leftarrow, x[\; = \bigcup_{y<x} \;]y, x[.$$

Hence in this case the order topology also has the base

$$\{ \,]x, y[\mid x, y \in X \}.$$

It follows from 2.49 (4) and the preceding paragraph that the order topology on R (for the usual total ordering of R) is just the usual topology on R.

(2) Consider the extended real line $R^* = R \cup \{ -\infty, +\infty \}$ with its total ordering $\leq$ and its metric D (Example 1.37). Then the order topology $\mathfrak{I}$ on R^* defined by $\leq$ is the same as the topology $\mathfrak{I}(D)$ on R^* induced by the metric D.

To prove our assertion it suffices to show that each $\mathfrak{I}$-neighborhood of a point $x \in R^*$ is a $\mathfrak{I}(D)$-neighborhood of x, and vice versa. Let $x \in R^*$.

Case (i): $x \in R$. Let V be a $\mathfrak{I}(D)$-neighborhood of x. Choose $\epsilon > 0$ with

$$\epsilon \leq \min \{ |\varphi(x) - 1|, |1 + \varphi(x)| \}$$

so that $B_\epsilon(x; D) \subset V$. In our earlier proof (1.38) that the metric on R induced by D is equivalent to the Euclidean metric on R, we actually showed that for such an ϵ the D-ball $B_\epsilon(x; D)$ is an open interval in R containing x. Hence V is a $\mathfrak{I}$-neighborhood of x in R^*.

Conversely, let V be a $\mathfrak{I}$-neighborhood of x. Choose $-\infty < a < x < b < +\infty$ with $]a, b[\subset V$. In the proof just cited we showed that $B_\epsilon(x; D) \subset]a, b[$ for a suitable $\epsilon > 0$. Hence V is a $\mathfrak{I}(D)$-neighborhood of x.

Case (ii): $x = +\infty$. If V is a $\mathfrak{I}(D)$-neighborhood of $+\infty$, then

$$B_\epsilon(+\infty; D) \subset V$$

for some $0 < \epsilon < 1$,

$$B_\epsilon(+\infty; D) = \left] \frac{1}{\epsilon} - 1, +\infty \right]$$

$$= \left] \frac{1}{\epsilon} - 1, \rightarrow \right[$$

by 1.39, and so V is a $\mathfrak{I}$-neighborhood of $+\infty$. Conversely, if V is a $\mathfrak{I}$-neighborhood of $+\infty$, then

$$]u, \rightarrow [\; = \;]u, +\infty] \subset V$$

for some $u > 0$,

$$]u, +\infty] = B_{1/(u+1)}(+\infty; D)$$

by 1.39, and so V is a $\mathfrak{I}(D)$-neighborhood of $+\infty$.

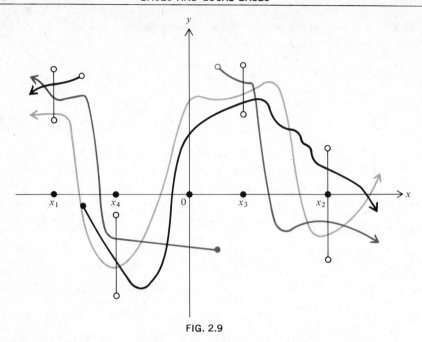

FIG. 2.9

Case (iii): $x = -\infty$. This case is treated in the same way as Case (ii).

(3) Let $\mathfrak{F}$ be the set of all functions $f\colon \mathsf{R} \to \mathsf{R}$. Given any integer $k \geq 1$, any k-tuple $(x_1, \ldots, x_k)$ of real numbers, and any k-tuple $(I_1, \ldots, I_k)$ of open intervals in R, define

$$B(x_1, \ldots, x_k; I_1, \ldots, I_k) = \{f \in \mathfrak{F} \mid f(x_j) \in I_j \ (j = 1, \ldots, k)\}.$$

Thus a function $f\colon \mathsf{R} \to \mathsf{R}$ belongs to $B(x_1, \ldots, x_k; I_1, \ldots, I_k)$ if and only if the graph of f passes through each of the vertical segments $\{x_1\} \times I_1, \ldots, \{x_k\} \times I_k$ in the plane $\mathsf{R} \times \mathsf{R}$. Fig. 2.9 depicts three members of a set of the form $B(x_1, x_2, x_3, x_4; I_1, I_2, I_3, I_4)$.

Define $\mathfrak{B}$ to be the collection of all such subsets $B(x_1, \ldots, x_k; I_1, \ldots, I_k)$ for all possible choices of k, $(x_1, \ldots, x_k)$, and $(I_1, \ldots, I_k)$. Property (B1) holds, for if $f \in \mathfrak{F}$, then $f \in B(x_1, I_1)$ where $x_1 \in \mathsf{R}$ and I_1 is any open interval with $f(x_1) \in I_1$. Porperty (B2) follows from the relation

$$B(x_1, \ldots, x_k; I_1, \ldots, I_k) \cap B(t_1, \ldots, t_m; J_1, \ldots, J_m)$$
$$= B(x_1, \ldots, x_k, t_1, \ldots, t_m; I_1, \ldots, I_k, J_1, \ldots, J_m).$$

From Theorem 2.51 we obtain a unique topology $\mathfrak{I}$ on $\mathfrak{F}$ having $\mathfrak{B}$ as a base. We call $\mathfrak{I}$ the *topology of pointwise convergence* on the function space $\mathfrak{F}$ (this terminology will be explained in Section 5 of Chapter 3).

We show that $(\mathfrak{F}, \mathfrak{I})$ is a Hausdorff space. Let $f, g \in \mathfrak{F}$ with $f \neq g$. Then $f(x) \neq g(x)$ for some $x \in \mathsf{R}$. Choose disjoint open intervals I and J containing the real numbers $f(x)$ and $g(x)$, respectively. Then $B(x; I)$ and $B(x; J)$ are disjoint open neighborhoods of f and g, respectively.

We show that $(\mathfrak{F}, \mathfrak{I})$ is not first-countable by showing that at *no* $f \in \mathfrak{F}$ is there a countable local base. Just suppose there is a countable local base at some $f \in \mathfrak{F}$. Since $\{B \mid f \in B \in \mathfrak{B}\}$ is a local base at f, there is a sequence $(B_n \mid n \in \mathsf{N})$ of members of $\mathfrak{B}$ such that $\{B_n \mid n \in \mathsf{N}\}$ is also a local base at f. For each n the set B_n has the form

$$B_n = B(x_{1n}, \ldots, x_{k_n n}; I_{1n}, \ldots, I_{k_n n}).$$

The set $\{x_{jn} \mid n \in \mathsf{N}, 1 \leq j \leq k_n\}$ of real numbers used to form these sets B_n is only countable, whereas R is uncountable. Hence we may choose $x_0 \in \mathsf{R}$ with $x_0 \neq x_{jn}$ for all $n \in \mathsf{N}$ and all $1 \leq j \leq k_n$. Let I_0 be any open interval containing $f(x_0)$. Then $B(x_0; I_0)$ is a neighborhood of f. Hence

$$B_n \subset B(x_0; I_0)$$

for some n.

We now obtain a contradiction by constructing a $g \in B_n$ with $g \notin B(x_0; I_0)$. For each $1 \leq j \leq k_n$, choose any $y_{jn} \in I_{jn}$. Choose any $y_0 \in \mathsf{R} \backslash I_0$. Define $g: \mathsf{R} \to \mathsf{R}$ by

$$g(x) = \begin{cases} y_{jn} & \text{if } x = x_{jn} \quad \text{for } 1 \leq j \leq k_n, \\ y_0 & \text{if } x = x_0, \\ 0 & \text{otherwise.} \end{cases}$$

In view of what we just proved and 2.47 (1), the space $(\mathfrak{F}, \mathfrak{I})$ is not metrizable.

The next countability property is defined in terms of bases.

2.53 Definition. A topological space is said to be *second-countable* if it has some countable base.

2.54 Examples

(1) The metrizable space R^n is second-countable. To prove this we use the max metric d_∞ on R^n (1.6). Let

$$\mathfrak{B} = \{B_r(z; d_\infty) \mid 0 < r \in \mathsf{Q}, z_i \in \mathsf{Q} \ (i = 1, \ldots, n)\},$$

the collection of all cubes whose sides have rational lengths and whose centers have all coordinates rational. Since both Q^n and Q are countable, so is $\mathfrak{B}$. Since d_∞ induces the usual topology of R^n, each member of $\mathfrak{B}$ is open.

To prove that $\mathfrak{B}$ is a base of R^n, we verify condition (2) of Proposition 2.50. Let $x \in \mathsf{R}^n$ and let V be any neighborhood of x. We construct a $B \in \mathfrak{B}$ with $x \in B \subset V$. Choose $\epsilon > 0$ with $B_\epsilon(x; d_\infty) \subset V$. For each $i = 1, \ldots, n$, there is a rational number z_i with $|z_i - x_i| < \epsilon/2$. Set $z = (z_1, \ldots, z_n)$, so

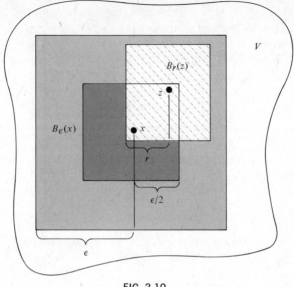

FIG. 2.10

that $d_\infty(z, x) < \epsilon/2$. Choose a rational number r with

$$d_\infty(z, x) < r < \frac{\epsilon}{2}$$

(see Fig. 2.10). Then $x \in B_r(z; d_\infty) \in \mathcal{B}$. Finally, $B_r(z; d_\infty) \subset B_\epsilon(x; d_\infty)$, for $y \in B_r(z; d_\infty)$ implies

$$d_\infty(y, x) \leq d_\infty(y, z) + d_\infty(z, x) < r + r < \frac{\epsilon}{2} + \frac{\epsilon}{2} = \epsilon.$$

(2) Let X be a discrete space. Then the space X is second-countable if and only if the underlying set X is countable. To see this, note first that X is countable if and only if the collection

$$\mathcal{S} = \{\{x\} \mid x \in X\}$$

is countable. If X is countable, then $\mathcal{S}$ is a countable base of X. Now suppose X is uncountable and let $\mathcal{B}$ be a base of X. If $x \in X$, then $x \in B \subset \{x\}$ for some $B \in \mathcal{B}$, so $\{x\} = B \in \mathcal{B}$. Hence $\mathcal{S} \subset \mathcal{B}$, and $\mathcal{B}$ is uncountable.

By putting the discrete topology on an uncountable set, we obtain a metrizable space that is not second-countable.

The preceding example shows that a first-countable space need not be second-countable. By Proposition 2.50, however, *every second-countable space is first-countable.*

Our third countability property uses the notion of a dense set (2.35).

2.55 Definition. A topological space X is said to be *separable* if some countable subset of X is dense in X.

2.56 Examples

(1) According to 2.36 (1) the real line R is separable.

(2) By 2.36 (2), a discrete space X is separable if and only if the underlying set X is countable. In view of 2.54 (2), a discrete space is separable if and only if it is second-countable.

(3) While proving in 2.54 (1) that R^n is second-countable, we showed that the countable set Q^n is dense in R^n. Hence Euclidean n-space R^n is separable.

(4) The Hilbert sequence space H (Example 1.9) is separable. Indeed, by analogy with the preceding example we would expect the set

$$\{y \in \mathsf{H} \mid y_i \in \mathsf{Q} \text{ for } i = 1, 2, \ldots\}$$

of all "rational points" in H to be dense in H—and it is—but this set is unfortunately uncountable. We must consider instead the smaller set D consisting solely of those rational points in H that are eventually zero:

$$D = \bigcup_{n=1}^{\infty} D_n$$

where for $n = 1, 2, \ldots$,

$$D_n = \{y \in \mathsf{H} \mid y_1, \ldots, y_n \in \mathsf{Q}, y_{n+1} = y_{n+2} = \cdots = 0\}.$$

For each n there is an obvious one-to-one correspondence between D_n and the denumerable set Q^n, so that D_n is denumerable. Hence D is denumerable.

To see that D is dense in H, let $x = (x_i \mid i = 1, 2, \ldots) \in \mathsf{H}$ and let $\epsilon > 0$. Since the series $\sum_{n=1}^{\infty} x_i^2$ converges, there exists $n \geq 1$ for which

$$\sum_{i=n+1}^{\infty} x_i^2 < \frac{\epsilon^2}{2}.$$

For each $i = 1, \ldots, n$ there is a rational number y_i with

$$|x_i - y_i|^2 < \frac{\epsilon^2}{2n}.$$

Then the point $y = (y_i \mid i = 1, 2, \ldots)$ with $y_i = 0$ for all $i > n$ belongs to D_n, and

$$[d_2(x, y)]^2 = \sum_{i=1}^{\infty} |x_i - y_i|^2$$

$$= \sum_{i=1}^{n} |x_i - y_i|^2 + \sum_{i=n+1}^{\infty} x_i^2$$

$$< n \cdot \frac{\epsilon^2}{2n} + \frac{\epsilon^2}{2} = \epsilon^2.$$

(5) The space $\mathfrak{C}([0, 1])$ of all continuous real-valued functions on $[0, 1,]$ with the topology induced by the max metric (1.7), is separable. One countable dense set in $\mathfrak{C}([0, 1])$ consists of all functions $f: [0, 1] \to \mathbb{R}$ having the property that for some positive integer n, f takes rational values at each of the points $0, 1/n, 2/n, \ldots, 1$ and is linear on each of the intervals $[0, 1/n]$, $[1/n, 2/n], \ldots, [(n-1)/n, 1]$. The proof requires ideas developed in Section 2 of Chapter 4 (see Example 4.44).

(6) The space $\mathfrak{B}([0, 1])$ of all bounded real-valued functions on $[0, 1]$, with its topology induced by the sup metric d_∞ (see 1.69), is not separable: Consider for each $x \in [0, 1]$ the characteristic function f_x of $\{x\}$ given by

$$f_x(x) = 1, \qquad f_x(y) = 0 \qquad (y \neq x).$$

Any dense set D in $\mathfrak{B}([0, 1])$ contains for each $x \in [0, 1]$ a function g_x with

$$d_\infty(f_x, g_x) < \tfrac{1}{2}.$$

Now $x, y \in [0, 1]$ with $x \neq y$ implies $d_\infty(f_x, f_y) = 1$ and hence $g_x \neq g_y$. Thus D contains the uncountable set $\{g_x \mid x \in [0, 1]\}$.

Examples (2) and (3) suggest a general theorem.

2.57 Theorem. A metrizable space X is separable if and only if it is second-countable.

Proof. Choose a metric d inducing the topology of X.

Assume X is second-countable. Let $(B_n \mid n \in \mathbb{N})$ be a sequence of non-empty open sets such that $\{B_n \mid n \in \mathbb{N}\}$ is a base of X. For each $n \in \mathbb{N}$, choose some $x_n \in B_n$. Then $\{x_n \mid n \in \mathbb{N}\}$ is a countable set which is easily seen to be dense in X. Hence X is separable.

Assume X is separable. Let D be a countable dense subset of X. Define a collection $\mathfrak{B}$ of open subsets of X by

$$\mathfrak{B} = \{B_r(z; d) \mid z \in D, 0 < r \in \mathbb{Q}\}.$$

Since D and $\mathbb{Q}$ are both countable, so is $\mathfrak{B}$.

To complete the proof that X is second-countable, we show that $\mathfrak{B}$ is a base of X by verifying condition (2) of Proposition 2.50. Let $x \in X$ and let V be a neighborhood of x. We must show $x \in B \subset V$ for some $B \in \mathfrak{B}$. Choose $\epsilon > 0$ with $B_\epsilon(x; d) \subset V$. Since D is dense in X, there is some $z \in D \cap B_{\epsilon/2}(x; d)$. Choose a rational number r with $d(z, x) < r < \epsilon/2$. Then $x \in B_r(z; d) \in \mathfrak{B}$ and $B_r(z; d) \subset B_\epsilon(x; d) \subset V$. $\square$

The next theorem states how bases and local bases in a subspace are obtained from the corresponding objects in the entire space.

2.58 Theorem. Let Y be a subspace of a topological space X. Then:
(1) If $y \in Y$ and if $\mathfrak{M}$ is a local base at y in X, then

$$\{V \cap Y \mid V \in \mathfrak{M}\}$$

is a local base at y in Y.
(2) If $\mathfrak{B}$ is a base of X, then

$$\{B \cap Y \mid B \in \mathfrak{B}\}$$

is a base of Y.

Proof. (1) Let $y \in Y$ and let $\mathfrak{M}$ be a local base at y in Y. Recall Proposition 2.14 which says that the neighborhoods of y in Y are precisely the sets $V \cap Y$ for all neighborhoods V of y in X. Then $V \cap Y$ is a neighborhood of y in Y for each $V \in \mathfrak{M}$. Now let W be any neighborhood of y in Y. Then $W = U \cap Y$ for some neighborhood U of y in X. Since $\mathfrak{M}$ is a local base at y in X, there is some $V \in \mathfrak{M}$ with $V \subset U$. Then $V \cap Y \subset U \cap Y = W$.
(2) Use (1) and Proposition 2.50. $\Box$

2.59 Corollary. Let Y be a subspace of a topological space X. Then:

(1) If X is first-countable, so is Y.
(2) If X is second-countable, so is Y.

Unlike first-countability and second-countability, separability is *not* always inherited from a space to its subspaces. For example, consider the half-disk space $X = L \cup H$ of Example 2.20 (3). The entire space X is separable, for its countable subset $\{(x, y) \in Q \times Q \mid y > 0\}$ is dense in X. Now the relative topology on the subspace $L = R \times \{0\}$ is the discrete topology since for each $x \in L$ the set

$$H_1(x) \cap L = \{x\}$$

is an open neighborhood of x in L. By Example 2.56 (2), an uncountable discrete space cannot be separable. Hence the subspace L of X is not separable.

EXERCISES

56. For each point x of the tangent disk space (Exercise 27) describe a local base at x different from the entire neighborhood system at x.

57. Let $\mathfrak{I}_1$ and $\mathfrak{I}_2$ be two topologies on the same set X and let $x \in X$. Suppose $\mathfrak{M}_1$ and $\mathfrak{M}_2$ are local bases at x for $\mathfrak{I}_1$ and $\mathfrak{I}_2$, respectively. In terms of $\mathfrak{M}_1$ and $\mathfrak{M}_2$, when is the $\mathfrak{I}_1$-neighborhood system at x the same as the $\mathfrak{I}_2$-neighborhood system at x?

58. (**a**) Is the Fort topology [Example 2.47 (3)] on a countable set first-countable?
(**b**) Is every topological space whose underlying set is countable a first-countable space?

59. Let X be a first-countable space and let $x \in X$. Prove that there is a sequence $(V_n \mid n \in N)$ such that $\{V_n \mid n \in N\}$ is a local base at x and $V_{n+1} \subset V_n$ for every

$n \in \mathbb{N}$. Can the sets V_n all be chosen to be open? Can they be chosen so that $\bigcap_{n=0}^{\infty} V_n = \{x\}$?

60. A topological space is said to be *regular* (and is called a T_3-*space*) if at each $x \in X$ there is some local base consisting solely of closed sets. [*Caution*: Some authors include as part of the definition of one of the terms 'regular' and 'T_3' the requirement of being a T_0-space (Exercise 32).] Prove that each of the following is a necessary and sufficient condition for X to be regular:
    (**a**) For each $x \in X$ and each neighborhood U of x there is a neighborhood V of x with cls $V \subset U$.
    (**b**) For each point $x \in X$ and each closed set $E \subset X$ with $x \notin E$ there exist disjoint open sets containing x and E.
    (**c**) For each point $x \in X$ and each closed set $E \subset X$ with $x \notin E$ there is a neighborhood V of x such that $E \cap$ cls $V = \varnothing$.

61. (Continuation of Exercise 60.)
    (**a**) Prove that every metrizable space is regular.
    (**b**) Verify that an uncountable set provided with its Fort topology [Example 2.47 (3)], which is a Hausdorff space, is regular but not metrizable.

62. (Continuation of Exercise 60.)
    (**a**) Prove: Every regular T_0-space (Exercise 32) is a T_2-space.
    (**b**) Find a T_2-space that is not regular and a regular space that is not a T_0-space (and hence not a T_2-space).

63. (Continuation of Exercise 60.) Is a subspace of a regular space necessarily regular?

64. For each point x in a topological space X, let $\mathcal{B}_x$ be a local base at x consisting of open sets. Establish that the following hold for all $x \in X$:
    (LB1) There exists at least one member of $\mathcal{B}_x$.
    (LB2) The point x belongs to each member of $\mathcal{B}_x$.
    (LB3) The intersection of any two members of $\mathcal{B}_x$ contains some member of $\mathcal{B}_x$.
    (LB4) If $B \in \mathcal{B}_x$, then for each $y \in B$ there is some $V \in \mathcal{B}_y$ with $V \subset B$.

65. Let X be a set. For each $x \in X$ let $\mathcal{B}_x$ be a given collection of subsets of X. Assume these collections have properties (LB1)–(LB4). Prove that there exists a unique topology $\mathfrak{I}$ on X such that for each $x \in X$, $\mathcal{B}_x$ is a local base at x consisting of $\mathfrak{I}$-open sets. (*Hint*: Use Theorem 2.17.)

66. Let $\mathcal{B}$ be a base of a topological space X.
    (**a**) Prove: A subset U of X is open in X if and only if for each $x \in U$ there is some $B \in \mathcal{B}$ with $x \in B \subset U$.
    (**b**) When, in terms of $\mathcal{B}$, is a subset E of X closed in X?
    (**c**) Do properties (B1) and (B2) of 2.51 necessarily hold for $\mathcal{B}$?

67. (**a**) Let $\mathcal{B}_1$ and $\mathcal{B}_2$ be bases of topologies $\mathfrak{I}_1$ and $\mathfrak{I}_2$, respectively, on the same set X. Prove that $\mathfrak{I}_1 \subset \mathfrak{I}_2$ if and only if for each $B_1 \in \mathcal{B}_1$ and each $x \in B_1$ there exists $B_2 \in \mathcal{B}_2$ with $x \in B_2 \subset B_1$.
    (**b**) By applying (a), show that two metrics d_1 and d_2 on a set X induce the same topology in case there are positive constants a and b with $a\,d_1(x, y) \leq d_2(x, y) \leq b\,d_1(x, y)$ for all $x, y \in X$.

68. Show that each of the following collections is a base of $\mathbb{R}^n$:
    (**a**) The collection of all subsets of $\mathbb{R}^n$ of the form $]a_1, b_1[\times \cdots \times]a_n, b_n[$.
    (**b**) The collection of all open sets in $\mathbb{R}^n$ that are convex.

69. Let $\leq$ be the total ordering of $\mathsf{R} \times \mathsf{R}$ defined in Exercise 29.
 (**a**) Describe explicitly the open intervals $]x, y[$, which form a base of the order topology on $\mathsf{R} \times \mathsf{R}$.
 (**b**) Compare the order topology with both the usual and the right-interval topologies on $\mathsf{R} \times \mathsf{R}$.

70. Find the boundary, interior, and closure of the set

$$\{ f \in \mathfrak{F} \mid 0 < f(x) < 1 \text{ for all } 0 \leq x < 1 \}$$

in the space $\mathfrak{F}$ of Example 2.52 (3).

71. Generalize the construction in Example 2.52 (3) as follows. Let X be a set, let Y be a topological space, and let $\mathfrak{F}$ be the set of all functions $f : X \to Y$. Define $\mathfrak{B}$ to be the collection of all subsets of $\mathfrak{F}$ of the form

$$B(x_1, \ldots, x_k; V_1, \ldots, V_k) = \{ f \in \mathfrak{F} \mid f(x_j) \in V_j \, (j = 1, \ldots, k) \},$$

where k is a positive integer, $x_1, \ldots, x_k \in X$, and $V_1, \ldots, V_k$ are open subsets of Y.
 (**a**) Verify that properties (B1) and (B2) of Theorem 2.51 are satisfied. Thus $\mathfrak{B}$ is a base of a unique topology on $\mathfrak{F}$, called the *topology of pointwise convergence*.
 (**b**) Show that $\mathfrak{F}$, with this topology, is a Hausdorff space whenever Y is.
 (**c**) When, if ever, will $\mathfrak{F}$ be first-countable?
 (**d**) If $X = \mathsf{N}$ and Y is a finite discrete space, show that this topology is separable and is induced by the metric of Example 1.14.

72. Is the extended real line (Example 1.37) separable?

73. Prove:
 (**a**) Every countable first-countable space is second-countable.
 (**b**) Every second-countable space is separable.

74. Which of the countability properties 'first-countable', 'second-countable', 'separable' is true of each of the following spaces?
 (**a**) A denumerable set with its finite-complement topology [2.3 (7)].
 (**b**) An uncountable set with its countable-complement topology (Exercise 6).
 (**c**) The half-disk space [2.20 (3)].
 (**d**) The set R with its right-interval topology [2.18(1)].

75. Is a subspace of a separable metrizable space necessarily separable?

76. Let X be the sum (Exercise 20) of a family $(X_i \mid i \in I)$ of pairwise disjoint spaces. Discuss the relationship of each of the countability properties for X to the corresponding property for the spaces X_i.

77. Let X be a totally ordered set and let Y be a subset of X having at least two elements. Then on Y we have the relative topology induced by the order topology [Example 2.52(1)] on X, and the order topology defined by the restriction of the total ordering of X to Y. Show that these two topologies may be distinct, even when $X = \mathsf{R}$ with its usual ordering.

78. A topological space X is said to be *zero-dimensional* if at each point of X there is a local base consisting of sets whose boundaries are empty.
 (**a**) Verify that any discrete space is zero-dimensional, but that R is not.
 (**b**) Let Y be a finite discrete space containing at least two points, let $\mathfrak{F}$ be the space of all functions $f : \mathsf{N} \to Y$, and provide $\mathfrak{F}$ with its topology of pointwise convergence (Exercise 71). Show that $\mathfrak{F}$ is nondiscrete and zero-dimensional.

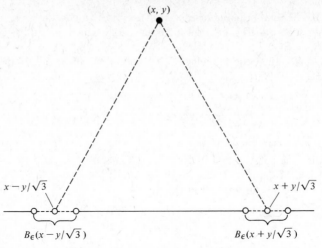

FIG. 2.11

79. Let $\mathcal{C}$ be any collection of subsets of a set X. Define $\mathcal{B}$ to be the collection of all subsets of X that are the intersection of finitely many members of $\mathcal{C}$.

(**a**) Show that $\mathcal{B}$ is a base of a topology $\mathfrak{I}$ on X containing $\mathcal{C}$.

(**b**) Prove that $\mathfrak{I} \subseteq \mathcal{S}$ for each topology $\mathcal{S}$ on X containing $\mathcal{C}$. We say that $\mathfrak{I}$ is *generated by* $\mathcal{C}$.

(**c**) Describe in terms of $\mathcal{C}$ the $\mathfrak{I}$-neighborhood system of an $x \in X$.

(**d**) Show that the topology of Example 2.52(3) is generated by the collection of all sets $B(x; I)$ for $x \in \mathsf{R}$ and I an open interval in R.

80. In the plane $\mathsf{R} \times \mathsf{R}$, let L be the x-axis $\mathsf{R} \times \{0\}$; below we denote each point $(x, 0) \in L$ simply by x. Let X be the countable subset

$$X = \{(x, y) \in \mathsf{Q} \times \mathsf{Q} \mid y \geq 0\}.$$

of $\mathsf{R} \times \mathsf{R}$. A remarkable topology on X will be defined with the aid of Exercise 79.

Given $\epsilon > 0$ and $(x, y) \in X$ with $y > 0$, define a set $U_\epsilon(x, y)$ as follows. The three points (x, y), $(x - y/\sqrt{3}, 0)$, $(x + y/\sqrt{3}, 0)$ are the vertices of an equilateral triangle $T(x, y)$ whose base is on L and whose sides have slopes $\sqrt{3}$ and $-\sqrt{3}$ and intersect L at the irrational points $x - y/\sqrt{3}$ and $x + y/\sqrt{3}$. Form the sets

$$B_\epsilon\left(x - \frac{y}{\sqrt{3}}\right) = \mathsf{Q} \cap \left]x - \frac{y}{\sqrt{3}} - \epsilon, x - \frac{y}{\sqrt{3}} + \epsilon\right[,$$

$$B_\epsilon\left(x + \frac{y}{\sqrt{3}}\right) = \mathsf{Q} \cap \left]x + \frac{y}{\sqrt{3}} - \epsilon, x + \frac{y}{\sqrt{3}} + \epsilon\right[$$

of rational points on L belonging to the symmetric intervals of length 2ϵ about these two points (see Fig. 2.11). Set

$$U_\epsilon(x, y) = \{(x, y)\} \cup B_\epsilon\left(x - \frac{y}{\sqrt{3}}\right) \cup B_\epsilon\left(x + \frac{y}{\sqrt{3}}\right).$$

Provide X with the topology generated by all the sets $U_\epsilon(x, y)$. The resulting topological space is the *Bing triangle space*, named after R. H. Bing. You are to show;

(**a**) Each $x \in X \cap L$ has a local base consisting of intervals in **Q** of the form **Q** ∪ $]x - \epsilon, x + \epsilon[$.

(**b**) For (x, y), $(x', y') \in X$ with $y > 0, y' > 0$, and $(x, y) \neq (x', y')$, the triangles $T(x, y)$ and $T(x', y')$ cannot have a vertex in common, nor can they intersect along an entire line segment not lying on L.

(**c**) The space X is a T_2-space.

(**d**) The space X is not regular (Exercise 60) and hence not metrizable.

(**e**) The space X is second-countable.

Continuity and Convergence

Our program of abstracting topological ideas from metric spaces is continued in this chapter with the study of continuous maps between topological spaces. Homeomorphisms are a special class of continuous maps examined in detail; when there is a homeomorphism from one space onto another, the two spaces are topologically the same in that they have precisely the same topological properties. With the aid of continuous maps we define appropriate topologies on product and quotient sets of topological spaces, thereby providing new methods for constructing topological spaces. Finally, after defining sequential convergence in topological spaces and seeing that it is not adequate for treating topological questions, we present a more general theory of convergence which is.

1. Continuous Maps

To define continuity of a map $f: X \to Y$ between arbitrary topological spaces we cannot simply mimic the definition (1.44) in the case of metric spaces, for that definition directly employed metrics. According to Theorem 1.46, however, if the topologies 3 and S on X and Y are induced by metrics d and d', then f is (d, d')-continuous at a point $x \in X$ precisely when each d'-neighborhood N of $f(x)$ contains the image $f(M)$ of some d-neighborhood M of x. Now a d'-neighborhood of $f(x)$ is just an S-neighborhood of $f(x)$ in Y, and a d-neighborhood of x is just a 3-neighborhood of x in X. Hence f is (d, d')-continuous at x precisely when each S-neighborhood N of $f(x)$ in Y contains the image $f(M)$ of some 3-neighborhood M of x in X. The latter condition no longer refers directly to metrics, only to topologies.

3.1 Definition. Let $f\colon X \to Y$ be a map from a topological space X to a topological space Y. For $x \in X$, the map f is *continuous at* x if for each neighborhood N of $f(x)$ in Y there exists some neighborhood M of x in X such that $f(M) \subset N$. The map f is *continuous* if it is continuous at each $x \in X$.

Sometimes it is convenient to work with the following condition (compare 1.46) instead of directly with the definition.

3.2 Lemma. A necessary and sufficient condition for a map $f\colon X \to Y$ between topological spaces to be continuous at a point $x \in X$ is that $f^{-1}(N)$ be a neighborhood of x in X for each neighborhood N of $f(x)$ in Y.

Proof. Let N be any neighborhood of $f(x)$ in Y. If $f(M) \subset N$ for a neighborhood M of x in X, then $f^{-1}(N)$ is a neighborhood of x because $M \subset f^{-1}(N)$. Conversely, if $f^{-1}(N)$ is a neighborhood of x in X, then $f(M) \subset N$ where $M = f^{-1}(N)$. □

In the same way that we deduced 1.48 from 1.46 we may deduce from 3.2 a topological criterion for continuity.

3.3 Theorem. A map $f\colon X \to Y$ from a topological space X to a topological space Y is continuous if and only if the inverse image $f^{-1}(V)$ of each open subset V of Y is open in X.

In terms of the topologies $\mathfrak{I}$ and $\mathfrak{S}$ on X and Y, respectively, the criterion 3.3 says simply that

$$V \in \mathfrak{S} \quad \Rightarrow \quad f^{-1}(V) \in \mathfrak{I}.$$

When using 3.3 to establish continuity of a map $f\colon X \to Y$, we can sometimes reduce the number of open subsets of Y whose inverse images in X must be examined. For example, if a metric d induces the topology of Y, we need only show that the inverse image of each d-ball $B_\epsilon(y; d)$ is open in X. This follows from the following corollary.

3.4 Corollary. A map $f\colon X \to Y$ from a topological space X to a topological space Y is continuous if there is some base $\mathfrak{B}$ of Y such that $f^{-1}(B)$ is open in X for each $B \in \mathfrak{B}$.

Proof. Assume $\mathfrak{B}$ is such a base. Let V be any open subset of Y. There is a family $(B_i \mid i \in I)$ of members of $\mathfrak{B}$ with

$$V = \bigcup_{i \in I} B_i.$$

Then

$$f^{-1}(V) = f^{-1}\left(\bigcup_{i \in I} B_i\right) = \bigcup_{i \in I} f^{-1}(B_i)$$

is a union of open subsets of X and hence is open in X. It follows from 3.3 that f is continuous.   ▯

Since open sets determine closed sets and vice versa, criterion 3.3 may be formulated in terms of closed sets.

3.5 Corollary. The map $f: X \to Y$ is continuous if and only if the inverse image $f^{-1}(E)$ of each closed subset E of Y is closed in X.

Proof. Assume f is continuous. Let E be closed in Y. Then $Y \setminus E$ is open in Y, by 3.3 the set

$$X \setminus f^{-1}(E) = f^{-1}(Y \setminus E)$$

is open in X, and hence $f^{-1}(E)$ is closed in X.

Conversely, assume $f^{-1}(E)$ is closed in X for each closed set E in Y. If V is open in Y, then $Y \setminus V$ is closed in Y, by assumption the set

$$X \setminus f^{-1}(V) = f^{-1}(Y \setminus V)$$

is closed in X, and hence $f^{-1}(V)$ is open in X. According to 3.3, then, f is continuous.   ▯

3.6 Corollary. Suppose $f: X \to Y$ is continuous. Then

$$f(\operatorname{cls} A) \subset \operatorname{cls} f(A)$$

for each set $A \subset X$. [Here, of course, $\operatorname{cls} A$ is the closure of A in X, and $\operatorname{cls} f(A)$ is the closure of the subset $f(A)$ of Y in Y.]

Proof. Let $A \subset X$. The set $\operatorname{cls} f(A)$ is closed in Y, so by 3.5 its inverse image $f^{-1}(\operatorname{cls} f(A))$ is closed in X. Now

$$A \subset f^{-1}(f(A)) \subset f^{-1}(\operatorname{cls} f(A))$$

because $f(A) \subset \operatorname{cls} f(A)$. Since $\operatorname{cls} A$ is the least closed subset of X containing A,

$$\operatorname{cls} A \subset f^{-1}(\operatorname{cls} f(A)).$$

Hence $f(\operatorname{cls} A) \subset \operatorname{cls} f(A)$.   ▯

Despite its trivial proof, the next theorem expresses the single most important property of continuity. It assures us that the most general way of combining continuous maps always leads to continuous maps. Moreover, it is a tool for establishing the continuity of maps built up from simpler maps. For example, from the continuity of the sine and square-root functions it allows us to conclude at once the continuity of the real-valued function $x \mapsto \sin \sqrt{x}$ on the set of nonnegative real numbers.

3.7 Theorem. Let $f\colon X \to Y$ and $g\colon Y \to Z$ be continuous maps between topological spaces. Then the composite map $g \circ f\colon X \to Z$ is continuous.

Proof. We use 3.3. Let W be any open subset of Z. By continuity of g the set $g^{-1}(W)$ is open in Y, and then by continuity of f the set

$$(g \circ f)^{-1}(W) = f^{-1}(g^{-1}(W))$$

is open in X. □

3.8 Examples

(1) Let X and Y be metrizable topological spaces. Then the discussion leading to Definition 3.1 shows that a map $f\colon X \to Y$ is continuous precisely when it is (d, d')-continuous for any metrics d and d' inducing the topologies on X and Y.

All the continuous maps between metric spaces discussed in Chapter 1, and in particular all the continuous functions encountered in calculus, are thus examples of continuous maps between topological spaces.

This example demonstrates once again that continuity of a map between metric spaces depends not on the particular metrics, but only on the topologies they induce (compare 1.47). Hence it supports our contention that topological spaces provide the proper setting in which to study continuity even of maps between metric spaces.

(2) A constant map $f\colon X \to Y$ with constant value c is continuous, for

$$f^{-1}(V) = \begin{cases} \varnothing & \text{if } c \notin V, \\ X & \text{if } c \in V \end{cases}$$

for each subset V of Y.

(3) Since every subset of a discrete space is open, a map $f\colon X \to Y$ from a discrete space X to an arbitrary topological space Y is continuous.

(4) Suppose A is a subspace of a topological space X. Then the inclusion map

$$j\colon A \to X$$

$$x \mapsto x$$

is continuous since by definition of the relative topology on A the set $j^{-1}(V) = V \cap A$ is open in A for each open subset V of X. In particular (for $A = X$), the identity map $i\colon X \to X$ of any topological space X is continuous.

(5) Let $f\colon X \to Y$ be a map from a topological space X to a topological space Y. If f is continuous, then for each set $A \subset X$ the restriction

$$f\,|_A\colon A \to Y$$

$$x \mapsto f(x)$$

of f to A is continuous, being the composite of the inclusion map $j\colon A \to X$

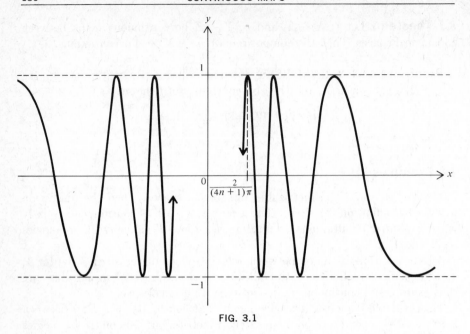

FIG. 3.1

and $f: X \to Y$; here we have used 3.7. Thus cutting down the domain X of f does not affect continuity.

Conversely, suppose $A \subset X$ and $g = f|_A: A \to Y$ is continuous. Then the extension f of g to X need not be continuous. For example, take $X = Y = \mathbf{R}$, $A = \mathbf{R}\backslash\{0\}$, and $f: X \to Y$ the map defined by

$$
f(x) = \begin{cases} \sin \dfrac{1}{x} & \text{if } x \neq 0, \\[2mm] 0 & \text{if } x = 0. \end{cases}
$$

Then $g = f|_A$ is continuous, being the composite of the continuous maps $x \mapsto 1/x$ of $A \to \mathbf{R}$ and $y \mapsto \sin y$ of $\mathbf{R} \to \mathbf{R}$. However, f cannot be continuous at $0 \in X$ because its graph oscillates between the lines $y = -1$ and $y = 1$ arbitrarily close to the origin (see Fig. 3.1). More precisely, if V is any neighborhood of $0 = f(0)$, then $]-\epsilon, \epsilon[\subset V$ for some $0 < \epsilon < 1$; any neighborhood U of 0 in X contains $2/(4n+1)\pi$ for some sufficiently large positive integer n, the value $f(2/(4n+1)\pi) = \sin (4n+1)\pi/2 = 1 \notin V$, and so $f(U) \not\subset V$.

Even had $f(0)$ been defined to be some number other than 0, a slight refinement of the preceding argument would show that f still would not be continuous at 0. In the jargon of advanced calculus, f has a "nonremovable discontinuity" at $x = 0$; in the language of elementary claculus, f cannot be made continuous at $x = 0$ because $\lim_{x\to 0} f(x)$ does not exist. Thus the continuous map $g: A \to Y$ does not have any continuous extension to X.

When does a continuous map $g: A \to Y$ on a subspace A of a space X have a continuous extension $f: X \to Y$ to X? This is one of the central questions of topology. A general situation where continuous extensions always exist is given below [3.11 (2)]. One instance where they do not is

$$A = Y = \{x \in \mathsf{R}^2 : ||x|| = 1\}, \qquad X = \{x \in \mathsf{R}^2 : ||x|| \leq 1\},$$

and $g: A \to Y$ is the identity map of the circle A. A continuous extension $f: X \to Y$ of g would be a continuous map of the disk X onto its bounding circle Y leaving each point of this circle fixed [that is, $f(x) = x$ for each $x \in A = Y$]. The reader is hereby challenged to prove no such f exists before he sees the machinery developed in Sections 4 and 5 of Chapter 5.

(6) The map $x \mapsto 1/x$ with domain $\mathsf{R} \backslash \{0\}$ is continuous whether considered as a map into R or into $\mathsf{R} \backslash \{0\}$. In general, cutting down or enlarging the codomain of a map does not affect its continuity: Let

$$g: X \to Z, \qquad f: X \to Y$$

be maps on a topological space X with Z a subspace of the topological space Y and

$$f(x) = g(x) \qquad\qquad (x \in X).$$

[Necessarily $f(X) \subset Z$; the two maps differ when $Z \neq Y$.] Then f is continuous if and only if g is. In fact, $f: X \to Y$ is the composite of $g: X \to Z$ and the inclusion map $j: Z \to Y$, so f is continuous if g is. Conversely, if $f: X \to Y$ is continuous, then an arbitrary open subset V of Z has the form $V = W \cap Z$ for some open subset W of Y, and $g^{-1}(V) = f^{-1}(W)$ is open in X.

If above we take $Z = f(X)$, the range of f, then we obtain a surjection $g: X \to f(X)$ having the same graph as $f: X \to Y$.

(7) Let $\mathfrak{I}$ and $\mathfrak{S}$ be two topologies on the same underlying set X. Consider the identity map $i: X \to X$ as a map $i: (X, \mathfrak{I}) \to (X, \mathfrak{S})$ from the topological space $(X, \mathfrak{I})$ to the topological space $(X, \mathfrak{S})$. [Generally, given a map $f: X \to Y$ between sets and topologies $\mathfrak{I}$ on X and $\mathfrak{S}$ on Y, we write

$$f: (X, \mathfrak{I}) \to (Y, \mathfrak{S})$$

when continuity of f with respect to these topologies is at issue.] Since $i^{-1}(V) = V$ for each $V \subset X$,

$$i: (X, \mathfrak{I}) \to (X, \mathfrak{S}) \text{ is continuous} \qquad \Leftrightarrow \qquad \mathfrak{S} \subset \mathfrak{I}.$$

We express the relation $\mathfrak{S} \subset \mathfrak{I}$ by saying that $\mathfrak{I}$ is *finer* or *stronger* or *larger* than $\mathfrak{S}$ and that $\mathfrak{S}$ is *coarser* or *weaker* or *smaller* than $\mathfrak{I}$. For example, the discrete topology is finer than, and the indiscrete topology weaker than, every topology on X. Example 1.34 concerning the set X of all continuous functions $x: [0, 1] \to \mathsf{R}$ shows that each d_1-ball at a point $x \in X$ contains a d_∞-ball at x, but not conversely; hence the topology on X induced by the metric d_∞ is *strictly* finer than the topology induced by d_1, in other words, the d_∞-topology is finer than but distinct from the d_1-topology.

An arbitrary pair of topologies on a set X need not, of course, be comparable with one another; that is, neither need be finer than the other (see Exercise 2.7).

The finer the topology on X, the more open sets there are and hence the more neighborhoods of each point there are. Moreover, the finer the topology on X, the more continuous maps on X there are: If $\mathcal{S} \subset \mathcal{I}$ and $(Y, \mathcal{U})$ is any topological space, then each continuous map $f \colon (X, \mathcal{S}) \to (Y, \mathcal{U})$ is also continuous as a map $f \colon (X, \mathcal{I}) \to (Y, \mathcal{U})$.

In connection with Example (5), observe that even when a continuous map $g \colon A \to Y$ on a subspace A of a space X has a continuous extension to X, such an extension need not be unique. This will be the case, for example, when X is discrete, $A \neq X$, and Y contains at least two points (the map can be extended to X in two different ways by assigning distinct constant values on $X \setminus A$). In one important case, however, such an extension, if it exists at all, must be unique.

3.9 Theorem (extension of identities). Let $f \colon X \to Y$ and $g \colon X \to Y$ be continuous maps from a topological space X into a Hausdorff space Y. Suppose $f|_A = g|_A$ for some dense subset A of X. Then $f = g$.

Proof. Assume $f(x) \neq g(x)$ for some $x \in X$. Choose disjoint open neighborhoods V of $f(x)$ and W of $g(x)$ in Y. The set

$$U = f^{-1}(V) \cap g^{-1}(W)$$

is open in X and, since it contains x, nonempty. Then there is some $a \in A \cap U$, so $f(a) \in V$ and $g(a) \in W$. Now $g(a) = f(a)$ since $a \in A$. Hence $f(a) \in V \cap W$, which is impossible by the choice of V and W.   ☐

Thus under the hypotheses of the theorem, the identity $f(x) = g(x)$ holds for all x in the entire space if it holds for all x in a dense subset. A simple application is that a continuous function $f \colon \mathbb{R} \to \mathbb{R}$ taking the value 0 at each rational number must take the value 0 at every real number.

A map f on a space X need not be continuous just because its restriction to a given subspace A of X is continuous. This is hardly surprising, for the behavior of f on the single subspace A need tell us nothing about its behavior at points not belonging to A. What happens if the restrictions of f to each of a whole family $(A_i \mid i \in I)$ of subspaces of X is continuous? To have any hope of inferring continuity of f on the entire space X, we should require that $(A_i \mid i \in I)$ *covers* X in the sense that

$$X = \bigcup_{i \in I} A_i.$$

3.10 Theorem (gluing lemma). Let $(A_i \mid i \in I)$ be a family of subspaces of a topological space X that covers X. Suppose either each A_i is open in X, or else $(A_i \mid i \in I)$ is finite and each A_i is closed in X. Then a map $f \colon X \to Y$ into a topological space Y is continuous if its restriction $f \mid A_i \colon A_i \to Y$ is continuous for each $i \in I$.

Proof. Assume the restriction of f to each A_i is continuous. Suppose first each A_i is open in X. Let V be any open set in Y. For each $i \in I$ the set

$$f^{-1}(V) \cap A_i = (f \mid A_i)^{-1}(V)$$

is open in A_i by 3.3 and hence is open in X. Then the union

$$f^{-1}(V) = f^{-1}(V) \cap X = f^{-1}(V) \cap \bigcup_{i \in I} A_i = \bigcup_{i \in I} [f^{-1}(V) \cap A_i]$$

is open in X. By 3.3, f is continuous.

Suppose now $(A_i \mid i \in I)$ is finite and each A_i is closed in X. Let E be any closed set in Y. For each $i \in I$ the set

$$f^{-1}(E) \cap A_i = (f \mid A_i)^{-1}(E)$$

is closed in A_i by 3.5 and hence is closed in X. Then the union

$$f^{-1}(E) = \bigcup_{i \in I} [f^{-1}(E) \cap A_i]$$

of finitely many closed sets is closed in X. By 3.5, f is continuous. $\quad\square$

Typically in applications of this gluing lemma we do not start with a map f already defined on the entire space X. Rather, we are given a whole family of maps

$$f_i \colon A_i \to Y \qquad\qquad (i \in I),$$

one for each of the subspaces A_i of X. If these maps "match up" on the overlaps of their domains, that is, if

$$f_i \mid (A_i \cap A_j) = f_j \mid (A_i \cap A_j) \qquad\qquad (i, j \in I),$$

then they may be "glued together" to form a single map $f \colon X \to Y$ defined by the rule that for $x \in X$,

$$f(x) = f_i(x) \qquad \text{where } i \in I \text{ and } x \in A_i;$$

this map is well defined because $x \in A_i$ and $x \in A_j$ implies $f_i(x) = f_j(x)$. Thus $f \colon X \to Y$ is the unique map such that

$$f \mid A_i = f_i \qquad\qquad (i \in I).$$

According to the gluing lemma, f will be continuous if each f_i is continuous.

3.11 Examples

(1) Consider the absolute-value function $f \colon \mathsf{R} \to \mathsf{R}$ given by

$$f(x) = \begin{cases} x & \text{if } x \geq 0, \\ -x & \text{if } x \leq 0. \end{cases}$$

The gluing lemma may be used to establish continuity of f as follows: Take

$$A_1 = \{x \in \mathsf{R} \mid x \geq 0\}, \qquad A_2 = \{x \in \mathsf{R} \mid x \leq 0\}$$

so that $(A_i \mid i \in I)$ covers R and consists of two closed sets. The restriction $f \mid A_1 : A_1 \to \mathsf{R}$ is the inclusion map, which is continuous; the restriction $f \mid A_2 : A_2 \to \mathsf{R}$ is also continuous, being a restriction of the continuous map $x \mapsto -x$ from R into R.

Of course, continuity of f may also be established directly by using the standard δ-ϵ argument.

(2) Here is a nontrivial application of the gluing lemma. Any continuous function

$$g : A \to [0, 1]$$

on a closed subspace A of a metrizable space X has a continuous extension $f : X \to [0, 1]$ to X. This is a special case of *Tietze's extension theorem*, which asserts that the same thing is true more generally for any "normal" space X (see the discussion following 4.16).

For technical reasons it is simpler to extend not $g : A \to [0, 1]$ but instead the continuous function

$$G : A \to [1, 2]$$

given by $G(x) = g(x) + 1$ for all $x \in A$. Once a continuous extension $F : X \to [1, 2]$ of G has been constructed, the function $f : X \to [0, 1]$ defined by $f(x) = F(x) - 1$ for all $x \in X$ will be the desired continuous extension of g.

Choose a metric d inducing the topology of X. According to 2.34 the distance

$$d(x, A) = \inf_{a \in A} d(x, a)$$

from a point x to the closed set A is positive when $x \notin A$. Then it is meaningful to define

$$F(x) = \begin{cases} G(x) & \text{if } x \in A, \\[2ex] \dfrac{1}{d(x, A)} H(x) & \text{if } x \in X \setminus A, \end{cases}$$

where

$$H(x) = \inf_{a \in A} G(a) \, d(x, a) \qquad\qquad (x \in X \setminus A).$$

Since $1 \leq G(a) \leq 2$ for $a \in A$, then $1 \leq F(x) \leq 2$ for $x \in X \setminus A$. Thus an extension $F : X \to [1, 2]$ of G has been constructed.

To see that F is continuous, consider its restrictions to the closed subspaces

$$A_1 = A = \operatorname{cls} A, \qquad A_2 = \operatorname{cls}(X \setminus A) = (X \setminus A) \cup \operatorname{bdy} A$$

of X. Now $A_1 \cup A_2 = X$, and $F \mid A_1 = G$ is already known to be continuous.

It remains only to show that $F \mid A_2$ is continuous. Let $x \in A_2$ and consider two cases.

Case (i): $x \in X \backslash A$. According to Exercise 1.56 the real-valued function $u \mapsto d(u, A)$ is continuous at x. If we can establish continuity of $H : X \backslash A \to \mathsf{R}$ at x, then continuity of $F \mid A_2$ at x will follow by the standard argument used to establish continuity of the quotient of a real-valued function by a positive-valued function on R (compare Exercise 66(a)).

Let $\epsilon > 0$. Let $u \in X \backslash A$ with

$$d(x, u) < \frac{\epsilon}{2}.$$

Then $a \in A$ implies

$$d(x, a) \leq d(x, u) + d(u, a) < \frac{\epsilon}{2} + d(u, a)$$

and, since $1 \leq G(a) \leq 2$,

$$G(a) \, d(x, a) < \epsilon + G(a) \, d(u, a).$$

Hence

$$H(x) \leq \epsilon + H(u).$$

Similarly, $H(u) \leq \epsilon + H(x)$. Then $|H(u) - H(x)| \leq \epsilon$.

Case (ii): $x \in \mathrm{bdy}\, A$. Let $\epsilon > 0$. By continuity of G at x there is a $\delta > 0$ such that

(*) $\qquad a \in A, \quad d(x, a) < \delta \quad \Rightarrow \quad |G(a) - G(x)| < \epsilon.$

Already, then, $|F(u) - F(x)| < \epsilon$ for $d(x, u) < \delta/4$ in case $u \in A$; we shall show the same thing is true in case $u \in X \backslash A$. Let $u \in X \backslash A$ with

$$d(x, u) < \frac{\delta}{4}.$$

The infimum

$$H(u) = \inf_{a \in A} G(a) \, d(u, a)$$

will be unchanged if we let a vary just over the subset

$$B = A \cap B_\delta(x; d)$$

of A, that is,

(**) $\qquad\qquad H(u) = \inf_{a \in B} G(a) \, d(u, a).$

In fact, $a \in A \backslash B$ implies

$$d(u, a) \geq d(x, a) - d(x, u) > \delta - \frac{\delta}{4} = \frac{3}{4}\delta$$

so that

$$\inf_{a \in A \setminus B} G(a)\, d(u, a) \geq \frac{3}{4}\, \delta,$$

whereas $x \in B$ and

$$G(x)\, d(u, x) \leq 2d(u, x) < \frac{\delta}{2} < \frac{3}{4}\, \delta.$$

By the same kind of argument just used,

$$\inf_{a \in B} d(u, a) = d(u, A),$$

and from (*)

$$G(x) - \epsilon < G(a) < G(x) + \epsilon \qquad\qquad (a \in B).$$

Hence,

$$[G(x) - \epsilon]\, d(u, A) \leq \inf_{a \in B} G(a)\, d(u, a) \leq [G(x) + \epsilon]\, d(u, A).$$

Using (**) we conclude

$$|F(u) - F(x)| = |F(u) - G(x)| \leq \epsilon.$$

Theorem 3.3 and Corollary 3.5 say that a map $f: X \to Y$ is continuous precisely when the inverse image $f^{-1}(B)$ of each open or closed set in Y is open or closed, respectively, in X; they say nothing about the direct image $f(A)$ of an open or closed subset A of X.

3.12 Definition. A map $f: X \to Y$ between topological spaces is *open* if the image $f(U)$ of each open subset U of X is open in Y, and *closed* if the image $f(E)$ of each closed subset E of X is closed in Y.

3.13 Examples

(1) Let X be R with its usual topology and let Y be R with its discrete topology. Then the identity map $i: X \to Y$ is not continuous but is both open and closed. The identity map $i: Y \to X$, on the other hand, is continuous but neither open nor closed.

(2) The projection

$$p_1: \mathsf{R}^2 \to \mathsf{R}$$

$$(x_1, x_2) \mapsto x_1$$

of the plane onto the horizontal axis is continuous. To see this we use the max metric d_∞ on R^2 and the Euclidean metric d on R. Given $x = (x_1, x_2) \in \mathsf{R}^2$ and $\epsilon > 0$, any $u = (u_1, u_2) \in \mathsf{R}^2$ with

$$d_\infty(x, u) = \max \{d(x_1, u_1), d(x_2, u_2)\} < \epsilon$$

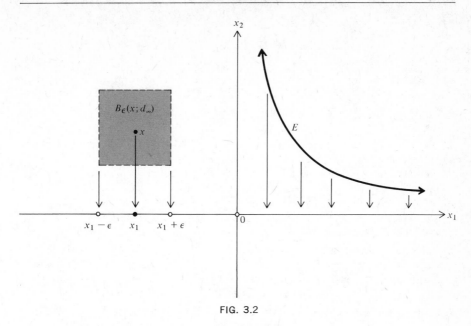

FIG. 3.2

will satisfy

$$d(p_1(x), p_1(u)) = d(x_1, u_1) < \epsilon.$$

The map p_1 is also open. Let U be an open set in R^2. To see that $p_1(U)$ is open in R, let $x_1 \in p_1(U)$. Then $x = (x_1, x_2) \in U$ for some x_2. For some $\epsilon > 0$ the square region

$$]x_1 - \epsilon, x_1 + \epsilon[\times]x_2 - \epsilon, x_2 + \epsilon[= B_\epsilon(x; d_\infty) \subset U.$$

Then $p_1(U)$ contains the neighborhood $]x_1 - \epsilon, x_1 + \epsilon[$ of x_1 (see Fig. 3.2).

The map p_1 is not closed, however. In fact, the hyperbolic arc

$$E = \left\{ (x_1, x_2) \in \mathsf{R}^2 \;\middle|\; x_1 > 0, x_2 = \frac{1}{x_1} \right\}$$

is closed in R^2, but its image

$$p_1(E) = \{ x_1 \in \mathsf{R} \mid x_1 > 0 \}$$

is certainly not closed in R (see Fig. 3.2).

Two facts about open maps will be useful later. First, a map $f\colon X \to Y$ is open if there is some base $\mathfrak{B}$ of X with $f(B)$ open in Y for each $B \in \mathfrak{B}$; this analog of 3.4 generalizes the technique used in Example (2) to show p_1 is open. Second, a bijection $f\colon X \to Y$ is an open map if and only if it is a closed map, owing to the identity

$$Y \backslash f(A) = f(X \backslash A)$$

which holds for all subsets A of X (compare the proof of 3.5).

EXERCISES

1. If $f: X \to Y$ is continuous at $x \in X$ and if $g: Y \to Z$ is continuous at $f(x)$, show that $g \circ f$ is continuous at x.

2. Prove that a map $f: X \to Y$ is continuous at a point $x \in X$ if and only if there is a local base $\mathcal{B}$ at $f(x)$ in Y such that $f^{-1}(B)$ is a neighborhood of x for each $B \in \mathcal{B}$.

3. Is the map $f: \mathsf{Q} \to \mathsf{R}$ such that $f(x) = 0$ if $x^2 < 2$, and $f(x) = 1$ otherwise, a continuous map?

4. Let $X = \mathsf{R} \cup \{z\}$ be the line with two origins [2.18(3)]. Which of the following maps are continuous?
    (**a**) The map $f: X \to \mathsf{R}$ with $f(x) =$ if $x \in \mathsf{R}$ and $f(z) = 0$.
    (**b**) The map $\tau: [-1, 1] \to X$ with $\tau(x) = x$ for all x.
    (**c**) The map $\tau': [-1, 1] \to X$ with $\tau'(x) = x$ for $x \neq 0$ and $\tau'(0) = z$.
    (**d**) The map $\sigma: [0, 2] \to X$ with $\sigma(t) = t$ for $0 \leq t \leq 1$, $\sigma(t) = 2 - t$ for $1 \leq t < 2$, and $\sigma(2) = z$.

5. (**a**) If $X = H \cup L$ is the half-disk space [2.20(3)], is the map $(x, y) \mapsto (x, 0)$ from X to its subspace L continuous? Is it open? Do your answers remain the same if L is given its usual topology instead of its relative topology as a subspace of L?
    (**b**) Repeat (a) if X is instead the tangent disk space (Exercise 2.27).

6. Show that a map $f: X \to Y$ will be continuous if the topology of Y is generated by some collection $\mathcal{Q}$ (Exercise 2.79) such that $f^{-1}(A)$ is open in X for each $A \in \mathcal{Q}$.

7. (**a**) Strengthen 3.6 by showing that a map $f: X \to Y$ is continuous if and only if $f(\mathrm{cls}\, A) \subset \mathrm{cls}\, f(A)$ for each subset A of X.
    (**b**) Is there an analogous characterization of continuity involving interiors?

8. If $f: X \to Y$ is a continuous map from a separable space X onto a topological space Y, must Y be separable?

9. Does the property that every map $f: X \to Y$ on a discrete space X to an arbitrary topological space is continuous actually characterize discrete spaces?

10. Determine all topological spaces Y such that any map $f: X \to Y$ from any topological space X is continuous.

11. Given a map $f: X \to Y$ from a space X to a space Y and a subspace A of X, show that continuity of $f \mid A$ is not necessarily equivalent to the continuity of f at each point of A. Find conditions on A sufficient to ensure it is equivalent.

12. Show directly, without using 3.11(2), that a continuous map $g: [a, b] \to [0, 1]$ on a closed interval $[a, b]$ in R can always be extended continuously to R. Can $[0,1]$ be replaced by other spaces?

13. Let $\mathfrak{I}$ and $\mathfrak{S}$ be topologies on a set X with $\mathfrak{S}$ stronger than $\mathfrak{I}$.
    (**a**) Given any subset A of X, compare the boundary, interior, and closure of A for $\mathfrak{S}$ with the boundary, interior, and closure for $\mathfrak{I}$.
    (**b**) Discuss the relation between the various countability properties and the Hausdorff separation property holding for $(X, \mathfrak{I})$ and their holding for $(X, \mathfrak{S})$.

14. Prove that the relative topology on a subset A of a topological space X is the weakest topology making the inclusion map $j: A \to X$ continuous.

15. Given two topologies $\mathfrak{I}$ and $\mathfrak{S}$ on the same set X, show that there is a largest topology smaller than both $\mathfrak{I}$ and $\mathfrak{S}$, and a smallest topology larger than both $\mathfrak{I}$ and $\mathfrak{S}$. (Compare Exercise 2.8.)

16. Show that a sufficient condition for a map $f: X \rightarrow Y$ to be continuous is that each $x \in X$ have some neighborhood V_x such that $f \mid V_x$ is continuous.

17. Verify that the conclusion of Theorem 3.10 still holds if the family $(A_i \mid i \in I)$ covering X is only a locally finite family (Exercise 2.49) of closed sets.

18. In the notation of the paragraph following 3.10, show that the graph of $f: X \rightarrow Y$ is the union of the graphs of the maps $f_i: A_i \rightarrow Y$ for all $i \in I$.

19. Using 3.11(2) to extend an appropriate function, show that for disjoint closed sets A and B in a metrizable space X there is a continuous function $f: X \rightarrow [0, 1]$ with $f(a) = 0$ for all $a \in A$ and $f(b) = 1$ for all $b \in B$. (*Note*: A direct proof of this result is indicated in Exercise 4.17.)

20. Prove: A map $f: X \rightarrow Y$ is open precisely when for each $x \in X$ the image $f(M)$ of each neighborhood M of x is a neighborhood of $f(x)$ in Y.

21. When is the inclusion map $j: A \rightarrow X$ of a subspace A of a space X an open map? When is it a closed map?

22. Let $f: R \times R \rightarrow R \times [0, \rightarrow [$ be the map given by $f(x, y) = (x, |y|)$, which folds the plane along the x-axis onto the upper half plane. Is f continuous? Is it open? Is it closed?

23. A map $f: X \rightarrow Y$ can be continuous or not continuous, open or not open, closed or not closed. Show by example that all of the $8 = 2^3$ resulting logical possibilities do actually occur. (Begin with Examples 3.13 and construct surjections wherever possible as your additional examples.)

24. Does an analog of Theorem 3.7 hold for open maps? for closed maps?

25. Let $f: X \rightarrow Y$ be a continuous open surjection. Prove:
  (**a**) If X is a Hausdorff space, and if f is injective, then Y is a Hausdorff space.
  (**b**) If X is second-countable, so is Y.
  (**c**) If X is first-countable, so is Y.

26. Prove that a map $f: X \rightarrow Y$ is both continuous and closed precisely when $f(\operatorname{cls} A) = \operatorname{cls} f(A)$ for each subset A of X. Give an analogous result for continuous open maps.

27. (**a**) If $f: X \rightarrow Y$ is a continuous surjection, show that a map $g: Y \rightarrow Z$ is open if $g \circ f$ is open.
  (**b**) If $g: Y \rightarrow Z$ is a continuous injection, show that a map $f: X \rightarrow Y$ is open if $g \circ f$ is open.
  (**c**) Do (a) and (b) remain true if 'open' is replaced by 'closed'?

28. A real-valued function f on a topological space X is said to be *lower semicontinuous* when $f^{-1}(]c, \rightarrow [)$ is open in X for each $c \in R$, and *upper semicontinuous* when $f^{-1}(] \leftarrow, c [)$ is open in X for each $c \in R$. Prove:
  (**a**) f is continuous if and only if f both lower semicontinuous and upper semicontinuous.
  (**b**) f is lower semicontinuous if and only if $-f$ is upper semicontinuous.
  (**c**) f is lower semicontinuous if and only if for each $x \in X$ and each $c < f(x)$ there is a neighborhood U of x with $c < f(u)$ for all $u \in U$. (This explains the terminology 'lower'.)
  (**d**) The analog of (c) for upper semicontinuity. [*Hint*: Use (b).]
  (**e**) The characteristic function $\chi_A: X \rightarrow R$ of a subset A of X is lower semicontinuous if A is open in X and is upper semicontinuous if A is closed in X.

2. Homeomorphisms

Earlier, in Section 3 of Chapter 1, we looked briefly at metric equivalence (isometry) and topological equivalence of metric spaces. Although metric equivalence depended on the particular metrics involved, topological equivalence depended only on the topologies induced by the metrics. Hence the latter notion may be generalized to a purely topological setting.

3.14 Definition. Let X and Y be topological spaces. If $f: X \to Y$ is a bijection such that the image $f(U)$ of each open subset U of X is open in Y and the inverse image $f^{-1}(V)$ of each open subset V of Y is open in X, then we call f a *homeomorphism from X to Y* and write

$$f: X \cong Y.$$

When some homeomorphism $f: X \cong Y$ exists, we say that X is *homeomorphic to Y* and write

$$X \cong Y.$$

Homeomorphisms are sometimes called *topological transformations*.
Any bijection $f: X \to Y$ induces a bijection

$$f^*: \mathcal{P}(X) \to \mathcal{P}(Y)$$
$$U \mapsto f(U)$$

from the collection of all subsets of X to the collection of all subsets of Y. Then to say f is a homeomorphism is to say f^* maps the topology of X onto the topology of Y. Thus a homeomorphism $f: X \cong Y$ is a one-to-one correspondence between the points of X and the points of Y which establishes a one-to-one correspondence between the open sets in X and the open sets in Y. For this reason a topologist regards two homeomorphic spaces as being essentially the same; more will be said about this shortly.

In view of Theorem 3.3 and Definition 3.12, evidently *a homeomorphism $f: X \cong Y$ is just a continuous open bijection*, or equivalently, a continuous closed bijection. Now a bijection $f: X \to Y$ is an open map precisely when $f^{-1}: Y \to X$ is continuous, because

$$(f^{-1})^{-1}(U) = f(U)$$

for each open subset U of X. Hence *a bijection $f: X \to Y$ is a homeomorphism if and only if both f and its inverse $f^{-1}: Y \to X$ are continuous.*

This last characterization, coupled with 3.7 and 3.8 (4), shows the following theorem.

3.15 Theorem. The identity map $i: X \to X$ of any topological space X is a homeomorphism

$$i: X \cong X.$$

If

$$f \colon X \cong Y$$

is a homeomorphism, then its inverse

$$f^{-1} \colon Y \cong X$$

is a homeomorphism. If

$$f \colon X \cong Y, \qquad g \colon Y \cong Z$$

are homeomorphisms, then their composite

$$g \circ f \colon X \cong Z$$

is a homeomorphism.

According to this theorem the relation 'is homeomorphic to' is an equivalence relation on the class of all topological spaces: For all topological spaces X, Y, and Z

$$X \cong X,$$
$$X \cong Y \quad \Rightarrow \quad Y \cong X,$$
$$X \cong Y \quad \text{and} \quad Y \cong Z \quad \Rightarrow \quad X \cong Z.$$

Thus homeomorphism is a means of classifying topological spaces as being alike or unalike one another, two spaces being classified as alike if and only if they are homeomorphic to one another.

3.16 Examples

(1) Two discrete spaces X and Y are homeomorphic to one another if and only if there is some bijection from X to Y, in other words, X has the same "number" of points as Y (technically, X has the same cardinality as Y). Then any denumerable discrete space is homeomorphic to Z, and in particular $\mathsf{N} \cong \mathsf{Z}$.

(2) The bijection

$$\varphi^* \colon \mathsf{R}^* \to [-1, 1]$$

used in Example 1.37 to construct the metric D on the extended real line R^* is a homeomorphism, for it is an isometry with respect to D and the Euclidean metric on $[-1, 1]$. The restriction

$$\varphi \colon \mathsf{R} \to \ ]-1, 1[$$

$$x \mapsto \frac{x}{1 + |x|}$$

of φ^* is a homeomorphism since by 2.9 (3) the real line R is a subspace of R^*.

(3) According to Example 1.43 (2) the bijection

$$f: [0, 1] \to [a, b]$$

$$x \mapsto (b - a)x + a$$

is a homeomorphism when $a < b$. Thus the *closed unit interval*

$$\mathsf{I} = [0, 1]$$

is a topological "representative" of all closed intervals $[a, b]$ in the real line containing more than one point. Since $\mathsf{I} \cong [-1, 1]$ and $\mathsf{R}^* \cong [-1, 1]$, from Theorem 3.15 we obtain

$$\mathsf{R}^* \cong \mathsf{I}.$$

(4) In Example (3) we have $f(0) = a$ and $f(b) = 1$, so by restricting both the domain and the codomain of f we obtain a homeomorphism

$$]0, 1[\,\cong\,]a, b[$$

whenever $a < b$. In particular, $]0, 1[\,\cong\,]-1, 1[$, so from Example (2) we obtain

$$\mathsf{R} \cong \,]0, 1[.$$

For any $a \in \mathsf{R}$ the map $x \mapsto x - a + 1$ gives a homeomorphism

$$]a, \to[\,\cong\,]1, \to[,$$

for any $a \geq 0$ the map $x \mapsto -x$ gives a homeomorphism

$$]a, \to[\,\cong\,]\leftarrow, -a[,$$

and the map $x \mapsto 1/x$ gives a homeomorphism

$$]1, \to[\,\cong\,]0, 1[.$$

Hence all nonempty open intervals in R—including all open rays and R itself—are homeomorphic to one another.

(5) The closed interval $[0, 1]$ is not homeomorphic to the open interval $]0, 1[$. Suppose, to the contrary, there exists some homeomorphism $h: [0, 1] \cong \,]0, 1[$. The function $f: \,]0, 1[\to \mathsf{R}$ given by $f(x) = 1/x$ is continuous, so the composite $f \circ h: [0, 1] \to \mathsf{R}$ is continuous. Then $f \circ h$ is bounded (once again the reader is asked to assume the boundedness of continuous real-valued functions on $[0, 1]$—see Chapter 4), so there is a constant c with

$$f(h(t)) \leq c \qquad\qquad (t \in [0, 1]).$$

Since h maps $[0, 1]$ onto $]0, 1[$,

$$f(x) \leq c \qquad\qquad (x \in \,]0, 1[),$$

which is patently absurd.

(6) Given two distinct points $x, y \in \mathsf{R}^n$, the line

$$L = \{ (1 - t)x + ty \mid t \in \mathsf{R} \}$$

in R^n passing through x and y is homeomorphic to the real line R. A suitable homeomorphism is

$$h: R \cong L$$

$$t \mapsto (1 - t)x + ty$$

which satisfies

$$\|h(s) - h(t)\| = |s - t| \cdot \|x - y\|.$$

(7) The homeomorphism from R to $]-1, 1[$ in Example (2) has an n-dimensional analog. Consider the *n-ball*

$$B_n = \{x \in R^n : \|x\| < 1\}$$

which is just the Euclidean ball of radius 1 at the origin $0 \in R^n$. For any $x \in R^n$,

$$\left\| \frac{1}{1 + \|x\|} x \right\| = \frac{1}{1 + \|x\|} \|x\| < 1,$$

so we have a map

$$f: R^n \to B_n$$

$$x \mapsto \frac{1}{1 + \|x\|} x.$$

Now for $x \in R^n$ and $y \in B_n$,

$$y = f(x) \qquad \Leftrightarrow \qquad x = (1 + \|x\|)y \quad \text{and} \quad \|x\| = (1 + \|x\|)\|y\|$$

$$\Leftrightarrow \qquad x = \frac{1}{1 - \|y\|} y,$$

so f is a bijection with inverse

$$g: B_n \to R^n$$

$$y \mapsto \frac{1}{1 - \|y\|} y.$$

Both f and g are continuous, so

$$f: R^n \cong B_n.$$

(8) Let

$$J = [-1, 1].$$

The bijection

$$k: I \to J$$

$$x \mapsto 2x - 1$$

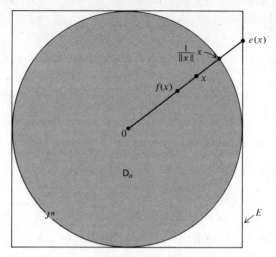

FIG. 3.3

induces a bijection

$$h: I^n \longrightarrow J^n$$

$$(x_1, \ldots , x_n) \mapsto (k(x_1), \ldots , k(x_n))$$

from the *n-cube*

$$I^n = \{x \in \mathbf{R}^n: 0 \leq x_i \leq 1 \text{ for } i = 1, \ldots , n\}$$

to the set

$$J^n = \{x \in \mathbf{R}^n: -1 \leq x_i \leq 1 \text{ for } i = 1, \ldots , n\}.$$

The map h is continuous since $x, u \in I^n$ implies

$$d_\infty (h(u), h(x)) = \max_{1 \leq i \leq n} |k(u_i) - k(x_i)|$$

$$= 2 \max_{1 \leq i \leq n} |u_i - x_i|$$

$$= 2 d_\infty (u, x)$$

for the max metric d_∞. Similarly, the inverse

$$h^{-1}: J^n \longrightarrow I^n$$

$$(y_1, \ldots , y_n) \mapsto (k^{-1}(y_1), \ldots , k^{-1}(y_n))$$

of h is continuous. Hence

$$h: I^n \cong J^n.$$

(9) We are going to show that the *n*-cube I^n is homeomorphic to the *n-disk*

$$\mathsf{D}_n = \{x \in \mathbf{R}^n: ||x|| \leq 1\}.$$

After Example (8), we need only show that $J^n \cong \mathsf{D}_n$. In terms of the Euclidean metric d and the max metric d_∞ on $\mathbf{R}^n$,

$$\mathbf{D}_n = \{x \in \mathbf{R}^n \mid d(x,0) \leq 1\}, \qquad J^n = \{x \in \mathbf{R}^n \mid d_\infty(x,0) \leq 1\}.$$

Then $\mathbf{D}_n \subset J^n$ because $d_\infty \leq d$. Let

$$E = \{x \in \mathbf{R}^n \mid d_\infty(x,0) = 1\}$$

which is, it so happens, the boundary of J^n in $\mathbf{R}^n$.

Suppose $x = (x_1, \ldots, x_n) \in J^n \backslash \{0\}$. Then the ray $\{\lambda x \mid \lambda \geq 0\}$ from the origin through x intersects E in exactly one point $e(x)$, as shown in Fig. 3.3. In fact, if $d_\infty(x,0) = |x_j|$, then $d_\infty(\lambda x, 0) = \lambda|x_j| = \lambda d_\infty(x,0)$ for $\lambda \geq 0$, so $\lambda x \in E$ if and only if $\lambda = 1/d_\infty(x,0)$. Hence

$$e(x) = \frac{1}{d_\infty(x,0)} x.$$

Since d_∞ induces the usual topology of $\mathbf{R}^n$, the function $x \mapsto d_\infty(x,0)$ is continuous. Hence the map

$$e \colon J^n \backslash \{0\} \to E$$

is continuous. For $x \in J^n \backslash \{0\}$ we have $d_\infty(x,0) \leq 1$ and $||x|| = d(x,0) \geq d_\infty(x,0)$, so that

$$||e(x)|| \geq ||x||, \qquad ||e(x)|| \geq 1.$$

We propose constructing a map $f \colon J^n \to \mathbf{D}_n$ that linearly contracts each line segment joining 0 to a point of E onto its part joining 0 to $\mathbf{S}_{n-1}$. Here $\mathbf{S}_{n-1}$ is the $(n-1)$-*sphere*

$$\mathbf{S}_{n-1} = \{x \in \mathbf{R}^n \colon ||x|| = 1\},$$

the boundary of $\mathbf{D}_n$ in $\mathbf{R}^n$ [see 2.23 (4)]. The ray from 0 through $x \in J^n \backslash \{0\}$ intersects E at $e(x)$ and $\mathbf{S}_{n-1}$ at $(1/||x||)x$; if $f(x) = \lambda x$, then we want the ratio

$$\frac{d(x,0)}{d(e(x),0)} = \frac{||x||}{||e(x)||}$$

of distances to be the same as the ratio

$$\frac{d(f(x),0)}{d((1/||x||)x,0)} = \frac{||f(x)||}{1} = \lambda||x||,$$

in other words, we want $\lambda = 1/||e(x)||$. Hence we define

$$f \colon J^n \to \mathbf{D}_n$$

by

$$f(x) = \begin{cases} \dfrac{1}{||e(x)||}\, x & \text{if } x \in J^n \backslash \{0\}, \\[4mm] 0 & \text{if } x = 0. \end{cases}$$

Continuity of f at each point of $J^n \setminus \{0\}$ follows from continuity of e there, and continuity of f at $x = 0$ follows from

$$\|f(u) - f(0)\| = \frac{\|u\|}{\|e(u)\|} \leq \|u\| = \|u - 0\| \qquad (u \in J^n \setminus \{0\}).$$

Whenever $x \in J^n \setminus \{0\}$ and $y \in D_n \setminus \{0\}$ are on the same ray through the origin, $e(x) = e(y)$, so

$$y = f(x) \qquad \Leftrightarrow \qquad x = \|e(x)\|y = \|e(y)\|y.$$

Hence f is a bijection having the inverse $g \colon D_n \to J^n$ given by

$$g(y) = \begin{cases} \|e(y)\|y & \text{if } y \in D_n \setminus \{0\}, \\ 0 & \text{if } y = 0. \end{cases}$$

The map g is continuous, and we conclude

$$f \colon J^n \cong D_n.$$

(10) Let d be the Euclidean metric on $\mathbf{R}^n$. Then for any point $y \in \mathbf{R}^n$ and any radius $\epsilon > 0$ there is a homeomorphism

$$B_\epsilon(y; d) \cong \mathbf{R}^n.$$

Such a homeomorphism can be obtained as the composite of three homeomorphisms: the translation

$$B_\epsilon(y; d) \cong B_\epsilon(0; d)$$

$$x \mapsto x - y,$$

the similarity

$$B_\epsilon(0; d) \cong B_1(0; d) = \mathbf{B}_n$$

$$x \mapsto \frac{1}{\epsilon} x,$$

and the homeomorphism

$$g \colon \mathbf{B}_n \cong \mathbf{R}^n$$

of Example (7).

(11) Consider the space obtained by removing from the n-sphere

$$\mathbf{S}_n = \{x \in \mathbf{R}^{n+1} \colon \|x\| = 1\}$$

its "north pole"

$$p = (0, \dots, 0, \dots, 0, 1).$$

We shall show

$$\mathbf{S}_n \setminus \{p\} \cong \mathbf{R}^n.$$

Now the "equatorial hyperplane"

$$H = \mathbf{R}^n \times \{0\} = \{x \in \mathbf{R}^{n+1} \mid x_{n+1} = 0\}$$

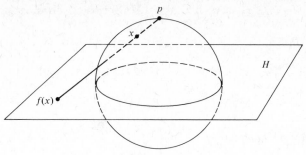

FIG. 3.4

in R^{n+1} is homeomorphic to R^n under the map

$$g: H \cong \mathsf{R}^n$$

$$(x_1, \ldots, x_n, 0) \mapsto (x_1, \ldots, x_n).$$

Hence we need only show $\mathsf{S}_n \backslash \{p\} \cong H$.

Let $x \in \mathsf{S}_n \backslash \{p\}$. Then the line through x and p intersects H in a unique point y (see Fig. 3.4). In fact, for $y = (y_1, \ldots, y_n, y_{n+1}) \in \mathsf{R}^{n+1}$ and $\lambda \in \mathsf{R}$ the conditions

(*) $$y_{n+1} = 0, \qquad y = \lambda x + (1 - \lambda)p$$

say

$$0 = y_{n+1} = \lambda x_{n+1} + (1 - \lambda)p_{n+1} = \lambda x_{n+1} + (1 - \lambda),$$

that is,

$$\lambda = \frac{1}{1 - x_{n+1}}$$

(since $x \in \mathsf{S}_n$ but $x \neq p$, then $x_{n+1} \neq 1$). Thus (*) has the unique solution

$$y = \frac{1}{1 - x_{n+1}} x + \frac{-x_{n+1}}{1 - x_{n+1}} p.$$

Call this solution $f(x)$. We have produced a continuous map

$$f: \mathsf{S}_n \backslash \{p\} \to H$$

with

$$f(x) = \frac{1}{1 - x_{n+1}} (x - x_{n+1}p) = \left(\frac{x_1}{1 - x_{n+1}}, \ldots, \frac{x_n}{1 - x_{n+1}}, 0 \right)$$

for each x.

To see that f is bijective and has a continuous inverse, for an arbitrary $y = (y_1, \ldots, y_n, 0) \in H$ we solve the equations

(**) $$y = f(x), \qquad ||x|| = 1$$

for $x = (x_1, \ldots, x_n, x_{n+1})$ with $x_{n+1} \neq 1$. In terms of coordinates, equations (**) say

$$y_i = \frac{x_i}{1 - x_{n+1}} \qquad (i = 1, \ldots, n)$$

$$\sum_{i=1}^{n} x_i^2 + x_{n+1}^2 = 1,$$

yielding the unique solution

$$x_i = \frac{2}{||y||^2 + 1} y_i \quad (i = 1, \ldots, n), \qquad x_{n+1} = \frac{||y||^2 - 1}{||y||^2 + 1}.$$

Thus

$$f^{-1}(y) = \frac{2}{||y||^2 + 1} y + \frac{||y||^2 - 1}{||y||^2 + 1} p \qquad (y \in H).$$

The homeomorphism

$$g \circ f \colon S_n \backslash \{p\} \cong R^n$$

is known as the *stereographic projection*. This map is of particular interest for dimension $n = 2$, because it allows us to represent the complex plane **C**, which as a topological space is just R^2, as the complement of a point in the sphere S_2 (see Example 4.56 for a further discussion).

(12) Let q be the "south pole" $(0, \ldots, 0, \ldots, 0, -1)$ of the n-sphere S_n. Composing the antipodal map

$$S_n \backslash \{q\} \cong S_n \backslash \{p\}$$

$$x \mapsto -x$$

with the stereographic projection $g \circ f \colon S_n \backslash \{p\} \cong R^n$ provides a homeomorphism

$$S_n \backslash \{q\} \cong R^n.$$

The two subsets $S_n \backslash \{p\}$ and $S_n \backslash \{q\}$ are open in S_n and homeomorphic to R^n, and each point of S_n belongs to at least one of these sets. Hence the n-sphere S_n is an "n-dimensional manifold" in accordance with the next definition.

3.17 Definition. Let n be a positive integer. A Hausdorff space X is called an *n-dimensional manifold* when each point of X has some open neighborhood that is homeomorphic to n-dimensional Euclidean space R^n.

The space R^n is itself an n-dimensional manifold, for $U = R^n$ is an open neighborhood of each point of R^n that is homeomorphic to R^n. More generally, any open subspace X of R^n is an n-dimensional manifold; in fact, if $x \in X$, there is a Euclidean ball $B_\epsilon(x; d) \subset X$, and $B_\epsilon(x; d) \cong R^n$ by 3.16 (10). Additional examples appear in the next two sections.

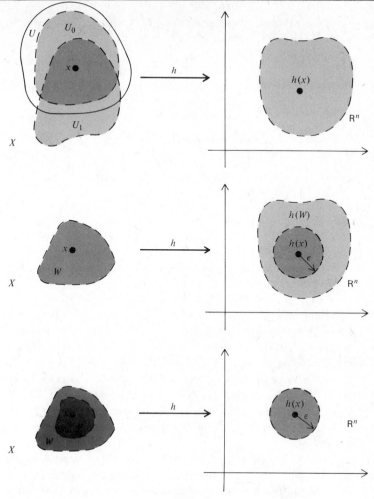

FIG. 3.5

Manifolds in the sense of Definition 3.17 are often called *topological manifolds* to distinguish them from "differentiable manifolds" which carry extra structure. The requirement in 3.17 that all manifolds be Hausdorff spaces is a standard one intended to exclude certain pathological examples.

An n-dimensional manifold is *locally Euclidean* in that each of its points has some sufficiently small neighborhood that looks like n-dimensional Euclidean space. Something more is true.

3.18 Proposition. Each point x in an n-dimensional manifold X has arbitrarily small Euclidean neighborhoods in the sense that each neighborhood U of x in X contains an open neighborhood V of x with $V \cong \mathsf{R}^n$.

Proof. We break the proof into three steps, corresponding to the three parts of Fig. 3.5.

Step 1. There is an open neighborhood U_0 of x with $U_0 \subset U$, and there is an open neighborhod U_1 of x and a homeomorphism

$$h \colon U_1 \cong \mathsf{R}^n.$$

Step 2. The set

$$W = U_0 \cap U_1$$

is an open neighborhood of x in the space U_1, so $h(W)$ is an open neighborhood of $h(x)$ in R^n (see Fig. 3.5). If d denotes the Euclidean metric on R^n, there is an $\epsilon > 0$ with

$$B_\epsilon(h(x); d) \subset h(W).$$

Step 3. Let

$$V = h^{-1}(B_\epsilon(h(x); d)).$$

From 3.16 (10) we have $V \cong \mathsf{R}^n$, and V is an open neighborhood of $x = h^{-1}(h(x))$ in W and hence in X.   ☐

We want to elaborate now on our earlier comment that two homeomorphic spaces should be regarded from a topological point of view as being essentially the same. Suppose X and Y are homeomorphic, so that there is a one-to-one correspondence between the points of X and the points of Y inducing a one-to-one correspondence between the open sets in X and the open sets in Y. Then regardless of the particular natures of their points, X and Y are indistinguishable purely as topological spaces. Hence anything we can correctly say about X with respect to its being a topological space, we can correctly say about Y, too.

Succinctly put, we are claiming that *homeomorphic topological spaces have the same topological properties*. The preceding was only a heuristic justification for this claim—and motivation for a precise definition of 'topological property'. Once we give a definition, our claim will be a tautology.

3.19 Definition. Suppose P is a property which it is meaningful, but not necessarily valid, to assert about an arbitrary topological space. Then P is said to be a *topological property* if each topological space homeomorphic to a topological space having property P also has property P.

We may say, then that a topological property is a property of topological spaces that is *preserved*, or *invariant*, under homeomorphisms. Some examples will clarify the definition.

3.20 Examples

(1) We begin with two "nonexamples". The property of being bounded with diameter at most 1 cannot qualify as a topological property, for it is

meaningless to assert it about a topological space (although it is meaningful for a metric space).

The property of being a subspace of the real line is not a topological property. Although it does make sense to say a given topological space is or is not a subspace of R, this property is not preserved under homeomorphism. For example, R itself is a subspace of R, the space $Y = \{ (x, 0) \mid x \in \mathsf{R} \}$ is homeomorphic to R, but Y is not a subspace of R because Y is not even a subset of R.

(2) Metrizability is a topological property. Certainly it is meaningful to ask of any topological space whether it is metrizable. Now suppose X is a metrizable topological space and Y is a topological space with $X \cong Y$. There is a metric d inducing the topology of X and homeomorphism $h: Y \cong X$. The formula

$$D(y, z) = d(h(y), h(z)) \qquad\qquad (y, z \in Y)$$

defines a metric on Y; symmetry of D, for example, follows from symmetry of d:

$$D(y, z) = d(h(y), h(z)) = d(h(z), h(y)) = D(z, y).$$

To complete the argument that Y is metrizable we show that D induces the given topology $\Im$ of Y. Suppose U is a D-open subset of Y. If $y \in U$, there is an $\epsilon > 0$ with $B_\epsilon(y; D) \subset U$, the set

$$h(B_\epsilon(y; D)) = B_\epsilon(h(y); d)$$

is open in X because d induces the topology of X, its inverse image

$$h^{-1}(B_\epsilon(h(y); d)) = B_\epsilon(y; D)$$

is $\Im$-open in Y because h is a homeomorphism, and so $B_\epsilon(y; D)$ is a $\Im$-neighborhood of y contained in U. Hence U is $\Im$-open. Conversely, suppose U is $\Im$-open. For each $y \in U$ the set $h(U)$ is an open neighborhood of $h(y)$ in X, there is an $\epsilon > 0$ with $B_\epsilon(h(y); d) \subset h(U)$, and

$$y \in B_\epsilon(y; D) = h^{-1}(B_\epsilon(h(y); d)) \subset U.$$

Hence U is D-open.

(3) Being a Hausdorff space is a topological property. Let X be a Hausdorff space and let $h: X \cong Y$ be a homeomorphism from X to a topological space Y. Let $y, z \in Y$ with $y \neq z$. Since $h^{-1}(y) \neq h^{-1}(z)$, there are disjoint open neighborhoods U and V of $h^{-1}(y)$ and $h^{-1}(z)$, respectively, in X. Then $h(U)$ and $h(V)$ are disjoint open neighborhoods of y and z in Y.

(4) Being a discrete space is a topological property. Let X be a discrete space and let $h: X \cong Y$ be a homeomorphism from X to a topological space Y. If V is any subset of Y, its inverse image $h^{-1}(V)$ is open in X, and then $V = h(h^{-1}(V))$ is open in Y.

(5) Second-countability is a topological property. Let X be a second-countable space and let $h: X \cong Y$ be a homeomorphism. Choose some countable base $\mathcal{B}$ of X. Then the countable collection $\{ h(B) \mid B \in \mathcal{B} \}$ is a base of

Y. In fact, each member $h(B)$ of this collection is open in Y. Now let $y \in V \subset Y$ with V open in Y. Then $h^{-1}(y) \in h^{-1}(V)$ with $h^{-1}(V)$ open in X, so there is some $B \in \mathcal{B}$ with $h^{-1}(y) \in B \subset h^{-1}(V)$. Hence $y \in h(B) \subset V$.

A similar argument shows that first-countability is a topological property. Separability, the third countability property considered in Chapter 2, is also a topological property (see Exercise 3.8).

The two most significant topological properties—compactness and connectedness—and related properties will be studied in depth in the next two chapters.

If we are given a topological space, then by listing enough of its topological properties we could hope to characterize it "up to homeomorphism" in the sense that any other space having the same properties is homeomorphic to the given one. Except for a few special spaces (see the note to Exercise 5.33 and Example 4.13), unfortunately, this hope cannot be fulfilled in any reasonable way.

Topological properties can, however, be used to distinguish between spaces that are not homeomorphic. If one of two spaces has a certain topological property and the other does not, then the two spaces cannot be homeomorphic to one another. For example, the metrizable subspace $\{(x, y) \in \mathsf{R} \times \mathsf{R} \mid y \geq 0\}$ of the plane is not homeomorphic to the nonmetrizable half-disk space [2.20 (3)]. The real line is not homeomorphic to the line with two origins [2.18 (3)] because the former is a Hausdorff space whereas the latter is not.

When a homeomorphic image of a space is a subspace of a larger space of interest, some special terminology is employed.

3.21 Definition. A map $f: X \to Y$ from a topological space X to a topological space Y is called an *embedding of X into Y* when the map

$$X \to f(X)$$
$$x \mapsto f(x)$$

obtained by restricting its codomain to its range is a homeomorphism from X to the subspace $f(X)$ of Y.

Thus an embedding of X into Y is in essence a homeomorphism from X to a subspace of Y, but regarded as a map into Y. Evidently a map $f: X \to Y$ is an embedding precisely when f is a continuous injection such that the image $f(U)$ of each open subset U of X is open in $f(X)$ (but not necessarily in Y).

3.22 Examples

(1) An isometric embedding of one metric space into another in the sense of Exercise 1.38 is an embedding of the associated topological spaces. To avoid ambiguity, we therefore sometimes call an embedding in the present sense a *homeomorphic embedding*.

(2) According to 3.16 (11) the map

$$\mathsf{R}^n \to \mathsf{R}^{n+1}$$

$$(x_1, \ldots, x_n) \mapsto (x_1, \ldots, x_n, 0)$$

is a homeomorphic embedding of R^n into R^{n+1} (in fact, an isometric embedding for the Euclidean metrics). In dimension $n = 1$ this is the embedding

$$h: \mathsf{R} \to \mathsf{R} \times \mathsf{R}$$

$$x \mapsto (x, 0)$$

of the real line into the plane with image the x-axis $\mathsf{R} \times \{0\}$.

The subspace $h(\mathsf{R}) = \mathsf{R} \times \{0\}$ of $\mathsf{R} \times \mathsf{R}$ is a homeomorphic copy of R that certainly is not identical to R. Nonetheless, by using h to "identify" each point $x \in \mathsf{R}$ with the corresponding point $h(x) = (x, 0) \in \mathsf{R} \times \mathsf{R}$ we may speak about R as if it actually were the subspace $\mathsf{R} \times \{0\}$ of $\mathsf{R} \times \mathsf{R}$. We may say, for example, that the subset $]0, 1[$ of the real line has empty interior in the plane, meaning really that $]0, 1[\times \{0\}$ has empty interior there.

This "abuse of language" of identifying the nonidentical objects R and $\mathsf{R} \times \{0\}$ is not peculiar to topology, but occurs in algebra as well. Recall that a typical complex number has the form $x + iy$ with x and y real, and when $y = 0$ such a complex number "is" the real number x. Now according to the formal definition of complex numbers, $x + iy$ is the ordered pair (x, y) of real numbers. In particular, the complex number $x + i \cdot 0$ is by definition the ordered pair $(x, 0)$ of real numbers, and this is *not* itself a real number. However, by identifying each real number x with the corresponding complex number $(x, 0) = h(x)$ we may treat the set R of all real numbers *as if* it were the subset $\mathsf{R} \times \{0\}$ of the set $\mathsf{C} = \mathsf{R} \times \mathsf{R}$ of all complex numbers. There is good reason to make this identification. Under the usual definitions

$$(x, y) + (u, v) = (x + u, y + v), \qquad (x, y) \cdot (u, v) = (xu - yv, xv + yu)$$

for adding and multiplying complex numbers, real numbers are manipulated algebraically the same way as the complex numbers with which they are identified by h:

$$h(x + u) = (x + u, 0) = (x, 0) + (u, 0) = h(x) + h(u),$$

$$h(x \cdot u) = (x \cdot u, 0) = (x, 0) \cdot (u, 0) = h(x) \cdot h(u).$$

Moreover, the complex number $i = (0, 1)$ satisfies $i \cdot i = (-1, 0) = h(-1)$ and thus, under the identification, is a solution of the equation $z^2 = -1$, which cannot be solved in R.

(3) Any two topological spaces X_1 and X_2 can simultaneously be embedded into the same topological space. Consider first the special case when X_1 is disjoint from X_2, and let $Z = X_1 \cup X_2$. The collection

$$\{U_1 \cup U_2 \mid U_1 \text{ open in } X_1, U_2 \text{ open in } X_2\}$$

of subsets of Z, which may also be described as

$$\{U \mid U \subset Z,\ U \cap X_1 \text{ open in } X_1,\ U \cap X_2 \text{ open in } X_2\},$$

is a topology on Z. Both X_1 and X_2 with their given topologies are then subspaces of Z with this topology [Z is the sum of the family $(X_i \mid i = 1, 2)$ as treated in Exercise 2.20].

Consider now the general case when X_1 need not be disjoint from X_2. The preceding construction need no longer work because X_1 and X_2 might induce distinct topologies on their overlap $X_1 \cap X_2$. To get around this obstacle we make X_1 disjoint from X_2 by labeling elements of X_1 with 1 and elements of X_2 with 2, obtaining the disjoint sets

$$X_1 \times \{1\} = \{(x, 1) \mid x \in X_1\}, \qquad X_2 \times \{2\} = \{(x, 2) \mid x \in X_2\}.$$

By endowing $X_1 \times \{1\}$, $X_2 \times \{2\}$ with the topologies

$$\{U_1 \times \{1\} \mid U_1 \text{ open in } X_1\}, \qquad \{U_2 \times \{2\} \mid U_2 \text{ open in } X_2\}$$

we obtain homeomorphisms

$$X_1 \cong X_1 \times \{1\}, \qquad X_2 \cong X_2 \times \{2\}.$$
$$x \mapsto (x, 1) \qquad\qquad x \mapsto (x, 2)$$

The construction in the preceding paragraph furnishes on

$$Z = (X_1 \times \{1\}) \cup (X_2 \times \{2\})$$

the topology

$$\{(U_1 \times \{1\}) \cup (U_2 \times \{2\}) \mid U_1 \text{ open in } X_1,\ U_2 \text{ open in } X_2\}$$

with respect to which $X_1 \times \{1\}$ and $X_2 \times \{2\}$ are subspaces of Z. Then the maps

$$X_1 \to Z \qquad , \qquad X_2 \to Z$$
$$x \mapsto (x, 1) \qquad\qquad x \mapsto (x, 2)$$

are embeddings. The space Z is called the *sum* of X_1 and X_2 and is denoted $X_1 + X_2$.

Sometimes a single space can be found into which all members of a wide class of spaces can simultaneously be embedded. One such space is the *Hilbert cube*

$$\mathsf{I}^\infty = \{(x_i \mid i = 1, 2, \ldots) : x_i \in \mathsf{R},\ |x_i| \leq \frac{1}{i} \quad (i = 1, 2, \ldots)\}$$

which according to Exercise 1.23 is a subset of the Hilbert sequence space H and therefore has a topology induced by the standard metric d_2 on H (Example 1.9). Recall that H is separable [2.56(4)]. Now a metrizable space is separable precisely when it is second-countable, and a subspace of a second-countable space is second-countable. Then each subspace of I^∞ is separable metrizable, and hence the same is true of each space that can be embedded into l^∞. This proves half of the following embedding theorem.

3.23 Theorem. A topological space X is separable and metrizable if and only if there is a homeomorphic embedding of X into the Hilbert cube l^∞.

Proof. Assume X is separable and metrizable. Choose a sequence $(x_i \mid i = 1, 2, \ldots)$ such that $\{x_i \mid i = 1, 2, \ldots\}$ is dense in X, and choose a metric d on X inducing the topology of X such that $d(x, y) \leq 1$ for all $x, y \in X$. Then the formula

$$h(x) = \left(\frac{1}{i} d(x, x_i) \,\middle|\, i = 1, 2, \ldots \right)$$

defines a map

$$h: X \to l^\infty.$$

Note that for $x, u \in X$ and for $j = 1, 2, \ldots$ we have

$$\frac{1}{j} |d(x, x_j) - d(u, x_j)| \leq \left[\sum_{i=1}^{\infty} \frac{1}{i^2} |d(x, x_i) - d(u, x_i)|^2 \right]^{1/2}$$

(*) $$= d_2(h(x), h(u))$$

$$\leq \left[\sum_{i=1}^{\infty} \frac{1}{i^2} [d(x, u)]^2 \right]^{1/2}.$$

We show that h is injective. Let $x, u \in X$ with $x \neq u$. There is a j with $d(x, x_j) < \frac{1}{2} d(x, u)$. Now $d(x, x_j) \neq d(u, x_j)$, for otherwise we would have

$$d(x, u) \leq d(x, x_j) + d(x_j, u) = 2d(x, x_j) < d(x, u).$$

Then from (*)

$$d_2(h(x), h(u)) \geq \frac{1}{j} |d(x, x_j) - d(u, x_j)| > 0,$$

and $h(x) \neq h(u)$.

We show next that h is continuous. Fix $x \in X$ and let $\epsilon > 0$. Since the series $\sum_{i=1}^{\infty} 1/i^2$ converges, we may choose n so large that

$$\sum_{i=n+1}^{\infty} \frac{1}{i^2} < \frac{\epsilon^2}{2}.$$

If $u \in X$ and

$$d(x, u) < \left[\frac{\epsilon^2}{2n} \right]^{1/2},$$

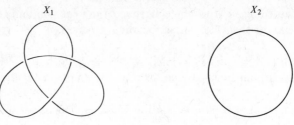

X_1 X_2

FIG. 3.6

then from (*)

$$[d_2(h(x), h(u))]^2 \leq \sum_{i=1}^{n} \frac{1}{i^2} [d(x, u)]^2 + \sum_{i=n+1}^{\infty} \frac{1}{i^2} [d(x, u)]^2$$

$$< n \cdot \frac{\epsilon^2}{2n} + \sum_{i=n+1}^{\infty} \frac{1}{i^2} < \frac{\epsilon^2}{2} + \frac{\epsilon^2}{2} = \epsilon^2,$$

and so $d_2(h(x), h(u)) < \epsilon$.

To conclude we show that h maps each open subset of X onto an open subset of $h(X) \subset I^\infty$. Fix $x \in X$ and let $\epsilon > 0$. It suffices to find a $\delta > 0$ with

$$B_\delta(h(x); d_2) \cap h(X) \subset h(B_\epsilon(x; d)).$$

There exists a j with $d(x, x_j) < \epsilon/3$. We claim that $\delta = \epsilon/3j$ will do. Suppose $y \in h(X)$ with $d_2(h(x), y) < \delta$. Choose $u \in X$ with $y = h(u)$. By (*),

$$|d(x, x_j) - d(u, x_j)| \leq j \cdot d_2(h(x), h(u)) < \frac{\epsilon}{3}.$$

Then

$$d(u, x_j) < d(x, x_j) + \frac{\epsilon}{3} < \frac{\epsilon}{3} + \frac{\epsilon}{3} = \frac{2\epsilon}{3},$$

$$d(u, x) \leq d(u, x_j) + d(x_j, x) < \frac{2\epsilon}{3} + \frac{\epsilon}{3} = \epsilon,$$

$u \in B_\epsilon(x; d)$, and hence $y = h(u) \in h(B_\epsilon(x; d))$. □

This is the place to correct a common misconception of topology which has been disseminated in certain popularizations of mathematics. This misconception is that two geometric objects are topologically the same only when one of them can be transformed into the other by "bending and stretching" but without "cutting and pasting". Consider two embeddings of the circle S_1 into Euclidean space R^3 whose images are the knotted circle X_1 and the unknotted circle X_2 depicted in Fig. 3.6. The spaces X_1 and X_2, being both homeomorphic to S_1, are topologically the same. However, X_1 cannot be transformed into X_2 without cutting it apart, unknotting it, and pasting it back together. (Mathematically, this means there is no homeomorphism of the ambient space R^3 onto itself that carries X_1 onto X_2. Although we do not prove that fact, a simple experiment with a loop of string should be convincing.)

EXERCISES

29. Give several examples of continuous bijections that are not homeomorphisms.

30. Show that a continuous map $f: X \to Y$ is a homeomorphism if there is a continuous map $g: Y \to X$ such that $g \circ f$ and $f \circ g$ are the identity maps on their respective domains.

31. When will two indiscrete spaces be homeomorphic?

32. Can there be more than one homeomorphism between two homeomorphic spaces? Can there be infinitely many?

33. How does a homeomorphism transform closures, interiors, and boundaries of subsets of its domain?

34. In terms of neighborhoods of points in X and in Y, when is a bijection $h: X \to Y$ between two topological spaces a homeomorphism?

35. Given a homeomorphism $h: X \cong Y$ and a set $A \subset X$, show that h restricts to homeomorphisms $A \cong h(A)$ and $X \backslash A \cong Y \backslash h(A)$.

36. Prove that each interval in R containing at least two points is homeomorphic to exactly one of the intervals $[0, 1]$, $[0, 1[$, and $]0, 1[$. (Remember that rays are intervals, too.)

37. Suppose spaces X_1 and X_2 have partitions $\{A_1, B_1\}$ and $\{A_2, B_2\}$, respectively, into closed sets with $A_1 \cong A_2$ and $B_1 \cong B_2$. Show that then $X_1 \cong X_2$. Is the supposition that the sets are closed really needed?

38. Construct a homeomorphism between the interval I and the U-shaped subset $(\{0\} \times I) \cup (I \times \{0\}) \cup (\{1\} \times I)$ of the plane.

39. (**a**) Construct a homeomorphism between each two of the three planar sets that are the graphs of

$$x^2 + y^2 = 1, \qquad \frac{x^2}{a^2} + \frac{y^2}{b^2} = 1, \qquad |x| + |y| = 1$$

(here a and b are nonzero constants).
(**b**) Do the same for the graphs of

$$x^2 + y^2 \leq 1, \qquad \frac{x^2}{a^2} + \frac{y^2}{b^2} \leq 1, \qquad |x| + |y| \leq 1.$$

(**c**) Do the same for the graphs of

$$x^2 + y^2 < 1, \qquad \frac{x^2}{a^2} + \frac{y^2}{b^2} < 1, \qquad |x| + |y| < 1.$$

40. Prove that the graph

$$G = \{(x, y) \in R^2 \mid y = f(x)\}$$

of a continuous function $f: R \to R$ is always homeomorphic to R. Does this result generalize to continuous functions $f: R^n \to R$?

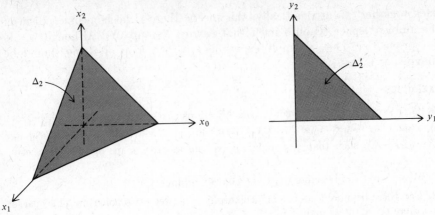

FIG. 3.7

41. Exhibit a homeomorphism from X to Y when:

   (**a**) X is the closed annulus $\{x \in \mathbf{R}^2 : 1 \le ||x|| \le 2\}$ and Y is the cylinder

$$S_1 \times I = \{x \in \mathbf{R}^3 \mid x_1^2 + x_2^2 = 1, 0 \le x_3 \le 1\}.$$

   (**b**) X is the "punctured plane" $\mathbf{R}^2 \setminus \{0\}$ and Y is the unbounded cylinder

$$S_1 \times \mathbf{R} = \{x \in \mathbf{R}^3 \mid x_1^2 + x_2^2 = 1\}.$$

42. Construct homeomorphisms between the two hemispheres

$$S_n^+ = \{x \in S_n \mid x_{n+1} \ge 0\}, \qquad S_n^- = \{x \in S_n \mid x_{n+1} \le 0\}$$

(on the n-sphere $S_n \subset \mathbf{R}^{n+1}$) and the n-disk D_n.

43. Show that the n-cube I^n is homeomorphic to the "standard n-simplex"

$$\Delta_n = \{(x_0, x_1, \ldots, x_n) \in \mathbf{R}^{n+1} \mid x_i \ge 0 \ (i = 0, \ldots, n), \sum_{i=0}^{n} x_i = 1\}$$

in $\mathbf{R}^{n+1}$. [You may wish to show first that Δ_n is homeomorphic to the subset

$$\Delta_n' = \{(y_1, \ldots, y_n) \in \mathbf{R}^n \mid y_i \ge 0 \ (i = 1, \ldots, n), \sum_{i=1}^{n} y_i \le 1\} .$$

Both Δ_n and Δ_n' are depicted in Fig. 3.7 for $n=2$.]

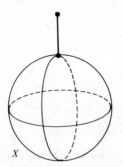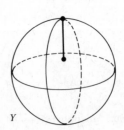

FIG. 3.8

44. Let X be the subspace of R^3 obtained by attaching to the sphere S_2 at its north pole a "whisker" pointing outward, and let Y be obtained by attaching to the same sphere at the same point a whisker pointing inward—see Fig. 3.8. Is X homeomorphic to Y?

45. Prove that an open subset of an n-dimensional manifold is itself an n-dimensional manifold.

46. Show that a topological space is an n-dimensional manifold if each of its points has some open neighborhood homeomorphic to some open subset of R^n.

47. A topological space is an *n-dimensional manifold-with-boundary* if each of its points has either an open neighborhood homeomorphic to an open set in R^n or else an open neighborhood homeomorphic to an open set in the subspace $\mathsf{R}_+{}^n = \{x \in \mathsf{R}^n \mid x_n \geq 0\}$ of R^n. (*Note*: That a point cannot have open neighborhoods of both kinds is guaranteed by the *invariance of domain theorem*, which says that a subset of R^n is open in R^n if it is homeomorphic to an open subset of R^n.)

Verify that the spaces below are n-dimensional manifolds-with-boundary:
  (**a**) An n-dimensional manifold.
  (**b**) The n-cube I^n.
  (**c**) The n-disk D_n.
  (**d**) The cylinder $\mathsf{S}_1 \times \mathsf{I}$ of Exercise 41 (a)—here $n = 2$.
  (**e**) A closed-halfspace in R^n (Exercise 1.55).

48. The *boundary* ∂M of an n-dimensional manifold-with-boundary M is the set of $x \in M$ not having any open neighborhood homeomorphic to an open subset of R^n.
  (**a**) What is ∂M when $M = \mathsf{D}_n$?
  (**b**) Give an example for $n = 2$ of such an M which is a subspace of R^n and for which ∂M is distinct from the boundary bdy M of M in R^n.
  (**c**) Prove that, in general, ∂M is closed in M.

49. Show that each of the properties 'T_0', 'T_1', and 'regular' is a topological property (see Exercises 2.32, 2.10, and 2.60).

50. Is the property of having a denumerable set of points a topological property? Must two spaces having this property be homeomorphic?

51. Is having some complete metric inducing the topology a topological property?

52. Show that being an n-dimensional manifold, and being an n-dimensional manifold-with-boundary (Exercise 47), are topological properties.

53. Use a suitable topological property to demonstrate that X is not homeomorphic to Y:
  (**a**) $X = \mathsf{R}$; $Y = \mathsf{Q}$.
  (**b**) $X = \mathsf{R}$ with its usual topology; $Y = \mathsf{R}$ with its right-interval topology [2.18(1)].
  (**c**) $X = \mathsf{R}$ with its usual topology; $Y = \mathsf{R}$ with its finite-complement topology [2.3(7)].
  (**d**) $X = \mathsf{H}$, the Hilbert sequence space (1.9); $Y =$ the set of all functions $f: \mathsf{R} \to \mathsf{R}$ with its topology of pointwise convergence [2.52(3)].

54. If each of two topological spaces can be homeomorphically embedded in the other, must they be homeomorphic to one another? (*Note*: There is a purely set-theoretic analog: If each of two sets has an injection into the other, must there be a bijection from one to the other? The answer, in the affirmative, is the content of the Schröder-Bernstein theorem—see Eisenberg [9, page 223].)

55. (**a**) Embed the Euclidean plane $\mathbb{R}^2$ in the half-disk space $H \cup L$ [2.20(3)].
    (**b**) Can $H \cup L$ be embedded in $\mathbb{R}^2$?

56. Let $h: X \to Y$ be an embedding. Construct a topological space Z containing X as a subspace and homeomorphic to Y. (Thus a space X that looks like a subspace of a space Y is actually a subspace of a space that looks like Y.) [*Hint*: Replace $h(X) \subset Y$ by X.]

57. Let $(X_i \mid i \in I)$ be a family of topological spaces. Construct a topological space X and embeddings $h_i: X_i \to X$, $i \in I$, such that $h_i(X_i)$ is open in X for each i and $X = \bigcup_{i \in I} h_i(X_i)$. Such a space X is called the *sum of* $(X_i \mid i \in I)$ and is denoted by

$$\underset{i \in I}{+} X_i.$$

[*Note*: This sum generalizes both the sum of a family of disjoint spaces as constructed in Exercise 2.20 and the sum of two spaces as constructed in Example 3.22(3).]

58. (Continuation of Exercise 57.) Prove that a map

$$f: \underset{i \in I}{+} X_i \to Y$$

into a topological space Y is continuous precisely when the maps $f \circ h_i: X_i \to Y$ are continuous for all $i \in I$.

59. (Continuation of Exercise 57.) Show that the sum of a family $(X_i \mid i \in I)$ is metrizable if each X_i is.

60. Discuss: "A topologist is somebody who cannot tell the difference between a coffee cup and a doughnut."

3. Product Spaces

The theme of this section and the next is the construction of new spaces from old. We look first at product spaces. Suppose we are given a number of topological spaces. How can we put a suitable topology on the product of their underlying sets? Just because this product is a set, several topologies are at our disposal—the discrete topology, the indiscrete topology, the finite-complement topology, and so forth—but these need not be related in any way to the given topologies on the various sets. To find a topology that is related to the given ones we turn once again to metric spaces for motivation.

We have met two general instances of metrics on products. The first concerns finitely many metric spaces $(X_1, d_1), \ldots, (X_n, d_n)$. Recall from Example 1.13 that on the product set $X = X_1 \times \cdots \times X_n$ the max metric d_∞ induced by $(d_1, \ldots, d_n)$ is defined by

$$d_\infty (x, y) = \max_{1 \le i \le n} d_i(x_i, y_i).$$

3.24 Proposition. The topology on $X = X_1 \times \cdots \times X_n$ induced by the max metric d_∞ has as a base the collection

$$\mathcal{B} = \{U_1 \times \cdots \times U_n \mid U_i \text{ is } d_i\text{-open in } X_i \ (i = 1, \ldots, n)\}$$

of all products of open sets in the factors $X_1, \ldots, X_n$.

Proof. We shall use the criterion of Proposition 2.50. Let $B \in \mathcal{B}$. We show B is d_∞-open by showing it contains a d_∞-ball at each of its points. Write

$$B = U_1 \times \cdots \times U_n$$

with each U_i being a d_i-open subset of X_i. Let $x = (x_1, \ldots, x_n) \in B$. For each $i = 1, \ldots, n$ the ith coordinate $x_i \in U_i$, so

$$B_{\epsilon_i}(x_i; d_i) \subset U_i$$

for some $\epsilon_i > 0$. Set

$$\epsilon = \min \{\epsilon_1, \ldots, \epsilon_n\}.$$

Then $B_\epsilon(x; d_\infty) \subset B$. In fact, if $y = (y_1, \ldots, y_n) \in B_\epsilon(x; d_\infty)$, then

$$d_i(x_i, y_i) \leq d_\infty(x, y) < \epsilon \leq \epsilon_i$$

for each $i = 1, \ldots, n$ and hence

$$y = (y_1, \ldots, y_n) \in B_{\epsilon_1}(x_1; d_1) \times \cdots \times B_{\epsilon_n}(x_n; d_n) \subset U_1 \times \cdots \times U_n = B.$$

Now let U be any d_∞-open set in X and let $x \in U$. We show $x \in B \subset U$ for some $B \in \mathcal{B}$. Choose $\epsilon > 0$ with

$$B_\epsilon(x; d_\infty) \subset U.$$

For each $i = 1, \ldots, n$ the ball $B_\epsilon(x_i; d_i)$ is a d_i-open subset of X_i, so the set

$$B = B_\epsilon(x_1; d_1) \times \cdots \times B_\epsilon(x_n; d_n)$$

belongs to $\mathcal{B}$. Clearly $x \in B$. Finally, $B \subset U$ because $y = (y_1, \ldots, y_n) \in X$ with $d_i(x_i, y_i) < \epsilon$ for each i implies $d_\infty(x, y) < \epsilon$. $\quad\square$

The second instance of a metric on a product concerns denumerably many metric spaces $(X_1, d_1), (X_2, d_2), \ldots$ each of which is bounded and of diameter of at most 1. Recall from Exercise 1.34 that the formula

$$d'(x, y) = \sum_{i=1}^{\infty} \frac{d_i(x_i, y_i)}{2^i}$$

defines a metric d' on the product

$$X = \underset{i=1}{\overset{\infty}{\times}} X_i.$$

Things are not quite as simple now as they were with finitely many factors, for even a product of balls need not be d'-open!

3.25 Proposition. The topology on

$$X = \underset{i=1}{\overset{\infty}{\times}} X_i$$

induced by the metric d' has as a base the collection

$$\mathcal{B} = \{U_1 \times \cdots \times U_n \times X_{n+1} \times X_{n+2} \times \cdots |$$

$$n \geq 1, \; U_i \text{ is } d_i\text{-open in } X_i \; (i = 1, \ldots, n)\}$$

of products.

Proof. Again we use Proposition 2.50. Let $x \in B \in \mathcal{B}$. We show $B_\epsilon(x; d') \subset B$ for some $\epsilon > 0$. Write

$$B = U_1 \times \cdots \times U_n \times X_{n+1} \times X_{n+2} \times \cdots$$

with $n \geq 1$ and U_i a d_i-open set in X_i for $i = 1, \ldots, n$. For each $i = 1, \ldots, n$ there is an $\epsilon_i > 0$ with

$$B_{\epsilon_i}(x_i; d_i) \subset U_i.$$

Set

$$\epsilon = \min \left\{ \frac{\epsilon_1}{2}, \frac{\epsilon_2}{2^2}, \ldots, \frac{\epsilon_n}{2^n} \right\}.$$

Then $B_\epsilon(x; d') \subset B$. In fact, if $y \in B_\epsilon(x; d')$, then

$$\frac{d_i(x_i, y_i)}{2^i} \leq d'(x, y) < \epsilon \leq \frac{\epsilon_i}{2^i}$$

and $d_i(x_i, y_i) < \epsilon_i$ for each $i = 1, \ldots, n$, so

$$y \in B_{\epsilon_1}(x_1; d_1) \times \cdots \times B_{\epsilon_n}(x_n; d_n) \times X_{n+1} \times X_{n+2} \times \cdots \subset B.$$

Now let U be any d'-open set in X and let $x \in U$. We show $x \in B \subset U$ for some $B \in \mathcal{B}$. Choose $\epsilon > 0$ with

$$B_\epsilon(x; d') \subset U$$

and then choose $n \geq 1$ with

$$\sum_{i=n+1}^{\infty} \frac{1}{2^i} < \frac{\epsilon}{2}.$$

The set

$$B = B_{\epsilon/2n}(x_1; d_1) \times \cdots \times B_{\epsilon/2n}(x_n; d_n) \times X_{n+1} \times X_{n+1} \times \cdots$$

belongs to $\mathcal{B}$. Clearly $x \in B$. Finally, $B \subset U$ because $y \in X$ with $d_i(x_i, y_i) <$

$\epsilon/2n$ for each $i = 1, \ldots , n$ implies

$$d'(x, y) = \sum_{i=1}^{n} \frac{d_i(x_i, y_i)}{2^i} + \sum_{i=n+1}^{\infty} \frac{d_i(x_i, y_i)}{2^i}$$

$$\leq \sum_{i=1}^{n} d_i(x_i, y_i) + \sum_{i=n+1}^{\infty} \frac{1}{2^i}$$

$$< n \cdot \frac{\epsilon}{2n} + \frac{\epsilon}{2} = \epsilon. \quad \square$$

The bases $\mathscr{B}$ in both 3.24 and 3.25 can be described in the same way with the aid of some unifying terminology.

3.26 Convention. For the remainder of this section, unless indicated otherwise, I will denote a nonempty set of natural numbers (usually I will be either $\{1, \ldots , n\}$ for some $n \geq 1$ or else $\{1, 2, \ldots\}$). A property will be said to hold for *almost all* $i \in I$ when it holds for all except a finite number (possibly zero) of values of $i \in I$, in other words, when the set of $i \in I$ for which it does *not* hold is finite (and possibly empty).

In both the finite case 3.24 and the denumerable case 3.25 the product set X has the form

$$X = \underset{i \in I}{\times} X_i,$$

where either $I = \{1, \ldots , n\}$ or else $I = \{1, 2, \ldots\}$. In both cases the base $\mathscr{B}$ of the metric topology on X has the description

$$\mathscr{B} = \left\{ \underset{i \in I}{\times} U_i \,\middle|\, U_i \text{ open in } X_i \text{ for all } i \in I, \right.$$

$$\left. U_i = X_i \text{ for almost all } i \in I \right\}.$$

Of course in the finite case $I = \{1, \ldots , n\}$, the 'almost all' clause is superfluous.

Suppose now we are given a family $(X_i \mid i \in I)$ of topological spaces, but with the topologies on the X_i not necessarily induced by metrics. If we can show that the collection $\mathscr{B}$ as described above is actually a base of some topology on $X = \times_{i \in I} X_i$, then it will be perfectly natural to endow the product X with this topology.

3.27 Lemma. Let $(X_i \mid i \in I)$ be a family of topological spaces. Then the collection

$$\mathcal{B} = \left\{ \underset{i \in I}{\times} U_i \,\middle|\, U_i \text{ open in } X_i \text{ for all } i \in I, \right.$$

$$\left. U_i = X_i \text{ for almost all } i \in I \right\}$$

is a base of a topology on the product set $X = \times_{i \in I} X_i$.

Proof. We shall verify hypotheses (B1) and (B2) of Theorem 2.51.

(B1) The entire product set X belongs to $\mathcal{B}$, so certainly each point of X belongs to some $B \in \mathcal{B}$.

(B2) Let $B_1, B_2 \in \mathcal{B}$. We show $B_1 \cap B_2 \in \mathcal{B}$. Write

$$B_1 = \underset{i \in I}{\times} U_i, \qquad B_2 = \underset{i \in I}{\times} V_i$$

with U_i and V_i open subsets of X_i for each $i \in I$ and with

$$U_i = X_i \qquad (i \notin J), \qquad V_i = X_i \qquad (i \notin K)$$

for some finite subsets J, K of the index set I. Then

$$B_1 \cap B_2 = \underset{i \in I}{\times} (U_i \cap V_i)$$

with $U_i \cap V_i$ open in X_i for each $i \in I$, and $U_i \cap V_i = X_i$ for all $i \in I$ not belonging to the finite subset $J \cup K$ of I. Hence $B_1 \cap B_2 \in \mathcal{B}$. $\square$

3.28 Definition. The *product* of a family $(X_i \mid i \in I)$ of topological spaces is the topological space whose underlying set is $\times_{i \in I} X_i$ and whose topology has the collection $\mathcal{B}$ described above as a base. This topology is called the *product topology*, and members of $\mathcal{B}$ are called *basic* open sets.

Unless otherwise indicated, any future reference to a topology on a product of topological spaces is to the product topology.

The most prominent product space is n-dimensional Euclidean space R^n with its usual topology. As a set, R^n is the product of n copies of the real line R. According to 3.24 the product topology on R^n (arising from the usual topology on each of these copies of R) is induced by the max metric, and we know that the max metric induces the usual topology on R^n.

When the index set I is finite, say $I = \{1, \ldots, n\}$ for some $n \geq 1$, the basic open sets in a product space

$$\underset{i \in I}{\times} X_i = X_1 \times \cdots \times X_n$$

are all the products

$$\overset{n}{\underset{i=1}{\times}} U_i = U_1 \times \cdots \times U_n$$

of open sets $U_1, \ldots, U_n$ in the respective factor spaces $X_1, \ldots, X_n$. In particular, when the index set consists of two elements, say $I = \{1, 2\}$, we are dealing with the product $X_1 \times X_2$ of two topological spaces X_1 and X_2, and then the basic open sets are all the products $U_1 \times U_2$ with U_1 open in X_1 and U_2 open in X_2.

Even when there are only two factors, *an open set in a product space need not be a product of open sets* in the factors—that is, need not be a basic open set. For example, the ball B_2 in $\mathsf{R}^2 = \mathsf{R} \times \mathsf{R}$ is open, but $\mathsf{B}_2 \neq U_1 \times U_2$ for any sets $U_1 \subset \mathsf{R}$, $U_2 \subset \mathsf{R}$. In fact, $(\frac{3}{4}, 0) \in \mathsf{B}_2$ and $(0, \frac{3}{4}) \in \mathsf{B}_2$; if $\mathsf{B}_2 = U_1 \times U_2$, then $\frac{3}{4} \in U_1$ and $\frac{3}{4} \in U_2$ whence $(\frac{3}{4}, \frac{3}{4}) \in U_1 \times U_2$, but $(\frac{3}{4}, \frac{3}{4}) \notin \mathsf{B}_2$.

In a product space

$$\overset{\infty}{\underset{i=1}{\times}} X_i = X_1 \times X_2 \times \cdots$$

the basic open sets are those products

$$U_1 \times \cdots \times U_n \times X_{n+1} \times X_{n+2} \times \cdots$$

of open sets for which the ith factor is the entire ith space for all sufficiently large values of i. However, *an arbitrary product* $V = \times_{i=1}^{\infty} V_i$ *of open sets need not be open in* $\times_{i=1}^{\infty} X_i$. For example, take X_i to be the two-point discrete space $\{0, 1\}$ and take V_i to be the open subset $\{0\}$ of X_i for each $i = 1, 2, \ldots$. The product V consists of the single point all of whose coordinates are 0. If V were open it would contain a basic open set and hence would contain a point almost all of whose coordinates are 1.

The collection $\mathcal{B}$ of all basic open sets in a product space $\times_{i \in I} X_i$ is a base of the product topology. Then according to 2.50, for each point $x \in \times_{i \in I} X_i$ the collection $\{B \mid x \in B \in \mathcal{B}\}$ of all basic open sets containing x is a local base at x. Hence neighborhoods in a product space are related to neighborhoods in the factor spaces as follows.

3.29 Proposition. Let $x = (x_i \mid i \in I)$ be a point in the product of a family $(X_i \mid i \in I)$ of topological spaces. Then a subset V of $\times_{i \in I} X_i$ is a neighborhood of x if and only if V contains a product $\times_{i \in I} V_i$, where V_i is a neighborhood of x_i for all $i \in I$ and $V_i = X_i$ for almost all $i \in I$.

Proof. If V is a neighborhood of x, there is a basic open set B with $x \in B \subset V$, and B is such a product $\times_{i \in I} V_i$. Conversely, suppose such a product $\times_{i \in I} V_i \subset V$. Set $J = \{i \in I \mid V_i \neq X_i\}$, a finite set. For each $i \in J$ the set V_i is a neighborhood of x_i in X_i and therefore contains an open neighbor-

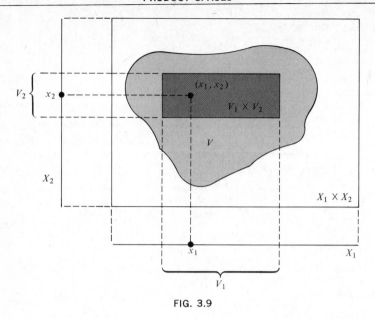

FIG. 3.9

hood U_i of x_i; let $U_i = X_i$ for each $i \in I \setminus J$. Then

$$B = \underset{i \in I}{\mathsf{X}} U_i$$

is a basic open set with

$$x \in B \subset \underset{i \in I}{\mathsf{X}} V_i \subset V,$$

so V is a neighborhood of x.   ☐

Nothing is changed above if we require in addition that each V_i be open in X_i. As usual, the 'almost all' clause may be omitted when the index set I is finite. In particular, then, a neighborhood of a point (x_1, x_2) in a product $X_1 \times X_2$ of two spaces is any subset of $X_1 \times X_2$ containing a product $V_1 \times V_2$ of a neighborhood V_1 of x_1 in X_1 with a neighborhood V_2 of x_2 in X_2 (see Fig. 3.9).

Product topologies mesh nicely with relative topologies. Consider, for example, the square $\mathsf{I} \times \mathsf{I}$. Each of the factors I has a topology as a subspace of the real line R, so $\mathsf{I} \times \mathsf{I}$ may be given the product topology obtained from these factors. At the same time $\mathsf{I} \times \mathsf{I}$ is a subset of the plane $\mathsf{R} \times \mathsf{R}$ and may be given its relative topology induced by the product topology on $\mathsf{R} \times \mathsf{R}$. These two topologies on $\mathsf{I} \times \mathsf{I}$ coincide (both are induced by the Euclidean metric). The same thing holds, more generally, for any product that is a subset of a larger product.

3.30 Proposition. Let $(Y_i \mid i \in I)$ and $(X_i \mid i \in I)$ be families of topological spaces with Y_i a subspace of X_i for each $i \in I$, so that

$$\underset{i \in I}{\times} Y_i \subset \underset{i \in I}{\times} X_i.$$

Then the product topology on $\times_{i \in I} Y_i$ equals the relative topology induced by the product topology on $\times_{i \in I} X_i$.

Proof. Let

$$X = \underset{i \in I}{\times} X_i, \qquad Y = \underset{i \in I}{\times} Y_i.$$

The collection

$$\mathcal{Q} = \{B \cap Y \mid B \text{ is a basic open set in } X\}$$

is a base of the relative topology on Y induced by the product topology on X. We need only show that $\mathcal{Q}$ is the collection of basic open sets in the product space Y.

Let $B = \times_{i \in I} U_i$ be a basic open set in X. Then

$$B \cap Y = \underset{i \in I}{\times} (U_i \cap Y_i)$$

with $U_i \cap Y_i$ open in Y_i for all $i \in I$, and with $U_i \cap Y_i = Y_i$ whenever $U_i = X_i$ and hence for almost all $i \in I$. Thus the member $B \cap Y$ of $\mathcal{Q}$ is a basic open set in Y. Conversely, let $V = \times_{i \in I} V_i$ be a basic open set in Y. For each i belonging to the finite set

$$J = \{i \in I \mid V_i \neq Y_i\}$$

there is an open subset U_i of X_i with $U_i \cap Y_i = V_i$; set $U_i = X_i$ for each $i \in I \setminus J$. Then $B = \times_{i \in I} U_i$ is a basic open set in X such that $B \cap Y = V$. Hence $V \in \mathcal{Q}$. $\quad \square$

From now on *we make the standing assumption that each product $\times_{i \in I} X_i$ we consider is nonempty*. This assumption implies that each factor of such a product is nonempty, for if some X_j were empty there certainly could not exist any family indexed by I whose jth coordinate belonged to X_j. (Conversely, if each factor $X_j \neq \varnothing$ then the product $\times_{i \in I} X_i \neq \varnothing$. The proof of this when I is infinite involves the "axiom of choice" of set theory—see Eisenberg [9, page 205].)

The next several results concern continuity of maps with respect to product topologies. Recall that if $(X_i \mid i \in I)$ is a family of sets, then for each $j \in I$ the jth projection is the map

$$p_j \colon \underset{i \in I}{\times} X_i \to X_j$$

$$x = (x_i \mid i \in I) \mapsto x_j$$

sending each point of the product to its jth coordinate. If $j \in I$ and $U \subset X_j$,

then

$$p_j^{-1}(U) = \{x \in \underset{i \in I}{\times} X_i \mid x_j \in U\},$$

or in other terms,

$$p_j^{-1}(U) = \underset{i \in I}{\times} U_i$$

where

$$U_i = \begin{cases} U & \text{if } i = j, \\ X_i & \text{if } i \neq j. \end{cases}$$

In Example 3.13 (2) we saw that the projection $p_1 \colon \mathsf{R}^2 \to \mathsf{R}$ is a continuous open map by using the max metric d_∞ on R^2. Since d_∞ induces the product topology on R^2, that example is a special case of the next theorem.

3.31 Theorem. Let $(X_i \mid i \in I)$ be a family of topological spaces. Then for each $j \in I$ the projection

$$p_j \colon \underset{i \in I}{\times} X_i \to X_j$$

is a continuous open surjection.

Proof. Let $X = \times_{i \in I} X_i$ and fix $j \in I$. We show that p_j is surjective. Let $y \in X_j$. By our standing assumption there exists some $z = (z_i \mid i \in I) \in X$. Define $x = (x_i \mid i \in I) \in X$ by

$$x_i = \begin{cases} y & \text{if } i = j, \\ z_i & \text{if } i \neq j. \end{cases}$$

Then $p_j(x) = x_j = y$.

The map p_j is continuous, for if U is an open set in X_j, then the formula above shows that $p_j^{-1}(U)$ is a basic open set in X. The map p_j is also open, for if $U = \times_{i \in I} U_i$ is a basic open set in X, then $p_j(U) = \varnothing$ if $U = \varnothing$ and $p_j(U) = U_j$ if $U \neq \varnothing$. □

Basic open sets can be represented by using the projections. Let $B = \times_{i \in I} U_i$ be a basic open set in $\times_{i \in I} X_i$. Suppose $B \neq \times_{i \in I} X_i$. Then the set

$$J = \{j \in I \mid U_j \neq X_j\}$$

is nonempty and finite. Hence

$$B = \{x \in \underset{i \in I}{\times} X_i \mid x_j \in U_j \, (j \in J)\}$$

$$= \bigcap_{j \in J} \{x \in \underset{i \in I}{\times} X_i \mid x_j \in U_j\}$$

$$= \bigcap_{j \in J} p_j^{-1}(U_j).$$

We motivated the definition of the product topology by means of metrics. Using the preceding representation, we can now furnish a purely topological reason for preferring the product topology over all other topologies on a product of spaces.

3.32 Corollary. Let $(X_i \mid i \in I)$ be a family of topological spaces. Then the product topology is the least topology on the set $X = \bigtimes_{i \in I} X_i$ making the projections $p_j \colon X \to X_j$ continuous for all $j \in I$.

Proof. Theorem 3.31 says that the product topology does make all the projections continuous. Suppose $\mathfrak{I}$ is any topology on X making all the projections continuous. To show the product topology is contained in $\mathfrak{I}$ it suffices to show each basic open set belongs to $\mathfrak{I}$. Let $B = \bigtimes_{i \in I} U_i$ be a basic open set. If $B = X$, certainly $B \in \mathfrak{I}$; so assume $B \neq X$. Then

$$B = \bigcap_{j \in J} p_j^{-1}(U_j)$$

where, as above, J is the nonempty finite set $\{j \in I \mid U_j \neq X_j\}$. For each $j \in J$ the map p_j is continuous for $\mathfrak{I}$ and U_j is open in X_j, so $p_j^{-1}(U_j) \in \mathfrak{I}$. Hence the finite intersection $B \in \mathfrak{I}$ also. □

In calculus we learn that a map $f \colon \mathsf{R}^m \to \mathsf{R}^n$ is continuous if and only if its component functions $f_1 \colon \mathsf{R}^m \to \mathsf{R}, \ldots, f_n \colon \mathsf{R}^m \to \mathsf{R}$ are continuous, where

$$f(x) = (f_1(x), \ldots, f_n(x)) \qquad\qquad (x \in \mathsf{R}^m).$$

This fact can be generalized to any product space once we note that $f_1 = p_1 \circ f, \ldots, f_n = p_n \circ f$, where $p_1 \colon \mathsf{R}^n \to \mathsf{R}, \ldots, p_n \colon \mathsf{R}^n \to \mathsf{R}$ are the projections.

3.33 Theorem. Let

$$f \colon Y \to \bigtimes_{i \in I} X_i$$

be a map from a topological space Y into the product of a family $(X_i \mid i \in I)$ of topological spaces. Then f is continuous if and only if the composites

$$p_j \circ f \colon Y \to X_j$$

of f with the jth projections are continuous for all $j \in I$.

Proof. Since the projections p_j are continuous, the composites $p_j \circ f$ are all continuous if f is. Conversely, assume $p_j \circ f$ is continuous for all $j \in I$. Let $B = \bigtimes_{i \in I} U_i$ be a basic open set in $\bigtimes_{i \in I} X_i$. If $B = \bigtimes_{i \in I} X_i$, then $f^{-1}(B) = Y$ is open in Y. Now suppose $B \neq \bigtimes_{i \in I} X_i$. As before we have

$$B = \bigcap_{j \in J} p_j^{-1}(U_j)$$

for a nonempty finite subset J of I. Then

$$f^{-1}(B) = \bigcap_{j \in J} f^{-1}(p_j^{-1}(U_j)) = \bigcap_{j \in J} (p_j \circ f)^{-1}(U_j)$$

is open in Y because, by the assumed continuity of $p_j \circ f$, the set $(p_j \circ f)^{-1}(U_j)$ is open in Y for each $j \in J$.   □

This theorem is most often applied in the following form. Suppose we are given a whole family of continuous maps

$$f_i \colon Y \to X_i,$$

one for each $i \in I$, having the common domain Y. These may be combined into the single map

$$f \colon Y \to \underset{i \in I}{\times} X_i$$

$$y \mapsto (f_i(y) \mid i \in I)$$

which is the unique map f satisfying

$$p_i \circ f = f_i \qquad\qquad (i \in I).$$

Then Theorem 3.33 tells us that f is continuous. The maps f_i are called the *components* of f.

Theorem 3.33 can also be used if we are given maps having different domains.

3.34 Corollary. Let $(Y_i \mid i \in I)$ and $(X_i \mid i \in I)$ be families of topological spaces indexed by the same set I, and for each $i \in I$ let

$$f_i \colon Y_i \to X_i$$

be a continuous map. Then the map

$$f \colon \underset{i \in I}{\times} Y_i \to \underset{i \in I}{\times} X_i$$

defined by

$$y = (y_i \mid i \in I) \in \underset{i \in I}{\times} Y_i \quad\Rightarrow\quad f(y) = (f_i(y_i) \mid i \in I)$$

is continuous. Moreover, if

$$f_i \colon Y_i \cong X_i$$

for each $i \in I$, then

$$f \colon \underset{i \in I}{\times} Y_i \cong \underset{i \in I}{\times} X_i.$$

Proof. Set

$$Y = \underset{i \in I}{\times} Y_i, \qquad X = \underset{i \in I}{\times} X_i,$$

and for each $j \in I$ let

$$q_j \colon Y \to Y_j, \qquad p_j \colon X \to X_j$$

be the projections. To show f is continuous, it suffices by 3.33 to show that the composites
$$p_j \circ f: Y \to X_j$$
are continuous for all $j \in I$. However, for $j \in I$ we have
$$p_j(f(y)) = p_j((f_i(y_i) \mid i \in I)) = f_j(y_j) = f_j(q_j(y))$$
for each $y = (y_i \mid i \in I) \in Y$,
$$p_j \circ f = f_j \circ q_j,$$
and $f_j \circ q_j$ is a continuous composite of continuous maps.

Suppose now $f_i: Y_i \cong X_i$ for each $i \in I$. Define
$$g_i = f_i^{-1}: X_i \cong Y_i \qquad\qquad (i \in I).$$
The map $g: X \to Y$ defined by
$$x = (x_i \mid i \in I) \in X \qquad \Rightarrow \qquad g(x) = (g_i(x_i) \mid i \in I)$$
satisfies
$$f(g(x)) = x \qquad (x \in X), \qquad g(f(y)) = y \qquad (y \in Y).$$
Hence f is bijective and $g = f^{-1}$. Since each g_i is continuous, the same reasoning as above, but with the roles of X and Y reversed, shows that g is continuous. $\square$

In the notation of 3.34, the maps $f_i: Y_i \to X_i$ are called the *components* of the map
$$f: \underset{i \in I}{\times} Y_i \to \underset{i \in I}{\times} X_i.$$
When the index set I consists of only two elements, say $I = \{1, 2\}$, the map $f: Y_1 \times Y_2 \to X_1 \times X_2$ has two components $f_1: Y_1 \to X_1$ and $f_2: Y_2 \to X_2$; in this case f is denoted by $f_1 \times f_2$, so that
$$(f_1 \times f_2)(y_1, y_2) = (f_1(y_1), f_2(y_2))$$
for $(y_1, y_2) \in Y_1 \times Y_2$.

The examples that follow illustrate the use of 3.33 and 3.34. Included are two earlier examples that can now be handled more cleanly, without special estimates involving metrics.

3.35 Examples

(1) Vector addition in $\mathbf{R}^n$ is the map
$$\alpha: \mathbf{R}^n \times \mathbf{R}^n \to \mathbf{R}^n$$
$$(x, y) \mapsto x + y$$
which has the components
$$\alpha_i: \mathbf{R}^n \times \mathbf{R}^n \to \mathbf{R} \qquad\qquad (i = 1, \ldots, n).$$
$$(x, y) \mapsto x_i + y_i$$

For $i = 1, \ldots, n$, the component α_i is the composite of the map

$$p_i \times p_i \colon \mathsf{R}^n \times \mathsf{R}^n \to \mathsf{R} \times \mathsf{R}$$

$$(x, y) \mapsto (x_i, y_i)$$

which is continuous by 3.34, and the map

$$\mathsf{R} \times \mathsf{R} \to \mathsf{R}$$

$$(u, v) \mapsto u + v$$

which is continuous by Exercise 1.48, so α_i is continuous. From 3.33 it follows that α is continuous.

Similarly, scalar multiplication

$$\mu \colon \mathsf{R} \times \mathsf{R}^n \to \mathsf{R}^n$$

$$(\lambda, x) \mapsto \lambda x$$

is continuous.

(2) Let $f_1, f_2 \colon X \to \mathsf{R}$ be continuous functions on a topological space X. The sum

$$f_1 + f_2 \colon X \to \mathsf{R}$$

$$x \mapsto f_1(x) + f_2(x)$$

is the composite of the two maps

$$X \to \mathsf{R} \times \mathsf{R}, \qquad\qquad \mathsf{R} \times \mathsf{R} \to \mathsf{R}.$$

$$x \mapsto (f_1(x), f_2(x)) \qquad\quad (u, v) \mapsto u + v$$

The first of these two maps is continuous by 3.33, the second by Exercise 1.48. Hence $f_1 + f_2$ is continuous. Similarly, the product

$$f_1 \cdot f_2 \colon X \to \mathsf{R}$$

$$x \mapsto f_1(x) \cdot f_2(x)$$

is continuous.

(3) The map

$$f \colon \mathsf{R}^n \to \mathsf{B}_n$$

$$x \mapsto \frac{1}{1 + ||x||} \, x$$

considered in Example 3.16 (7) is the composite of the maps

$$\mathsf{R}^n \to \mathsf{R} \times \mathsf{R}^n, \qquad\qquad \mathsf{R} \times \mathsf{R}^n \to \mathsf{R}^n.$$

$$x \mapsto \left(\frac{1}{1 + ||x||}, x \right) \qquad\quad (\lambda, u) \mapsto \lambda u$$

The former of these is continuous by 3.33, the latter by Example (1) above. Hence f is continuous.

(4) The map

$$h: I^n \to J^n$$

$$(x_1, \ldots, x_n) \mapsto (k(x_1), \ldots, k(x_n))$$

constructed in Example 3.16 (8) is obtained from n component functions all equal to the homeomorphism $k: I \cong J = [-1, 1]$. From 3.34 it follows at once that h is a homeomorphism.

(5) The Hilbert cube

$$I^\infty = \left\{ (x_i \mid i = 1, 2, \ldots) : x_i \in \mathsf{R}, |x_i| \leq \frac{1}{i} \, (i = 1, 2, \ldots) \right\}$$

can be topologized in two natural ways. First, as a subset of the Hilbert sequence space H, I^∞ has the topology induced by the usual metric d_2 on H. Second, as a product

$$I^\infty = \underset{i=1}{\overset{\infty}{\mathsf{X}}} \left[-\frac{1}{i}, \frac{1}{i} \right]$$

of intervals, I^∞ has the product topology. These two topologies are the same, as we shall demonstrate in a moment. Then since $[-1/i, 1/i] \cong [0, 1] = I$ for each $i = 1, 2, \ldots$, we can conclude from 3.34 that

$$I^\infty \cong I \times I \times \cdots \times I \times \cdots$$

(this explains the notation I^∞ for the Hilbert cube).

Let $X_i = [-1/i, 1/i]$ for $i = 1, 2, \ldots$. To see that d_2 does in fact induce the product topology on $I^\infty = X_1 \times X_2 \times \cdots$, fix a point $x = (x_1, x_2, \ldots) \in I^\infty$. The collections

$$\{ U \mid U \text{ a basic open set in } I^\infty, x \in U \},$$

$$\{ B_\epsilon(x; d_2) \cap I^\infty \mid \epsilon > 0 \}$$

are local bases at x for the product topology and for the topology induced by d_2, respectively. We show that these collections define the same neighborhoods of x.

Suppose

$$U = U_1 \times \cdots \times U_n \times X_{n+1} \times X_{n+2} \times \cdots$$

is a basic open set in I^∞ with $x \in U$. Then $U_1 \times \cdots \times U_n$ is a neighborhood of the point $(x_1, \ldots, x_n)$ in the product space $X_1 \times \cdots \times X_n$. Because the Euclidean metric induces the product topology on $X_1 \times \cdots \times X_n$, there is an $\epsilon > 0$ such that for $(y_1, \ldots, y_n) \in X_1 \times \cdots \times X_n$,

$$\sum_{i=1}^{n} (x_i - y_i)^2 < \epsilon^2 \quad \Rightarrow \quad (y_1, \ldots, y_n) \in U_1 \times \cdots \times U_n.$$

Then $y = (y_1, y_2, \ldots) \in I^\infty$ with $d_2(x, y) < \epsilon$ implies

$$\sum_{i=1}^{n} (x_i - y_i)^2 \le \sum_{i=1}^{\infty} (x_i - y_i)^2 = [d_2(x, y)]^2 < \epsilon^2$$

so that $y \in U$. Hence $B_\epsilon(x; d_2) \cap I^\infty \subset U$.

Now suppose $\epsilon > 0$. We show that $x \in U \subset B_\epsilon(x; d_2)$ for some basic open set U in I^∞. The series $\sum_{i=1}^{\infty} (2/i)^2$ converges, so we may choose n so large that

$$\sum_{i=n+1}^{\infty} \left(\frac{2}{i}\right)^2 < \frac{\epsilon^2}{2}.$$

Since the Euclidean metric induces the product topology on $X_1 \times \cdots \times X_n$, there are open neighborhoods U_i of x_i in X_i for $i = 1, \ldots, n$ such that

$$(y_1, \ldots, y_n) \in U_1 \times \cdots \times U_n \quad \Longrightarrow \quad \sum_{i=1}^{n} (x_i - y_i)^2 < \frac{\epsilon^2}{2}.$$

The basic open set

$$U = U_1 \times \cdots \times U_n \times X_{n+1} \times X_{n+2} \times \cdots$$

in I^∞ contains x. Moreover, $U \subset B_\epsilon(x; d_2)$ because $y = (y_1, \ldots, y_n, y_{n+1}, \ldots) \in U$ implies $(y_1, \ldots, y_n) \in U_1 \times \cdots \times U_n$ and $|x_i - y_i| < 2/i$ for $i = n + 1, n + 2, \ldots$, so that

$$[d_2(x, y)]^2 = \sum_{i=1}^{n} (x_i - y_i)^2 + \sum_{i=n+1}^{\infty} (x_i - y_i)^2 < \frac{\epsilon^2}{2} + \frac{\epsilon^2}{2} = \epsilon^2.$$

(6) A fact related to Example (5), proved by R. D. Anderson in 1966, is that the Hilbert sequence space H is homeomorphic to the product of denumerably many lines:

$$\mathsf{H} \cong \mathsf{R} \times \mathsf{R} \times \cdots \times \mathsf{R} \times \cdots.$$

Another application of Theorem 3.33 is the commutative law for products, which says that a product space is unchanged topologically if its factors are permuted. We consider here only the special case of two factors, leaving the general case to the exercises. The product spaces $X \times Y$ and $Y \times X$ obtained from two spaces X and Y need not be the same, for even as sets $X \times Y \ne Y \times X$ unless $X = Y$. Now the map $(x, y) \mapsto (y, x)$ which interchanges coordinates is a bijection from $X \times Y$ to $Y \times X$. The commutative law tells us this map makes $X \times Y$ topologically the same as $Y \times X$.

3.36 Proposition. Let X and Y be topological spaces. Then the bijection

$$h \colon X \times Y \to Y \times X$$

$$(x, y) \mapsto (y, x)$$

is a homeomorphism. Thus

$$X \times Y \cong Y \times X.$$

Proof. The projections

$$p_1 \colon X \times Y \to X, \qquad p_2 \colon X \times Y \to Y$$

$$(x, y) \mapsto x \qquad\qquad (x, y) \mapsto y$$

and

$$q_1 \colon Y \times X \to Y, \qquad q_2 \colon Y \times X \to X$$

$$(y, x) \mapsto y \qquad\qquad (y, x) \mapsto x$$

are all continuous. Then the components $q_1 \circ h \colon X \times Y \to Y$, $q_2 \circ h \colon X \times Y \to X$ of h are continuous because

$$q_1 \circ h = p_2, \qquad q_2 \circ h = p_1.$$

Hence h is continuous. Similarly, h^{-1} is continuous. $\quad\square$

There is also an associative law for product spaces, which says that a product space is topologically unchanged if its factors are grouped in any way. Again we formulate here only a special case.

3.37 Proposition. Let $X_1, \ldots, X_n$ be topological spaces. Then

$$X_1 \times \cdots \times X_n \cong (X_1 \times \cdots \times X_{n-1}) \times X_n.$$

Proof. Set

$$X = X_1 \times \cdots \times X_n, \qquad Y = X_1 \times \cdots \times X_{n-1}.$$

An element of X is an n-tuple $(x_1, \ldots, x_n)$, while an element of $Y \times Y_n$ is a pair (y, x) with y itself an $(n-1)$-tuple $(x_1, \ldots, x_{n-1})$. We have a natural map

$$h \colon X \to Y \times X_n$$

$$(x_1, \ldots, x_n) \mapsto ((x_1, \ldots, x_{n-1}), x_n)$$

which is a bijection whose inverse is

$$h^{-1} \colon Y \times X_n \to X$$

$$((y_1, \ldots, y_{n-1}), x) \mapsto (y_1, \ldots, y_{n-1}, x).$$

The map h has the two components

$$X \to Y \qquad\qquad\qquad\qquad X \to X_n$$

$$(x_1, \ldots, x_n) \mapsto (x_1, \ldots, x_{n-1}), \qquad (x_1, \ldots, x_n) \mapsto x_n .$$

The latter of these two maps is continuous because it is just the nth projection of the product X. The former of these is a map into the product Y and by 3.33 is continuous because its components are the 1st through $(n-1)$st projections of the product Y. Hence by 3.33 the map h is continuous. Similarly, h^{-1} is continuous. $\quad\square$

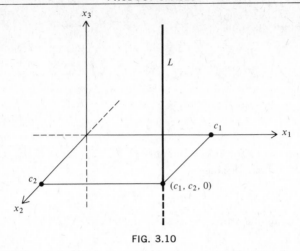

FIG. 3.10

Proposition 3.37 was already used implicitly in Exercise 3.41 when we wrote

(*) $$S_1 \times I = \{x \in R^3 \mid x_1^2 + x_2^2 = 1, 0 \leq x_3 \leq 1\}.$$

The space $S_1 \times I$ on the left is, of course, not a subspace of $R^3 = R \times R \times R$, but of $R^2 \times R = (R \times R) \times R$. It is, however, the subspace of $R^2 \times R$ corresponding to the set on the right under the homeomorphism $R^3 \cong R^2 \times R$ provided by the proof of 3.37. Thus the equality sign $=$ in (*) should really be $\cong$, but we have implicitly made an identification of the two spaces in (*). We reserve the right to make similar identifications in the future which make implicit use of 3.37.

If we choose points $c_1 \in R, c_2 \in R$, then the subspace $L = \{(x_1, x_2, x_3) \in R^3 \mid x_1 = c_1, x_2 = c_2\}$ of R^3 obtained by fixing the first two coordinates of points is a line that is homeomorphic to the third factor R of $R^3 = R \times R \times R$ (see Fig. 3.10). In general, a product space contains copies of each of its factors.

3.38 Theorem. Let $(X_i \mid i \in I)$ be a family of topological spaces, let $c = (c_i \mid i \in I) \in \bigtimes_{i \in I} X_i$, and let $j \in I$. Then the map

$$f: X_j \to \bigtimes_{i \in I} X_i$$

defined by

$$f(y) = (x_i \mid i \in I)$$

where

$$x_i = \begin{cases} y & \text{if } i = j, \\ c_i & \text{if } i \neq j \end{cases}$$

is an embedding.

Proof. The range $f(X_j)$ of f is the subspace $A = \times_{i \in I} A_i$ of $\times_{i \in I} X_i$ with

$$A_i = \begin{cases} X_j & \text{if } i = j, \\ \{c_i\} & \text{if } i \neq j. \end{cases}$$

We show that f, considered as a map from X_j onto A, is a homeomorphism. Note that by 3.30 the relative topology on A is the product topology on A.

Let $p_i \colon A \to A_i$ be the ith projection for each $i \in I$. Then $y \in X_j$ implies

$$p_j \circ f(y) = y, \qquad p_i \circ f(y) = c_i \qquad (i \neq j).$$

This means that the components $p_i \circ f \colon X_j \to A_i$ are all constant maps except the jth, which is the identity map of $X_j = A_j$, so these components are all continuous. Hence $f \colon X_j \to A$ is continuous. The relation $p_j \circ f(y) = y$ above coupled with

$$f \circ p_j(a) = a \qquad\qquad (a \in A)$$

shows that f is bijective and

$$f^{-1} = p_j \colon A \to A_j = X_j.$$

Hence f^{-1} is continuous, too. $\Box$

This theorem can sometimes be used to show that if a product space has a certain topological property, then each of its factors has the same property. Of course, it is more interesting to know, conversely, that a topological property is "productive" in the sense that if each of a family of spaces has the property, then their product has the same property.

3.39 Theorem. A product space $\times_{i \in I} X_i$ is a Hausdorff space if and only if its factors X_i are Hausdorff spaces for all $i \in I$.

Proof. Assume $X = \times_{i \in I} X_i$ is a Hausdorff space. Then each subspace of X is a Hausdorff space. If $i \in I$, then X_i is by 3.38 homeomorphic to a subspace of X, so X_i is a Hausdorff space.

Conversely, assume X_i is a Hausdorff space for each $i \in I$. Let $x = (x_i \mid i \in I), y = (y_i \mid i \in I) \in X$ with $x \neq y$. Then $x_j \neq y_j$ for some index $j \in I$. By assumption there are disjoint neighborhoods U_j and V_j of x_j and y_j in X_j. Define $U_i = V_i = X_i$ for all $i \in I$ with $i \neq j$. By 3.29 the sets $U = \times_{i \in I} U_i$ and $V = \times_{i \in I} V_i$ are neighborhoods of x and y, respectively, in X. The sets U and V are disjoint since no point of X can have its jth coordinate both in U_j and in V_j. $\Box$

3.40 Theorem. A product space $\times_{i \in I} X_i$ is metrizable if and only if its factors X_i are metrizable for all $i \in I$. (As usual, we assume that $I \subset \mathbb{N}$, so that I is countable.)

Proof. Since a subspace of a metrizable space is metrizable, 3.38 implies that each X_i is metrizable if $\bigtimes_{i \in I} X_i$ is. Conversely, assume each X_i is metrizable. Choose a metric d_i inducing the topology of X_i for each i.

Case (i): I is finite. By relabeling the factors if necessary, we may assume $I = \{1, \ldots, n\}$ for some n. Then by 3.24 the max metric induced by $(d_1, \ldots, d_n)$ induces the product topology on $\bigtimes_{i \in I} X_i$.

Case (ii): I is infinite. Then I is denumerable, so we may assume $I = \{1, 2, \ldots\}$. Replace each d_i if necessary by an equivalent metric for which diam $X_i \leq 1$. Then the metric d' considered in 3.25 induces the product topology on $\bigtimes_{i \in I} X_i$. $\square$

The 'if' parts of 3.39 and 3.40 say that the properties of being a Hausdorff space and of being metrizable are "preserved under the formation of products". In the next two chapters we shall meet other properties that behave the same way.

3.41 Note. On all the products considered above we have imposed the restriction that the index set be countable. One reason for this restriction is that a typical element of such a product is a familiar object—either an n-tuple for some positive integer n or else a sequence. Another reason is that we can avoid certain advanced set-theoretic tools (the axiom of choice and Zorn's lemma) which would otherwise be required to prove certain key theorems (in particular, Tychonoff's theorem on compactness). However, there is no real difficulty in defining products over more general index sets.

Suppose $(X_i \mid i \in I)$ is a family of topological spaces indexed by an arbitrary, not necessarily countable, nonempty set I. Then the product set $\bigtimes_{i \in I} X_i$ is defined to be the set of all families $(x_i \mid i \in I)$ such that $x_i \in X_i$ for each $i \in I$. Lemma 3.27 still holds with the same proof, so a product topology on $\bigtimes_{i \in I} X_i$ may be defined exactly as in 3.28. With the single exception of the metrizability theorem 3.40, all the results of this section remain valid in this more general setting.

3.42 Example. In Chapter 2 we have already met product spaces of the more general type just considered, but in a different guise. Let X be any set and let Y be any topological space. Consider the product space

$$\mathfrak{F} = \bigtimes_{x \in X} Y_x$$

where X is the index set and

$$Y_x = Y \qquad\qquad (x \in X).$$

A typical element of $\mathfrak{F}$ is by definition a family $f = (f_x \mid x \in X)$ indexed by

X such that $f_x \in Y_x = Y$ for each $x \in X$, that is, a map

$$f: X \to Y$$

$$x \mapsto f_x$$

from X to Y. Thus the *product set* $\mathfrak{F}$ *is the set of all maps from X to Y*. We shall revert to the usual functional notation, writing $f(x)$ for the xth coordinate f_x of an $f \in \mathfrak{F}$.

A typical basic open set U in the product space $\mathfrak{F}$ is obtained as follows: Choose finitely many indices $x_1, \ldots, x_k \in X$ and for each $i = 1, \ldots, k$ choose an open subset V_i of the x_ith factor $Y = Y_{x_i}$; then

$$U = \underset{x \in X}{\times} U_x$$

where

$$U_x = \begin{cases} V_i & \text{if } x = x_i, i = 1, \ldots, k, \\ Y_x = Y & \text{otherwise.} \end{cases}$$

Then for $f \in \mathfrak{F}$,

$$f \in U \quad \Leftrightarrow \quad f(x_i) \in V_i \quad (i = 1, \ldots, k).$$

In the notation of Exercise 2.71 we may therefore write

$$U = B(x_1, \ldots, x_k; V_1, \ldots, V_k).$$

The basic open sets U form a base of the product topology on $\mathfrak{F}$. According to Exercise 2.71—and according to Example 2.52 (3) in the case $X = Y = \mathbf{R}$—the sets $B(x_1, \ldots, x_k; V_1, \ldots, V_k)$ form a base of the topology of pointwise convergence on $\mathfrak{F}$. Hence *the product topology on $\mathfrak{F}$ is its topology of pointwise convergence*.

We shall look again at the topology of pointwise convergence and we shall explain its name in Section 5.

EXERCISES

61. Let A and B be subsets of topological spaces X and Y, respectively. Prove:
    (**a**) int $(A \times B) = (\text{int } A) \times (\text{int } B)$.
    (**b**) cls $(A \times B) = (\text{cls } A) \times (\text{cls } B)$.
    (**c**) bdy $(A \times B) = [(\text{bdy } A) \times (\text{cls } B)] \cup [(\text{cls } A) \times (\text{bdy } B)]$.

62. Do formulas (a) and (b) in Exercise 61 generalize to more than two factors?

63. Let A be a subset of a product set $X = \times_{i \in I} X_i$, and for $i \in I$ let $p_i: X \to X_i$ be the ith projection. Prove or disprove:
    (**a**) $A \subset \times_{i \in I} p_i(A)$.
    (**b**) $A = \times_{i \in I} p_i(A)$.
    (**c**) If $x = (x_i \mid i \in I) \notin A$, then $x_i \notin p_i(A)$ for some $i \in I$.

64. For each $i \in I$ let A_i be a subset of a topological space X_i.
    (**a**) Suppose A_i is nonempty and open in X_i for each $i \in I$. We know that $\times_{i \in I} A_i$ is open in $\times_{i \in I} X_i$ if $A_i = X_i$ for almost all $i \in I$. Show that the converse is true.

(**b**) Prove: If A_i is closed in X_i for each $i \in I$, then $\underset{i \in I}{\times} A_i$ is closed in $\underset{i \in I}{\times} X_i$. Is the converse true?

(**c**) When is $\underset{i \in I}{\times} A_i$ dense in $\underset{i \in I}{\times} X_i$?

65. (**a**) Prove that the product of finitely many discrete spaces is discrete.

(**b**) Show that the product of infinitely many discrete spaces cannot be discrete unless almost all of them consist of a single point.

66. Use the techniques of this section to establish the continuity of the following maps:

(**a**) The quotient $x \mapsto f(x)/g(x)$ of two real-valued functions f and g on a topological space X, where $g(x) \neq 0$ for all $x \in X$.

(**b**) The map $e: J^n\backslash\{0\} \to E$ of Example 3.16(9).

(**c**) The map $f: S_n\backslash\{p\} \to H = \mathsf{R}^n \times \{0\}$ of Example 3.16(11).

(**d**) The antipodal map $x \mapsto -x$ of $S_n \to S_n$.

67. If X is a topological space, show that the *diagonal*

$$\Delta = \{(x, x) \mid x \in X\}$$

of $X \times X$ is homeomorphic to X.

68. Prove that a topological space X is a Hausdorff space if and only if the diagonal Δ of $X \times X$ (Exercise 67) is closed in $X \times X$.

69. Let $f: X \to Y$ be a continuous map and let

$$G = \{(x, y) \in X \times Y \mid y = f(x)\}$$

be its graph. Prove:

(**a**) The domain X of f is homeomorphic to G. (Compare Exercise 3.40).

(**b**) If Y is a Hausdorff space, then G is closed in $X \times Y$.

70. Construct a homeomorphism $h: [0, 1] \times [0, 1[\cong [0, 1[\times [0, 1[$. (*Note:* Thus a product space does not uniquely determine its factor spaces. Moreover, there can be no cancellation law $X \times Y \cong Z \times Y \Rightarrow X \cong Z$.)

71. Establish the continuity of the vector operations

$$\mathsf{H} \times \mathsf{H} \to \mathsf{H} \qquad \mathsf{R} \times \mathsf{H} \to \mathsf{H}$$
$$(x, y) \mapsto x + y, \qquad (\lambda, x) \mapsto \lambda x$$

in the Hilbert sequence space H. [*Note:* The continuity of these operations in the vector space H means that H is a *topological vector space*. Likewise, Example 3.35 (1) says that R^n is a topological vector space.]

72. Prove the general commutative law for product spaces: Let σ be a permutation of the index set I of a family $(X_i \mid i \in I)$ of topological spaces (that is, σ is a bijection from I to I). Then

$$\underset{i \in I}{\times} X_i \cong \underset{i \in I}{\times} X_{\sigma(i)}.$$

73. If $(X_i \mid i \in I)$ and $(Y_i \mid i \in I)$ are two families of topological spaces indexed by the same set I, construct a homeomorphism

$$\underset{i \in I}{\times} (X_i \times Y_i) \cong \left(\underset{i \in I}{\times} X_i\right) \times \left(\underset{i \in I}{\times} Y_i\right).$$

Apply this result with $I = \{1, \ldots, n\}$ and $X_i = Y_i = \mathsf{R}$ to reprove continuity of vector addition $(x, y) \mapsto x + y$ in R^n.

74. (**a**) Prove the general associative law for product spaces: Let $(I_j \mid j \in J)$ be a partition of the index set I of a family $(X_i \mid i \in I)$ of topological spaces (so that each I_j is a nonempty subset of I, I_j is disjoint from I_k whenever $j \neq k$, and $\bigcup_{j \in J} I_j = I$). Then

$$\underset{i \in I}{\times} X_i \cong \underset{j \in J}{\times} \left(\underset{i \in I_j}{\times} X_i \right).$$

(**b**) Obtain Proposition 3.37 as a special case.

75. Generalize Theorem 3.38 by showing that $\times_{j \in J} X_j$ can be embedded into $\times_{i \in I} X_i$ for each nonempty subset J of I.

76. Is the property of being a T_0-space (Exercise 2.32) preserved under the formation of product spaces? The property of being a T_1-space (Exercise 2.10)?

77. Must a product space be regular (Exercise 2.60) if each of its factors is regular? Must the factors be regular if the product space is regular?

78. (**a**) Show that the product of an m-dimensional manifold and an n-dimensional manifold is an $(m + n)$-dimensional manifold.

(**b**) Use 3.37 to deduce from (a) that the product of finitely many manifolds is a manifold.

79. Let $(X_i \mid i \in I)$ be a family of topological spaces. Let $\mathcal{B}_i$ be a base of X_i for each $i \in I$. Show that the collection of all products $\times_{i \in I} B_i$ such that $B_i = X_i$ for almost all i and $B_i \in \mathcal{B}_i$ whenever $B_i \neq X_i$ is a base of the product topology. Describe this base by means of the projections.

80. Let $(X_i \mid i \in I)$ be a family of topological spaces, and let $x = (x_i \mid i \in I) \in \times_{i \in I} X_i$. Suppose $\mathfrak{M}_i$ is a local base at x_i in X_i for each $i \in I$. Use the collections $\mathfrak{M}_i$ to determine a local base at x in $\times_{i \in I} X_i$.

81. Prove that a product space is second-countable (respectively, first-countable) if and only if each of its factors is second-countable (respectively, first-countable). Here, as usual, the index set is assumed countable. [*Hint*: Use Exercise 79 (respectively, Exercise 80).]

82. Let $(X_i \mid i \in I)$ be a family of spaces whose index set I is, as in 3.41, not necessarily countable. Suppose X_i contains at least two points for uncountably many indices $i \in I$. Show that $\times_{i \in I} X_i$ cannot be first-countable even if each X_i is first-countable. [*Hint*: Compare Example 2.52 (3).]

83. When is the set of all functions from a set X to a topological space Y metrizable when provided with its topology of pointwise convergence?

Exercises 84–88 require a slight knowledge of groups.

84. A *topological group* is a group G provided with a topology making the maps

$$G \times G \to G, \qquad G \to G$$
$$(x, y) \mapsto x \cdot y \qquad x \mapsto x^{-1}$$

continuous (here $G \times G$ is given its product topology induced by the topology on G). Verify that the following are topological groups:

(**a**) The multiplicative group $\mathbb{R} \setminus \{0\}$ with its usual topology.

(**b**) Any group G with its discrete topology.

(**c**) The additive group R with its usual topology (here, of course, we write $x + y$ instead of $x \cdot y$ and $-x$ instead of x^{-1}.)

(**d**) The additive group R^n with its usual topology.

(**e**) The *circle group*—that is, the multiplicative group $\{z \in C : |z| = 1\}$—with its usual topology.

(**f**) The multiplicative group $GL(2, R)$ of all nonsingular 2×2 matrices with real coefficients with the following topology $\Im$. Let

$$\phi : GL(2, R) \to R^4$$

be the injection defined by

$$\begin{bmatrix} a_{11} & a_{12} \\ a_{21} & a_{22} \end{bmatrix} \mapsto (a_{11}, a_{12}, a_{21}, a_{22}).$$

Then $\Im$ is the unique topology on $GL(2, R)$ making ϕ an embedding. (*Hint*: Write out explicitly the formulas for the product of two matrices and for the inverse of a matrix.)

85. Fix an element a in a topological group G. Show that each of the maps from G to G below is a homeomorphism:

(**a**) Left translation $x \mapsto ax$ by a.

(**b**) Right translation $x \mapsto xa$ by a.

(**c**) Conjugation $x \mapsto axa^{-1}$ by a.

86. (**a**) Prove that a group G provided with a topology is a topological group if and only if the single map

$$G \times G \to G$$

$$(x, y) \mapsto xy^{-1}$$

is continuous.

(**b**) If U is a neighborhood of the identity element e of a topological group G, show that U contains a neighborhood V of e such that $V^2 \subset U$ and $V = V^{-1}$; here

$$V^2 = \{x \cdot y \mid x, y \in V\}, \qquad V^{-1} = \{x^{-1} \mid x \in V\}.$$

87. Let G be a topological group and let e be its identity element. Prove:

(**a**) If V is a neighborhood of e, then xV is a neighborhood of x for each $x \in G$.

(**b**) G is first-countable if there is some countable local base at e.

(**c**) G is separated if for each $x \neq e$ there is some neighborhood U of e with $x \notin U$.

88. Let H be a subgroup of a topological group G. Prove:

(**a**) Under its relative topology H is a topological group.

(**b**) The closure of H in G is also a subgroup of G.

(**c**) If H is open in G, then H is closed in G.

4. Quotient Spaces

To motivate the next general method for constructing new spaces from old, we pose the problem of representing mathematically a simple physical experiment. A straight piece of very thin, flexible wire is bent around until its ends touch. The shape that results is a circle or—in case the bending was not done uniformly or some kinks or corners were introduced—a shape topologically the same as a circle.

Let us represent a circular shape by the unit circle

$$S_1 = \{ (x_1, x_2) \in R^2 \mid x_1{}^2 + x_2{}^2 = 1 \}$$

and the original, unbent wire by the closed interval

$$X = [0, 2\pi]$$

whose length is the same as the circumference of the circle. (We could instead choose an interval whose length differs from the circle's circumference in order to represent the wire's being stretched while being bent, and this would just introduce a slight change in the analysis below.) Bending the wire until its ends touch may be represented by identifying the endpoints 0 and 2π of X—that is, by treating them as if they were the same point. In strictly mathematical terms we form the equivalence relation $\sim$ on X given by

$$t \sim s \quad \Leftrightarrow \quad (t = 0 \text{ and } s = 2\pi) \text{ or } (t = 2\pi \text{ and } s = 0) \text{ or } (t = s).$$

The bent piece of wire is then represented by the quotient set

$$Y = X/\sim$$

whose points are the equivalence classes

$$0/\sim \ = \ 2\pi/\sim \ = \ \{0, 2\pi\},$$

$$t/\sim \ = \ \{t\} \qquad\qquad (0 < t < 2\pi).$$

The mathematical problem, then, is to put a topology on the quotient set Y which makes Y homeomorphic to S_1.

To attack this problem we change the point of view and think of wrapping the wire around an existing circular shape (whose circumference equals the length of the wire). In mathematical terms we consider the mapping

$$f \colon X \to S_1$$

$$t \mapsto (\cos t, \sin t)$$

of the interval X onto the circle S_1. Here we rely on the familiar properties of the cosine and sine functions which guarantee that each point $(x_1, x_2) \in S_1$ has a parametric representation

$$\begin{cases} x_1 = \cos t \\ \\ x_2 = \sin t \end{cases}$$

for some $0 \le t \le 2\pi$. Each point of S_1 has a unique representation of this sort except the point $(1, 0)$, which has the two representations

$$\begin{cases} 1 = \cos 0 \\ 0 = \sin 0, \end{cases} \qquad \begin{cases} 1 = \cos 2\pi \\ 0 = \sin 2\pi. \end{cases}$$

In other words, for $t, s \in X$ we have

$$(*) \qquad\qquad f(t) = f(s) \qquad \Leftrightarrow \qquad t \sim s.$$

Property (*) allows us to relate $Y = X/\!\sim$ to S_1 by constructing a map

$$h\colon Y \to \mathsf{S}_1$$

according to the following rule: If $y \in Y$, then y is an equivalence class in X under $\sim$, and we define $h(y)$ to be $f(t)$ for any representative t of this equivalence class; $h(y)$ is uniquely determined by y alone, independently of the choice of t, because if t and s are two representatives of y, then $t \sim s$ and hence $f(t) = f(s)$ by (*).

The map h is a bijection: Suppose $y, z \in Y$ with $h(y) = h(z)$. Choose any representatives t of y and s of z. Then

$$f(t) = h(y) = h(z) = f(s),$$

$t \sim s$ by (*), and so $y = z$. Hence h is injective. Now suppose $x \in \mathsf{S}_1$. Since f is surjective, $f(t) = x$ for some $t \in X$. Then $y = t/\!\sim$ is an element of Y such that $h(y) = f(t) = x$. Hence h is surjective, too.

We now endow Y with its unique topology $\mathfrak{I}$ making h a homeomorphism— and hence making Y homeomorphic to S_1—namely,

$$\mathfrak{I} = \{U \subset Y \mid h(U) \text{ is open in } \mathsf{S}_1\}.$$

This solves our problem, but not in quite the form suitable for generalization. We seek a description of $\mathfrak{I}$ purely in terms of the topology of X and the equivalence relation $\sim$ on X.

Let

$$p\colon X \to Y = X/\!\sim$$
$$t \mapsto t/\!\sim$$

be the quotient map which sends each element of X to its equivalence class under $\sim$. Because each $t \in X$ is a representative of its equivalence class $p(t)$, the definition of h says that

$$h(p(t)) = f(t) \qquad\qquad (t \in X).$$

Then

$$h \circ p = f,$$

so we have the commutative diagram below

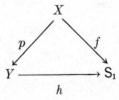

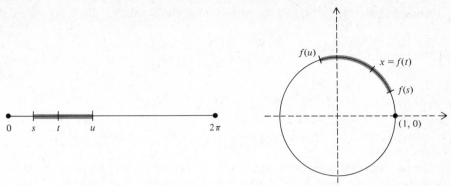

FIG. 3.11

with h being bijective. Hence

(**) $f^{-1}(h(U)) = p^{-1}(U)$ $(U \subset Y)$.

The crucial observation now concerns the special nature of f:

(***) V is open in S_1 $\Leftrightarrow$ $f^{-1}(V)$ is open in X $(V \subset S_1)$.

In fact, f is continuous because its component functions $t \mapsto \cos t$, $t \mapsto \sin t$ are continuous, so $f^{-1}(V)$ is open in X if V is open in S_1. Conversely, suppose V is a subset of S_1 such that $f^{-1}(V)$ is open in $X = [0, 2\pi]$. Let $x \in V$; we show V contains an open arc containing x.

Case (i). $x \neq (1, 0)$. Then $f(t) = x$ for a unique $t \neq 0, 2\pi$. Since $f^{-1}(V)$ is an open neighborhood of t in $[0, 2\pi]$, there exist $0 < s < t < u < 2\pi$ with $]s, u[\subset f^{-1}(V)$. Then $f(]s, u[)$ is the desired open arc in S_1 (see Fig. 3.11).

Case (ii). $x = (1, 0)$. Then $f(0) = x = f(1)$, the points 0 and 1 both belong to $f^{-1}(V)$, and there exist $0 < s < u < 2\pi$ with $[0, s[\cup]u, 2\pi] \subset f^{-1}(V)$. In this case $f([0, s[) \cup f(]u, 2\pi])$ is the desired open arc (see Fig. 3.12).

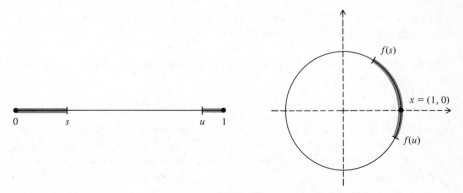

FIG. 3.12

Combining (**), (***), and the definition of the topology $\mathfrak{I}$ on Y, we obtain

$$U \in \mathfrak{I} \quad \Leftrightarrow \quad h(U) \text{ open in } \mathsf{S}_1$$

$$\Leftrightarrow \quad p^{-1}(U) = f^{-1}(h(U)) \text{ open in } X.$$

Thus

$$\mathfrak{I} = \{U \subset Y \mid p^{-1}(U) \text{ is open in } X\}.$$

This formula describes in terms of the topological space X and the equivalence relation $\sim$ on X a topology on the quotient set $Y = X/\sim$ that makes Y homeomorphic to S_1.

We now proceed rapidly to extract from this example the general definition of a quotient space and some general principles that will allow us to give additional concrete examples.

3.43 Lemma. Let $\sim$ be an equivalence relation on a topological space X, and let $p: X \to X/\sim$ be the quotient map. Then the collection

$$\mathfrak{I} = \{V \subset X/\sim \,:p^{-1}(V) \text{ open in } X\}$$

is a topology on the quotient set $X/\sim$.

Proof. Since $p^{-1}(\varnothing) = \varnothing$ and $p^{-1}(X/\sim) = X$, both $\varnothing$ and $X/\sim$ belong to $\mathfrak{I}$. If $(V_i \mid i \in I)$ is a family of sets belonging to $\mathfrak{I}$, then $p^{-1}(V_i)$ is open in X for each $i \in I$, so

$$p^{-1}\left(\bigcup_{i \in I} V_i\right) = \bigcup_{i \in I} p^{-1}(V_i)$$

is open in X. If $U \in \mathfrak{I}$ and $V \in \mathfrak{I}$, then $p^{-1}(U)$ and $p^{-1}(V)$ are open in X, so $p^{-1}(U \cap V) = p^{-1}(U) \cap p^{-1}(V)$ is open in X. □

3.44 Definition. The *quotient* of a topological space X under an equivalence relation $\sim$ is the topological space whose underlying set is $X/\sim$ and whose topology is the collection $\mathfrak{I}$ described above. This topology is called the *quotient topology*.

Unless otherwise indicated, any future reference to a topology on a quotient set of a topological space is to the quotient topology.

A set $V \subset X/\sim$ has as its points equivalence classes in X and has as its inverse image

$$p^{-1}(V) = \{x \in X: x/\sim \,\in V\} = \bigcup \{x/\sim\,: x \in X, x/\sim \,\in V\},$$

the union of those equivalence classes belonging to V (the relation between V and $p^{-1}(V)$ is indicated schematically in Fig. 3.13, where equivalence classes in X are represented by vertical line segments). Hence a *typical open set in the quotient space $X/\sim$ is just a collection of equivalence classes whose union is an open subset of X.*

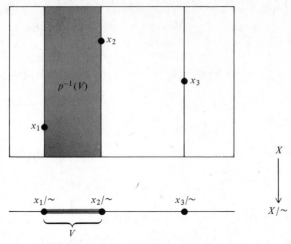

FIG. 3.13

Let us call a subset U of X *saturated* by the equivalence relation $\sim$ when U is a union of equivalence classes under $\sim$, that is, when

$$x \in U \quad \text{and} \quad y \sim x \quad \Rightarrow \quad y \in U.$$

If V is a subset of $X/\sim$, then $p^{-1}(V)$ is saturated by $\sim$; if U is any saturated subset of X, then $U = p^{-1}(V)$ for a subset of $X/\sim$, namely, $V = p(U)$.

The quotient topology may now be described as

$$\{p(U) \mid U \text{ open in } X, \ U \text{ saturated by } \sim\}.$$

Then each open set V in $X/\sim$ is the image under p of an open subset of X, namely, of the saturated open subset $p^{-1}(V)$ of X. However, the image $p(U)$ of an arbitrary open subset of X need not be open in $X/\sim$ (so p need not be an open map). For example, the set $p([0, \pi[)$ is not open in the quotient space $[0, 2\pi]/\sim$ considered at the beginning of this section, because $p^{-1}(p([0, \pi[)) = [0, \pi[\ \cup \ \{2\pi\}$ is not open in $[0, 2\pi]$.

By definition of the quotient topology, if V is a subset of $X/\sim$, then

$$V \text{ is open in } X/\sim \quad \Leftrightarrow \quad p^{-1}(V) \text{ is open in } X.$$

Dually, if E is a subset of $X/\sim$, then

$$E \text{ is closed in } X/\sim \quad \Leftrightarrow \quad p^{-1}(E) \text{ is closed in } X.$$

To see this, set $Y = X/\sim$. Then E is closed in Y if and only if $Y \backslash E$ is open in Y, that is, $p^{-1}(Y \backslash E)$ is open in X. Now $p^{-1}(Y \backslash E) = X \backslash p^{-1}(E)$, and the latter set is open in X if and only if $p^{-1}(E)$ is closed in X.

Our definition of the quotient topology was a generalization of a single example. Nonetheless, among all topologies on a quotient set, the quotient topology is the most natural one, as the following characterization demonstrates.

3.45 Proposition. Let $p\colon X \to X/\!\sim$ be the quotient map induced by an equivalence relation $\sim$ on a topological space X. Then the quotient topology is the greatest topology on $X/\!\sim$ making p continuous.

Proof. The quotient topology $\Im$ certainly makes p continuous by its very definition. Now let $\mathcal{S}$ be any topology on $X/\!\sim$ that makes p continuous. If $V \in \mathcal{S}$, then $p^{-1}(V)$ is open in X by assumption, so $V \in \Im$ by definition of $\Im$. Hence $\mathcal{S} \subset \Im$.   ☐

The continuous maps on a quotient of a space are determined by the continuous maps on that space.

3.46 Theorem. A map

$$g\colon X/\!\sim\, \to Y$$

from a quotient space $X/\!\sim$ into a topological space Y is continuous if and only if its composite

$$g \circ p\colon X \to Y$$

with the quotient map $p\colon X \to X/\!\sim$ is continuous.

Proof. Since p is continuous, $g \circ p$ is continuous if g is. Conversely, assume $g \circ p$ is continuous. To see that g is then continuous, let W be any open subset of Y. By assumption, the set

$$p^{-1}(g^{-1}(W)) = (g \circ p)^{-1}(W)$$

is open in X. Hence $g^{-1}(W)$ is open in $X/\!\sim$.   ☐

A composite

$$f = g \circ p\colon X \to Y$$

of the kind just considered behaves in a very special way with respect to the equivalence relation $\sim$. If x and t are elements of X belonging to the same equivalence class under $\sim$, that is, if $x \sim t$, then $p(x) = p(t)$ and hence $f(x) = g(p(x)) = g(p(t)) = f(t)$. In short, *f is constant on each equivalence class under* $\sim$. Conversely, a map on X that behaves this way determines a map on $X/\!\sim$ (the map $h\colon [0, 2\pi]/\!\sim\, \to S_1$ was constructed from $f\colon [0, 2\pi] \to S_1$ in this manner at the beginning of the section).

3.47 Theorem. Let $p\colon X \to X/\!\sim$ be the quotient map induced by an equivalence relation $\sim$ on a topological space X. Let $f\colon X \to Y$ be a continuous map from X into a topological space Y that is constant on each equivalence class under $\sim$, that is,

$$x \sim t \quad \Rightarrow \quad f(x) = f(t) \qquad\qquad (x, t \in X).$$

Then there is a unique continuous map

$$f^*\colon X/\!\sim\, \to Y$$

such that

$$f^* \circ p = f.$$

Moreover:

(1) The map f^* is surjective if f is surjective.

(2) The map f^* is injective if f takes distinct values at representatives of different equivalence classes under $\sim$.

(3) The map f^* is open if f^* is injective and the open subsets of Y are those subsets W of Y for which $f^{-1}(W)$ is open in X.

Proof. For convenience, set $Z = X/\sim$. The map f^* is defined by the following rule: If $z \in Z$, then z is an equivalence class under $\sim$, and we define $f^*(z)$ to be $f(x)$ for any representative x of this equivalence class; $f^*(z)$ is determined by z alone, independently of the choice of x, because if x and t are both representatives of z, then $f(x) = f(t)$ by hypothesis. Automatically $f^* \circ p = f$, for if $x \in X$, then x is a representative of the equivalence class $p(x)$ and so $f^*(p(x)) = f(x)$ by the preceding definition.

To establish uniqueness of f^*, suppose $g : Z \to Y$ is any map such that $g \circ p = f$. If z is an arbitrary element of Z, then $z = p(x)$ for some $x \in X$ and so

$$g(z) = g(p(x)) = f(x) = f^*(p(x)) = f^*(z).$$

Hence $g = f^*$. [Observe that everything so far, and the proofs of (1) and (2) below, are purely set theoretic and have nothing to do with topology.]

Continuity of f^* follows from that of f by 3.46.

(1) Assume f is surjective. Let $y \in Y$. Then $y = f(x)$ for some $x \in X$. Hence $z = p(x) \in Z$ with $f^*(z) = f^*(p(x)) = f(x) = y$.

(2) Assume f takes distinct values at representatives of different equivalence classes. Suppose $z_1, z_2 \in Z$ with $f^*(z_1) = f^*(z_2)$. Choose representatives x_1, x_2 of the equivalence classes z_1, z_2 respectively. Then $z_1 = p(x_1)$, $z_2 = p(x_2)$, so

$$f(x_1) = f^*(p(x_1)) = f^*(z_1) = f^*(z_2) = f^*(p(x_2)) = f(x_2).$$

By assumption, x_1 and x_2 must be representatives of the same equivalence class. Hence $z_1 = z_2$.

(3) Assume f^* is injective and the open sets in Y are as stated in (3). Let V be open in Z. To show $f^*(V)$ is open in Y we need therefore only check that $f^{-1}(f^*(V))$ is open in X. To do this we need only show

$$f^{-1}(f^*(V)) = p^{-1}(V),$$

for $p^{-1}(V)$ is open in X by continuity of p.

If $x \in p^{-1}(V)$, then $p(x) \in V$,

$$f(x) = f^*(p(x)) \in f^*(V),$$

and $x \in f^{-1}(f^*(V))$. Conversely, let $x \in f^{-1}(f^*(V))$. Then $f(x) = f^*(v)$

for some $v \in V$, so

$$f^*(p(x)) = f(x) = f^*(v).$$

Now $p(x) = v$ because f^* is injective. Hence $x \in p^{-1}(V)$.   ☐

The map f^* is said to be *induced by f* or to be obtained *by passing to the quotient*. The commutative diagram below summarizes the relationship between p, f, and the induced map f^*. If f is surjective, takes distinct values at

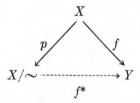

representatives of different equivalence classes, and has the topological behavior stated in (3), we may conclude $f^*\colon X/\!\sim \,\cong Y$. This conclusion suggests the possibility of representing a given quotient space by a "concrete" space homeomorphic to it by constructing a map f exhibiting this behavior. Hence we pause to look at such maps in general, without reference to quotient spaces.

A *quotient map* from a topological space X into a topological space Y is a surjection $f\colon X \to Y$ such that for each $V \subset Y$,

$$V \text{ is open in } Y \quad \Leftrightarrow \quad f^{-1}(V) \text{ is open in } X.$$

According to our definition of a quotient topology, the quotient map $p\colon X \to X/\!\sim$ induced by an equivalence relation $\sim$ on a topological space X is a quotient map in the above sense. For further examples we rely on the following proposition.

3.48 Proposition. Each continuous open surjection and each continuous closed surjection is a quotient map.

Proof. Let $f\colon X \to Y$ be a continuous surjection. Already $f^{-1}(V)$ is open in X for each open subset V of Y.

Assume f is an open map. Let V be a subset of Y with $f^{-1}(V)$ open in X. Then $V = f(f^{-1}(V))$ is open in Y.

Assume next f is a closed map. Again let V be a subset of Y with $f^{-1}(V)$ open in X. Then $X \backslash f^{-1}(V)$ is closed in X, so $f(X \backslash f^{-1}(V))$ is closed in Y. Now $X \backslash f^{-1}(V) = f^{-1}(Y \backslash V)$, so $Y \backslash V = f(f^{-1}(Y \backslash V))$ is closed in Y. Hence V is open in Y.   ☐

At last we are in a position to discuss additional examples of quotient spaces.

$S_1 \cong [0, 1]/\sim$

FIG. 3.14

3.49 Examples

(1) Let $X = [0, 1]$. Identify the endpoints 0 and 1 of X by means of the equivalence relation $\sim$ such that

$$t \sim s \quad \Leftrightarrow \quad (t = 0 \text{ and } s = 1) \quad \text{or} \quad (t = 1 \text{ and } s = 0) \quad \text{or} \quad (t = s).$$

It is geometrically evident that $X/\sim \; \cong \; S_1$ (see Fig. 3.14). To prove this we consider the continuous surjection

$$f \colon X \to S_1$$

$$t \mapsto (\cos 2\pi t, \sin 2\pi t).$$

For $t, s \in X$ we have

$$f(t) = f(x) \quad \Leftrightarrow \quad t \sim s,$$

so f is constant on each equivalence class under $\sim$ and takes distinct values at representatives of different equivalence classes. It remains only to show that f is a quotient map, for then by 3.47 f induces a homeomorphism $f^* \colon X/\sim \; \cong \; S_1$.

We show that f is a closed map; from 3.48 it will follow that f is a quotient map. Let E be a closed subset of X and let $z \in S_1$ with $z \notin f(E)$. We shall find an open arc in S_1 containing z and disjoint from $f(E)$.

Case (i). $z \neq (1, 0)$. Choose $t \in X$ with $f(t) = z$. Then $t \neq 0, 1$ and $t \notin E$, there is an open interval $U \subset \,]0, 1[$ disjoint from E and containing t, and $f(U)$ is the desired open arc.

Case (ii). $z = (1, 0)$. Then $f(0) = z$ and $0 \notin E$, so there is an interval $[0, a[\subset [0, 1]$ disjoint from E; similarly there is an interval $]b, 1] \subset [0, 1]$ disjoint from E. Then $f([0, a[) \cup f(]b, 1])$ is the desired open arc.

In this example and in some others below we resort to a special argument to verify that a continuous surjection $f \colon X \to Y$ is closed. Corollary 4.20 will often provide a shortcut for such a verification: A continuous map from a "compact" space onto a Hausdorff space is necessarily a closed map. Any subspace of $\mathbf{R}^n$ that is closed in $\mathbf{R}^n$ and d-bounded (where d is the Euclidean metric) is compact. The reader should feel free to use these results.

(2) The equivalence relation $\sim$ on $[0, 1]$ above may also be described by

$$t \sim s \quad \Leftrightarrow \quad t - s = -1 \quad \text{or} \quad t - s = 1 \quad \text{or} \quad t = s$$

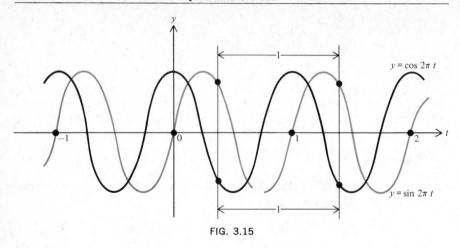

FIG. 3.15

and hence by

$$t \sim s \quad \Leftrightarrow \quad t - s \in \mathsf{Z}.$$

The latter rule also defines an equivalence relation on the real line R which identifies any two points an integral distance apart, and for which

$$t/\sim \; = \; \{. \; . \; . \; , \; t - 2, t - 1, t, t + 1, t + 2, \; . \; . \; .\}$$

for each $t \in \mathsf{R}$.

The quotient space

$$\mathsf{R}/\sim \; \cong \; \mathsf{S}_1.$$

To see this, consider the continuous surjection

$$f : \mathsf{R} \rightarrow \mathsf{S}_1.$$

$$t \mapsto (\cos 2\pi t, \sin 2\pi t).$$

The graphs of the functions $t \mapsto \cos 2\pi t$, $t \mapsto \sin 2\pi t$, shown in Fig. 3.15, are obtained from the graphs of $t \mapsto \cos t$, $t \mapsto \sin t$ by rescaling the t-axis so that they have period 1 instead of period 2π. Inspection of these graphs reveals

$$\cos 2\pi t = \cos 2\pi s \quad \& \quad \sin 2\pi t = \sin 2\pi s \quad \Leftrightarrow \quad t = s + n \text{ for some integer } n,$$

in other words,

$$f(t) = f(s) \quad \Leftrightarrow \quad t \sim s.$$

Thus f is constant on each equivalence class under $\sim$ and takes distinct values at representatives of different equivalence classes. Hence f induces a continuous bijection $f^* : \mathsf{R}/\sim \; \rightarrow \mathsf{S}_1$. The map f is an open map because it carries each open interval in R onto an open arc in S_1. Hence f is a quotient map, and f^* is a homeomorphism.

(3) Let

$$X = \mathsf{I} \times \mathsf{I} = [0, 1] \times [0, 1],$$

the unit square in the plane. Define $\sim$ to be the equivalence relation that identifies two points on the vertical edges of X having the same height.

Thus for (t, s), $(u, v) \in X$,

$$(t, s) \sim (u, v) \qquad \Leftrightarrow \qquad (t, s) = (u, v)$$
$$\text{or} \quad (t = 0, u = 1, s = v)$$
$$\text{or} \quad (t = 1, u = 0, s = v).$$

The simple experiment of rolling a square of paper into a tube leads us to predict that

$$X/\!\sim \; \cong \mathsf{S}_1 \times \mathsf{I},$$

a cylinder (see Fig. 3.16).

To confirm this prediction consider the map

$$g: X \to \mathsf{S}_1 \times \mathsf{I}$$
$$(t, s) \mapsto (f(t), s)$$

where $f: \mathsf{I} \to \mathsf{S}_1$ is the map used in Example (1). Then g is a continuous surjection such that

$$g(t, s) = g(u, v) \qquad \Leftrightarrow \qquad (t, s) \sim (u, v).$$

Hence g induces a continuous bijection $g^*: X/\!\sim \; \to \mathsf{S}_1 \times \mathsf{I}$. To show that g^* is a homeomorphism, it suffices to show that g is a closed map.

Let E be a closed subset of X and let $(z, s) \in \mathsf{S}_1 \times \mathsf{I}$ with $(z, s) \notin g(E)$. We shall find an open neighborhood of (z, s) in $\mathsf{S}_1 \times \mathsf{I}$ that is disjoint from $g(E)$.

Case (i). $(z, s) \notin \{(1, 0)\} \times \mathsf{I}$. Choose $t \in \mathsf{I}$ with $f(t) = z$, so that $g(t, s) = (z, s)$. Then $t \neq 0, 1$ and $(t, s) \notin E$. There is a neighborhood of (t, s) in X of the form $U \times V$, where U is an open interval contained in $]0, 1[$. Then $g(U \times V) = f(U) \times V$, the product of an open arc in S_1 and an open set in I, is the desired neighborhood of (z, s).

Case (ii). $(z, s) \in \{(1, 0)\} \times \mathsf{I}$. Then $f(0) = z = f(1)$, so that $g(0, s) = (z, s) = g(1, s)$. Since $(0, s) \notin E$ and $(1, s) \notin E$, there are intervals $[0, a[$ and $]b, 1]$ contained in $[0, 1]$ and an open neighborhood V of s in I such that $([0, a[\, \cup \,]b, 1]) \times V$ is disjoint from E. The image of this product under g is the desired neighborhood of (z, s).

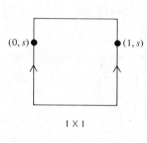

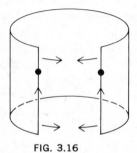

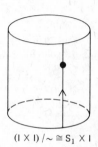

$(0, s) \qquad\qquad (1, s)$

$\mathsf{I} \times \mathsf{I}$

$(\mathsf{I} \times \mathsf{I})/\!\sim \; \cong \mathsf{S}_1 \times \mathsf{I}$

FIG. 3.16

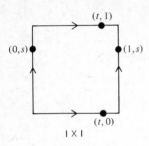
$I \times I$

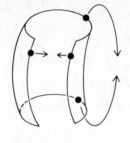

$(I \times I)/{\sim} \cong S_1 \times S_1$

FIG. 3.17

(4) Again let $X = I \times I$. This time the equivalence relation $\sim$ on X is to identify not only points on the vertical edges that were identified in Example (3), but also points on the horizontal edges having the same distance from the left-hand edge. Thus we define $\sim$ by

$$(t, s) \sim (u, v) \qquad \Leftrightarrow \qquad (t, s) = (u, v)$$

$$\text{or} \quad (t = 0, u = 1, s = v)$$

$$\text{or} \quad (t = 1, u = 0, s = v)$$

$$\text{or} \quad (t = u, s = 0, v = 1)$$

$$\text{or} \quad (t = u, s = 1, v = 0)$$

(see Fig. 3.17).

We have

$$X/{\sim} \cong S_1 \times S_1.$$

In fact, the map

$$g \colon X \to S_1 \times S_1,$$

$$(t, s) \mapsto (f(t), f(s)),$$

where $f \colon I \to S_1$ is the map considered in Example (1), is a continuous surjection such that

$$g(t, s) = g(u, v) \qquad \Leftrightarrow \qquad (t, s) \sim (u, v).$$

Hence g induces a continuous bijection $g^* \colon X/{\sim} \to S_1 \times S_1$. One shows that g is a closed map by modifying the argument used in Example (3) [this time, for $(z, w) \notin g(E)$, one must consider separately whether z is or is not $(1, 0)$ and whether w is or is not $(1, 0)$]. Hence g^* is a homeomorphism.

The product $S_1 \times S_1$ to which our quotient is homeomorphic is called the *torus* (or the 2-*torus*, the product of n circles being the n-*torus*). This product $S_1 \times S_1$ is a subspace of $R^2 \times R^2 \cong R^4$. Actually, the torus is homeomorphic to a subspace of R^3, namely, the doughnut-shaped surface T obtained by rotating the circle

$$C = \{ (0, y, z) \in R^3 \mid (y - 2)^2 + z^2 = 1 \}$$

in the yz-plane about the z-axis; the surface T and its generating circle C are

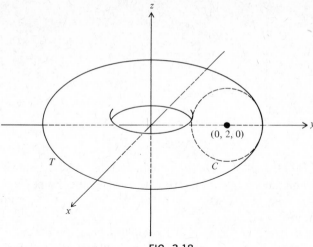

FIG. 3.18

shown in Fig. 3.18. A homeomorphism $h: \mathsf{S}_1 \times \mathsf{S}_1 \cong T$ may be constructed as follows (see Fig. 3.19). Let $(u, v) = ((u_1, u_2), (v_1, v_2)) \in \mathsf{S}_1 \times \mathsf{S}_1$. Identify the first factor S_1 with the circle C so that u is identified with the point $(0, u_1 + 2, u_2) \in C$. Draw the circle C_u passing through this point, parallel to the xy-plane, and having its center on the z-axis. Identify the second factor

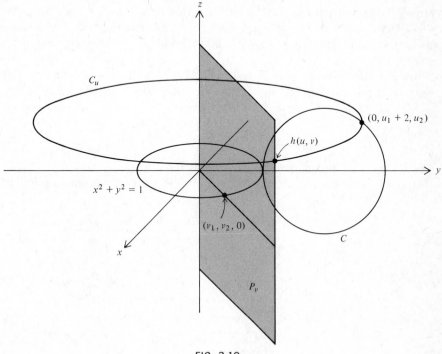

FIG. 3.19

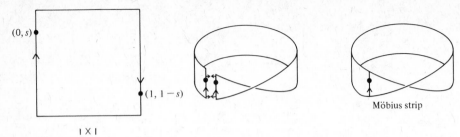

FIG. 3.20

S_1 with the circle $x^2 + y^2 = 1$ in the xy-plane so that v is identified with the point $(v_1, v_2, 0)$ on that circle. Construct the half-plane P_v passing through this point and having the z-axis as an edge. Then $h(u, v)$ is the unique point at which C_u intersects P_v.

(5) Once again let $X = I \times I$. As in Example (3) only points on the vertical edges will now be identified, but the left and right edges will this time be identified in *opposite* directions. Thus we define an equivalence relation $\sim$ on X by

$$(t, s) \sim (u, v) \quad \Leftrightarrow \quad (t, s) = (u, v)$$
$$\text{or} \quad (t = 0, u = 1, v = 1 - s)$$
$$\text{or} \quad (t = 1, u = 0, v = 1 - s).$$

The quotient space $X/\sim$, known as the *Möbius strip*, is homeomorphic to the surface in $\mathbb{R}^3$ shown in Fig. 3.20. A model of this surface can be made by giving a half-twist to a strip of paper and then gluing its ends together. Some of the fascinating and surprising properties of the Möbius strip are discussed in the exercises.

(6) For the final time let $X = I \times I$. This time the equivalence relation $\sim$ on X identifies the vertical edges in opposite directions and identifies the horizontal edges in the same direction, so that

$$(0, s) \sim (1, 1 - s) \qquad\qquad (0 \leq s \leq 1)$$
$$(t, 0) \sim (t, 1) \qquad\qquad (0 \leq t \leq 1)$$

(see Fig. 3.21). The quotient space $X/\sim$ is the *Klein bottle*.

It is a fact that although the Klein bottle is a two-dimensional manifold,

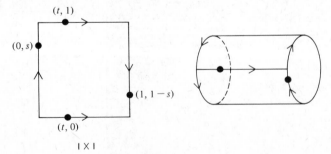

FIG. 3.21

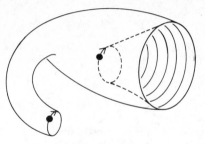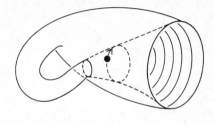

FIG. 3.22

it cannot be homeomorphically embedded in R^3! The reason for this may be understood by inspecting the cylinder in Fig. 3.21, which shows only the top and bottom edges identified; the circular ends of this cylinder must be identified in the directions shown in order to obtain the Klein bottle. We might imagine making the latter identification as suggested in Fig. 3.22 by pushing in the cylinder's right end, grabbing hold of the cylinder's left end, pushing it though the cylinder's side, and then joining the two ends. Of course each point on the small circle at which the cylinder intersects itself on its side represents an identification of two points in the interior of $I \times I$ which are not supposed to be identified under $\sim$. Hence the subspace of R^3 shown at the right in Fig. 3.22 is *not* the Klein bottle. Rather, it is the image of the Klein bottle $X/\sim$ under a "local homeomorphism"—a map that is a homeomorphism when restricted to a suitably small neighborhood of each point of $X/\sim$.

(7) For $n \geq 1$ let $\sim$ be the equivalence relation on the n-sphere $S_n \subset R^{n+1}$ which identifies any pair of antipodal (that is, diametrically opposite) points with each other:

$$x \sim y \quad \Leftrightarrow \quad x = y \quad \text{or} \quad x = -y.$$

The quotient space $S_n/\sim$ is *(real) projective n-space* RP_n. In particular, RP_1 is the *(real) projective line,* and RP_2 is the *(real) projective plane.*

The projective spaces RP_n arise in projective geometry, as we indicate for dimension $n = 2$. Let $p \colon S_2 \to RP_2$ be the quotient map induced by $\sim$. Call the image $p(x) = \{x, -x\}$ of a point $x \in S_2$ a "projective point" and the image $p(C)$ of a great circle $C \subset S_2$ a "projective line." Since any two points of S_2 that are not antipodal lie on a unique great circle, and since any two distinct great circles intersect in one pair of antipodal points (see Fig. 3.23),

FIG. 3.23

FIG. 3.24

we have:

(P1) Any two distinct projective points lie on a unique projective line.
(P2) Any two distinct projective lines intersect at a unique projective point.

Property (P1) is just like the corresponding property of "ordinary" points and lines in the plane R^2 (which, when only the incidence properties of its points and lines are under discussion, should be called the "affine plane"). However, (P2) exhibits a radical difference between projective lines and ordinary lines because two lines in R^2 intersect only if they are not parallel. To explain the difference we first derive another description of RP_2.

The equivalence relation $\sim$ on the sphere S_2 induces an equivalence relation $\simeq$ on its lower hemisphere

$$S_2^- = \{x \in S_2 \mid x_3 \leq 0\}$$

given by

$$x \simeq y \quad \Leftrightarrow \quad x = y \quad \text{or} \quad x = -y.$$

Set

$$P = S_2^- / \simeq.$$

Now two distinct points of S_2^- are antipodal if and only if they are diametrically opposite points on the equator

$$E = \{x \in S_2 \mid x_3 = 0\}$$

(see Fig. 3.24). Hence a point of P is either a point x of S_2^- lying below the equator (actually, a singleton $\{x\}$ for such an x) or else a pair $\{x, -x\}$ of diametrically opposite points on the equator.

A homeomorphism

$$g^*: RP_2 \simeq P$$

is obtained as follows: The map $k: S_2 \to S_2^-$ given by

$$k(x) = \begin{cases} x & \text{if } x \in S_2^-, \\ -x & \text{if } x \notin S_2^- \end{cases}$$

is continuous; its composite

$$g = q \circ k: S_2 \to P$$

with the quotient map $q: S_2^- \to P$ is continuous and is constant on each equivalence class under $\sim$. Hence g induces a continuous map

$$g^*: RP_2 = S_2/\sim \to S_2^-/\simeq = P.$$

Similarly, the composite $f = p \circ j \colon S_2^- \to RP_2$ of the inclusion map $j \colon S_2^- \to S_2$ with the quotient map $p \colon S_2 \to RP_2$ induces a continuous map $f^* \colon P \to RP_2$, and we can readily check that $g^* \circ f^*$ and $f^* \circ g^*$ are identity maps. Hence g^* is a homeomorphism (and f^* is its inverse).

Let us now refer to P itself as the projective plane, to the points of P as projective points, and to the images of projective lines in RP_2 under g^* as projective lines. Then each projective line in P has the form $q(C \cap S_2^-)$ for some great circle C in S_2. This follows from the commutativity of the diagram

$$
\begin{array}{ccc}
 & k & \\
S_2 & \longrightarrow & S_2^- \\
p \downarrow & & \downarrow q \\
 & g^* & \\
RP_2 & \longrightarrow & P
\end{array}
$$

and the fact that k maps each great circle C of the sphere S_2 onto its intersection $C \cap S_2^-$ with the lower hemisphere.

Let us also work not with R^2 itself, but instead with the plane

$$H = \{x \in R^3 \mid x_3 = -1\}$$

in R^3 to which it is homeomorphic under the map

$$(x_1, x_2) \qquad \longmapsto \qquad (x_1, x_2, -1).$$

This map carries lines in R^2 onto lines in H, so R^2 and H are geometrically, as well as topologically, the same.

To each $x \in H$ we associate the point $r(x) \in S_2^- \setminus E$ at which the line joining x to the origin $(0, 0, 0)$ meets S_2^- (see Fig. 3.25). The resulting map

$$r \colon H \to S_2^-$$

is a homeomorphic embedding. It is easy to see that q maps the subspace $S_2^- \setminus E = r(H)$ of S_2^- homeomorphically onto the subspace $P \setminus q(E)$ of P. Hence

$$h = q \circ r \colon H \to P$$

is a homeomorphic embedding of the plane H into the projective plane P.

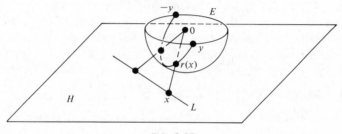

FIG. 3.25

What is really of interest here is the geometric, not the topological, behavior of h. A simple geometric argument shows that the image $r(L)$ of any line $L \subset H$ has the form

$$r(L) = (C \cap S_2^-)\backslash\{y, -y\}$$

where C is a great circle and y, $-y$ are the diametrically opposite points of C on E (see Fig. 3.25); hence $h(L)$ is a projective line in P with a single projective point $i(L) = \{y, -y\}$ removed. Thus the projective plane can be obtained from the ordinary (affine) plane by adjoining to each ordinary line L a single "ideal" projective point $i(L)$, and then $L \cup \{i(L)\}$ is a projective line. Besides these projective lines there is the single "ideal" projective line $q(E)$ which is the set of all ideal projective points.

For additional information about projective geometry, consult Birkhoff and MacLane [2, Chap. IX, Sec. 14] or Rosenbaum [29].

(8) In Example (1) we identified distinct points of the space $[0, 1]$ only when they belonged to its subset $\{0, 1\}$. We generalize that example.

Let A be a nonempty closed subset of a topological space X. Define $\sim$ to be the equivalence relation on X that identifies all points of A with each other:

$$x \sim y \quad \Leftrightarrow \quad (x = y) \quad \text{or} \quad (x \in A \quad \text{and} \quad y \in A).$$

Then the points of the quotient set $X/\sim$ are the singletons $\{x\}$ for $x \notin A$ together with the distinguished point A (note that $A = a/\sim \in X/\sim$ for each $a \in A$). The quotient space $X/\sim$, sometimes denoted by X/A, is said to be obtained by *collapsing A to a point*.

Let

$$q\colon X \to X/A$$

be the quotient map. Then q maps the open subspace $X \backslash A$ of X onto the complement $(X/A)\backslash\{A\}$ of the point A in the quotient space. This complement is open in X/A because $X \backslash A$, its inverse image under q, is open in X. Moreover, q induces a homeomorphism

$$X \backslash A \cong (X/A)\backslash\{A\}.$$

In fact, q is a continuous map whose restriction to $X \backslash A$ is injective; if U is open in $X \backslash A$, then $q^{-1}(q(U)) = U$ is open in X whence $q(U)$ is open in X/A. Thus X/A contains a point whose complement is topologically the same as the portion $X \backslash A$ of X in which no identifications are made.

As a specific example take $X = D_2$, the 2-disk in the plane, and $A = S_1$, its boundary in the plane. To see what D_2/S_1 looks like, imagine a circular piece of cloth with a drawstring around its edge; when the string is drawn tight, a spherical bag results (see Fig. 3.26). Thus we should expect

$$D^2/S_1 \cong S_2.$$

To find a homeomorphism $h\colon D_2/S_1 \cong S_2$ we need only construct a continu-

FIG. 3.26

ous closed surjection $f: D_2 \to S_2$ that sends each point of S_1 to the north pole p of S_2 and maps $D_2 \setminus S_1$ injectively onto $S_2 \setminus \{p\}$, for then we may take h to be the induced map f^*. One such f, indicated in Fig. 3.27, is the composite of two maps. The first is the homeomorphism $(x_1, x_2) \mapsto (\pi x_1, \pi x_2, -1)$ from D_2 onto the disk

$$D = \{x \in \mathsf{R}^3 \mid x_1{}^2 + x_2{}^2 \le \pi, x_3 = -1\}$$

tangent to S_2 at the south pole. The second wraps each radial segment of D upward onto a meridian of S_2 so that an $x \in D$ with cylindrical coordinates $(r, \theta, -1)$ is sent to the point on S_2 having spherical coordinates $(\rho, \phi, \theta) = (1, \pi - r, \theta)$. That f is actually a closed map is most easily established by the shortcut mentioned at the end of Example (1). (Another way of constructing the desired f, more amenable to generalization to higher dimensions, is indicated in Exercise 104.)

(9) Let X and Y be topological spaces and let $f: A \to Y$ be a continuous map on a nonempty closed subset A of X. Imagine joining X and Y together by gluing each point $a \in A$ to the point $f(a) \in Y$ (continuity of f means that nearby points in A are glued to nearby points in Y). The result should be a topological space Z which contains (homeomorphic copies of) $X \setminus A$ and Y and in which each $y \in f(A)$ represents an identification of all $a \in f^{-1}(y)$ with y. We shall show that such a space Z actually exists.

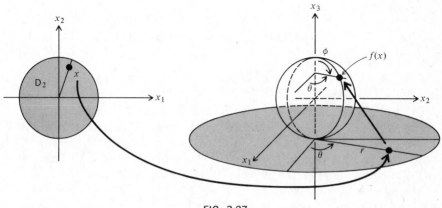

FIG. 3.27

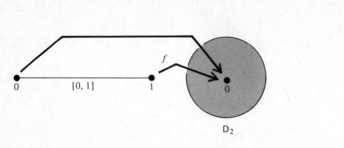

FIG. 3.28

Let us temporarily make the simplifying assumption that X is disjoint from Y. Then the set $X \cup Y$, endowed with the topology constructed in the first paragraph of Example 3.22 (3), contains both X and Y as open, closed subspaces. Let $\sim$ be the equivalence relation on $X \cup Y$ given by

$$u \sim v \quad \Leftrightarrow \quad (u = v) \quad \text{or} \quad (u \in A \ \& \ v = f(u)) \quad \text{or} \quad (v \in A \ \& \ u = f(v))$$

$$\text{or} \quad (u \in A \ \& \ v \in A \ \& \ f(u) = f(v)).$$

Finally, take Z to be $(X \cup Y)/\sim$. We use the notation

$$X \cup_f Y = (X \cup Y)/\sim$$

and say that $X \cup_f Y$ is *obtained by attaching X to Y by f; f* is the *attaching map*.

We show that $X \cup_f Y$ has the desired properties. Let

$$p : X \cup Y \to X \cup_f Y$$

be the quotient map. Clearly

$$p(a) = p(f(a)) \qquad\qquad (a \in A),$$

$$X \cup_f Y = p(X) \cup p(Y) = p(X \backslash A) \cup p(A) \cup p(Y),$$

$$p(X) \cap p(Y) = p(A) = p(f(A)).$$

We claim, moreover, that:

(i) $p(X \backslash A)$ is open in $X \cup_f Y$, and p maps $X \backslash A$ homeomorphically onto $p(X \backslash A)$.

(ii) $p(Y)$ is closed in $X \cup_f Y$, and p maps Y homeomorphically onto $p(Y)$.

The set $p(Y)$ is closed because $p^{-1}(p(Y)) = A \cup Y$ is closed in $X \cup Y$. The map $p \mid Y$ is continuous and injective; if E is closed in Y, then $p(E)$ is closed in $p(Y)$ because $p^{-1}(p(E)) = A \cup E$ is closed in $X \cup Y$. This proves (ii). The proof of (i) is similar—work with open sets.

As a specific example, take $X = [0, 1]$, $A = \{0, 1\}$, $Y = D_2$, and $f : A \to Y$ the map such that $f(0) = f(1) = 0 \in D_2$. Then $X \cup_f Y$ is homeomorphic to the subspace

$$W = \{x \in \mathbf{R}^3 \mid x_3 = 0, x_1{}^2 + x_2{}^2 \leq 1\} \ \cup$$

$$\{x \in \mathbf{R}^3 \mid x_1 = 0, x_2{}^2 + (x_3 - 1)^2 = 1\}$$

of $\mathbf{R}^3$ shown in Fig. 3.28. To prove this, consider the map $g \colon X \cup Y \to W$ such that

$$g(x) = (0, \sin 2\pi x, 1 - \cos 2\pi x) \qquad\qquad (x \in X)$$

$$g(y) = (y_1, y_2, 0) \qquad\qquad (y \in Y)$$

and pass to the quotient.

To construct $Z = X \cup_f Y$ in the general case that X is not necessarily disjoint from Y, the procedure above must be modified. Form the sum $X + Y$ and the embeddings

$$h \colon X \to X + Y, \qquad k \colon Y \to X + Y$$

as in 3.22(3). Let $\sim$ be the equivalence relation on $X + Y$ given by

$$u \sim v \qquad \Leftrightarrow \qquad u = v$$

$$\text{or} \quad [u = h(a) \text{ for some } a \in A, v = k(f(a))]$$

$$\text{or} \quad [v = h(a) \text{ for some } a \in A, u = k(f(a))]$$

$$\text{or} \quad [u = h(a), v = h(b) \text{ for } a, b \in A \text{ with } f(a) = f(b)].$$

Finally, set

$$X \cup_f Y = (X + Y)/\sim.$$

If $p \colon X + Y \to X \cup_f Y$ is the quotient map, then

$$X \cup_f Y = p \circ h(X) \cup p \circ k(Y) = p \circ h(X \backslash A) \cup p \circ h(A) \cup p \circ k(Y),$$

$$p \circ h(X) \cap p \circ k(Y) = p \circ h(A) = p \circ k(f(A)),$$

$p \circ h$ embeds $X \backslash A$ as an open subspace of $X \cup_f Y$, and $p \circ k$ embeds Y as a closed subspace of $X \cup_f Y$.

EXERCISES

89. Let $p \colon X \to X/\sim$ be the quotient map induced by an equivalence relation $\sim$ on a space X. Suppose $\mathfrak{I}$ is a topology on $X/\sim$ such that p is continuous with respect to $\mathfrak{I}$ and such that an arbitrary map $g \colon X/\sim \to Y$ is continuous with respect to $\mathfrak{I}$ precisely when its composite $g \circ p \colon X \to Y$ is continuous. Must $\mathfrak{I}$ be the quotient topology?

90. (**a**) Let $f \colon X \to Y$ be a continuous map. Let $\sim$ and $\simeq$ be equivalence relations on X and Y, respectively, such that $x \sim t$ always implies $f(x) \simeq f(t)$. Use f to construct a continuous map $f' \colon X/\sim \to X/\simeq$.

(**b**) If f is a quotient map, show that f' is a quotient map.

91. What can be said about a quotient space $X/\sim$ if $\sim$ is the equality relation $=$ on X?

92. Let X be the space $[0, 1]$, let Y be the Sierpinski space $[2.3(6)]$, and let $f: X \to Y$ be the characteristic function of $[\frac{1}{2}, 1]$. Verify that f is a quotient map that is neither open nor closed.

93. Let $f: X \to Y$ be an arbitrary quotient map. The *equivalence kernel* of f is the equivalence relation $\sim$ on X given by

$$x \sim t \iff f(x) = f(t).$$

Obviously f is constant on each equivalence class under $\sim$, so f induces a map $f^*: X/\!\!\sim \, \to Y$ such that $f^* \circ p = f$, where $p: X \to X/\!\!\sim$ is the quotient map induced by $\sim$. Prove that $f^*: X/\!\!\sim \, \cong Y$. (Thus any "abstract" quotient map is essentially the same as the "concrete" quotient map induced by an equivalence relation.)

94. Find a quotient space of the plane $\mathbf{R}^2$ that is homeomorphic to the torus.

95. Show that the torus is a 2-dimensional manifold.

96. (a) Let $\sim$ and $\sim^*$ be two equivalence relations on a space X such that $x \sim^* u$ always implies $x \sim u$. Prove that $\sim$ induces an equivalence relation $\simeq$ on the quotient $Y = X/\!\!\sim^*$ such that $Y/\!\!\simeq$ is homeomorphic to $X/\!\!\sim$.
(b) Figure 3.17 suggests that the torus can be obtained from a cylinder by identifying the ends of the cylinder. Show that this is indeed the case by applying (a) with $X = \mathsf{I} \times \mathsf{I}$ and $\sim$ the equivalence relation of 3.49 (4).
(c) Show similarly that the Klein bottle is a quotient space of a cylinder.

97. The product $\mathsf{S}_1 \times \mathsf{D}_2$ is the *solid torus*.
(a) Embed the solid torus as a solid in $\mathbf{R}^3$ which is bounded by the surface T described in 3.49(4).
(b) Represent the 3-sphere S_3 as the union of two subspaces X_1 and X_2 each of which is homeomorphic to the solid torus such that $X_1 \cap X_2$ is homeomorphic to the torus.

98. Let M be the Möbius strip. The *edge* of M is, in the notation of 3.49(5), the image E of the set $\mathsf{I} \times \{0, 1\}$ under the quotient map $p: \mathsf{I} \times \mathsf{I} \to (\mathsf{I} \times \mathsf{I})/\!\!\sim \, = M$.
(a) Show that $E \cong \mathsf{S}_1$.
(b) Prove that M is a 2-dimensional manifold-with-boundary (Exercise 3.47) with $E = \partial M$ (Exercise 3.48).

99. (Continuation of Exercise 98.) Given a space X and points $x, y \in X$, a *path in X from x to y* is a continuous map $\sigma: \mathsf{I} \to X$ such that $\sigma(0) = x$, $\sigma(1) = y$ [see Section 3 of Chapter 5 for more about paths].
(a) If $x, y \in M \setminus E$, construct a path σ in $M \setminus E$ from x to y. [Thus we can move continuously from any point not on the edge of M to any other such point without ever crossing the edge. *Hint*: Choose $a \in p^{-1}(x)$, $b \in p^{-1}(y)$, construct a suitable map $\tau: \mathsf{I} \to \mathsf{I} \times \mathsf{I}$, and pass to the quotient.]
(b) Let $K = p(\mathsf{I} \times \{\frac{1}{2}\})$. If x, y are arbitrary points in $M \setminus K$, must there exist a path in $M \setminus K$ from x to y? (*Project*: Represent M by a twisted strip of paper with its ends glued together. Draw the curve representing K. What happens when you cut the paper along this curve?)

100. Noting that each pair of antipodal points on S_n determines a unique line in $\mathbf{R}^{n+1}$ through the origin 0, show that $\mathsf{RP}_n \cong (\mathbf{R}^{n+1} \setminus \{0\})/\!\!\simeq$, where $x \simeq y$ exactly when x and y lie on the same line in $\mathbf{R}^{n+1}$ through the origin.

101. (**a**) Show that $RP_n \cong D_n/\!\sim$, where $\sim$ identifies each pair of antipodal points of $S_{n-1} \subset D_n$.

(**b**) Deduce that the projective line $RP_1 \cong S_1$.

102. Use the map

$$f: S_2 \to R^4$$

$$(x_1, x_2, x_3) \mapsto (x_1{}^2 - x_2{}^2, x_1 x_2, x_1 x_3, x_2 x_3)$$

to induce an embedding of the projective plane into R^4.

103. Starting with the embedding

$$f: S_n \to S_{n+1}$$

$$x \mapsto (x_1, \ldots, x_{n+1}, 0),$$

use Exercise 90 to construct an embedding $f': RP_n \to RP_{n+1}$ such that $RP_{n+1} \backslash f'(RP_n) \cong B_n$.

104. Let $g: B_2 \cong R^2$ be the homeomorphism of 3.16(7), and let $k: R^2 \cong S_2 \backslash \{p\}$ be the inverse of the stereographic projection $[3.16(11)]$, where p is the north pole. Then $k \circ g$ defines an embedding of B_2 into S_2 which may be extended to a continuous map $f: D_2 \to S_2$ by taking $f(x) = p$ for all $x \in S_1$. Using this data, demonstrate anew that $D_2/S_1 \cong S_2$. (See Exercise 4.72 for a generalization.)

105. (**a**) Show that the space obtained from the cylinder $S_1 \times I$ by collapsing its top $S_1 \times \{1\}$ to a point is homeomorphic to the conical surface K in R^3 having vertex $(0, 0, 1)$ and base $\{x \in R^3 \mid x_3 = 0, x_1{}^2 + x_2{}^2 = 1\}$.

(**b**) If, instead, the bottom $S_1 \times \{0\}$ is collapsed to a point, is the space obtained still homeomorphic to K?

106. Use the map $(x, t) \mapsto tx$ from $S_{n-1} \times I$ into D_n to induce a homeomorphism $(S_{n-1} \times I)/(S_{n-1} \times \{0\}) \cong D_n$.

107. Let A be a nonempty closed subset of a space X that is both regular (Exercise 2.60) and T_0 (Exercise 2.32). Prove that X/A is a Hausdorff space. [*Hint*: Let $x, y \in X$ with $q(x) \neq q(y)$, where $q: X \to X/A$ is the quotient map. Consider separately the cases (i) $x \notin A$ and $y \notin A$, (ii) $x \in A$ or $y \in A$.]

108. Show that the sphere with a whisker attached, shown in Fig. 3.8, is homeomorphic to the space $I \cup_f S_2$, where $f: \{0\} \to S_2$ sends 0 to the north pole.

109. Given points a and b in spaces X and Y, respectively, the *one-point union* of X with Y is the space $X \cup_f Y$, where $f: \{a\} \to Y$ takes the value b. Show that for any choices of $a, b \in S_1$, the one-point union of S_1 with itself is homeomorphic to a "figure eight".

110. Let A be a nonempty closed subset of a space X. Show that X/A can be obtained by attaching X to a one-point space Y by a constant map $f: A \to Y$.

111. If $f: S_{n-1} \to D_n$ is the inclusion map, to what familiar space is $D_n \cup_f D_n$ homeomorphic?

112. Let M be the Möbius strip and let ∂M be its edge (Exercise 98). If $f: \partial M \to M$ is the inclusion map, show that $M \cup_f M$ is homeomorphic to the Klein bottle.

113. The *suspension* $S(X)$ of a topological space X is the quotient $(X \times [-1, 1])/\!\sim$, where $\sim$ identifies all points of $X \times \{-1\}$ with one another and all points of $X \times \{1\}$ with one another.

(**a**) Verify that $S(S_n) \cong S_{n+1}$.

(**b**) Prove that, in general, $S(X)$ can be obtained by attaching $X \times [-1, 1]$ to a two-point space by means of a suitable map.

114. Represent the sphere as $S_2 = D^+ \cup C \cup D^-$, where D^+ is the part above and on the plane $x_3 = \frac{1}{2}$, D^- is the part below and on the plane $x_3 = -\frac{1}{2}$, and C is the part between and on these two planes. Let $p \colon S_2 \to RP_2$ be the quotient map. Verify:

(**a**) $D^+ \cong D_2$, $C \cong S_1 \times I$.

(**b**) $p \mid D^+$ is an embedding.

(**c**) $p(C) \cong M$, the Möbius strip, and $p(C^+) \cong \partial M$, the edge of M (Exercise 98), where C^+ is the intersection of C with the plane $x_3 = \frac{1}{2}$.

(**d**) $RP_2 = p(C) \cup p(D^+)$, $p(C) \cap p(D^+) \cong S_1$.

115. (Continuation of Exercise 114.) Deduce that the projective plane RP_2 can be obtained by attaching the disk D_2 to the Möbius strip by an embedding $f \colon S_1 \to M$ with $f(S_1) = \partial M$.

116. Let $\sim$ be an equivalence relation on a topological space X. For $E \subset X$, the *saturation of E by* $\sim$ is the union of the equivalence classes of all points of E. In each case below compute the saturations of the specified subsets E of $X = I \times I$ by the given relation $\sim$.

(**a**) Let $\sim$ collapse $\{0\} \times I$ to a point. Let E be an arbitrary subset of X, but consider separately the cases that E does or does not intersect $\{0\} \times I$.

(**b**) Let $\sim$ collapse $]0, 1] \times I$ to a point. Let E be an arbitrary subset of X, but consider separately the cases that E is or is not contained in $\{0\} \times I$.

(**c**) Let $x \sim u$ if and only if $x = u$ or $x_1 = u_1 \in R \backslash Q$. First let $E = I \times [0, \frac{1}{2}[$; then let $E = I \times [0, \frac{1}{2}]$.

117. (Continuation of Exercise 116.) The equivalence relation $\sim$ is said to be *open* (respectively, *closed*) when the saturation of each open (respectively, closed) subset of X by $\sim$ is itself open (respectively, closed) in X. Determine which of the equivalence relations in (a)–(c) above are open and which are closed.

118. (Continuation of Exercise 117.)

(**a**) Show that $\sim$ is open (respectively, closed) precisely when the quotient map $p \colon X \to X/\sim$ is an open (respectively, a closed) map. (*Hint*: Express the saturation of $E \subset X$ in terms of p.)

(**b**) Show that $\sim$ is closed if and only if for each equivalence class A under $\sim$ and each open set U in X with $A \subset U$, there is some saturated open set V in X with $A \subset V \subset U$.

119. (Continuation of Exercise 118.) Prove:

(**a**) The quotient $X/\sim$ is a T_1-space (Exercise 2.10) if and only if the equivalence class of each $x \in X$ is closed in X.

(**b**) The quotient $X/\sim$ is a Hausdorff space if the equivalence relation $\sim$ is open and its graph $\{(x, y) \in X \times X \mid x \sim y\}$ is a closed subset of $X \times X$.

(**c**) The quotient $X/\sim$ is a Hausdorff space if $\sim$ is both open and closed and the space X is both regular (Exercise 2.60) and T_0 (Exercise 2.32). [*Hint*: Use 119(b). *Note*: Exercise 107 provides a criterion for a quotient space to be Hausdorff even when the hypotheses of (b) and (c) need not hold.]

120. The space $X = \{x \in R \mid x > 0\}$ is first-countable, second-countable, and metrizable. Show, however, that its quotient space X/A, where $A = \{1, 2, 3, \ldots\}$, has none of these three properties. (*Note:* Topological properties that *are* preserved under the formation of quotient spaces are considered in the next two chapters.)

121. Let G be a topological group (Exercise 3.84) and let H be a subgroup of G. Let G/H denote the collection $\{gH \mid g \in G\}$ of left cosets of G in H, where $gH = \{gh \mid h \in H\}$ for each $g \in G$. Then G/H is a partition of G whose associated equivalence relation $\sim$ is given by

$$x \sim y \quad \Leftrightarrow \quad xH = yH.$$

Let $p: G \to G/H = G/\!\sim$ be the quotient map. Prove:

(**a**) The equivalence relation $\sim$ is open (Exercise 117).

(**b**) If $x \in G$, then the neighborhood system at $p(x) = xH$ in the quotient space G/H is

$$\{p(U) \mid U \text{ open in } G, \, x \in UH\}.$$

Here $UH = \{uh \mid u \in U, h \in H\}$.

(**c**) The space G/H is *homogeneous* in the sense that for any two points $z, w \in G/H$ there is some homeomorphism $f: G/H \cong G/H$ with $f(z) = w$.

122. (Continuation of Exercise 121.) What can be said about the space G/H if H is open in G? if H is closed in G? if H is dense in G?

123. (Continuation of Exercise 121.) Suppose H is a normal subgroup of G, so that G/H is a group under the operation $(xH) \cdot (yH) = (xy)H$. Show that the quotient topology makes G/H into a topological group. As an example, show that R/Z is, as a topological group, the same as the circle group [Exercise 3.84(e)] by constructing a homeomorphism from R/Z onto the circle group which is also a group isomorphism.

5. Convergence

Having generalized the notion of continuity from the setting of metric spaces to that of topological spaces, we now do the same with the notion of sequential convergence. Our definition is based on Theorem 1.51, which said that convergence of a sequence in a metric space (X, d) depends only on d-neighborhoods and hence only on neighborhoods defined by the topology induced by d.

3.50 Definition. Let $(x_n \mid n \in \mathsf{N})$ be a sequence of points of a topological space X. If $x \in X$, then we say that $(x_n \mid n \in \mathsf{N})$ *converges to x in X* when for each neighborhood V of x in X there exists some $m \in \mathsf{N}$ with

$$n \geq m \quad \Rightarrow \quad x_n \in V.$$

We say that $(x_n \mid n \in \mathsf{N})$ *converges in X* when it converges to some x in X.

Since a sequence converges to a point in a metric space precisely when it converges to the same point in the associated topological space, we already have an adequate supply of examples from Chapter 1. The first thing to do, then, is to generalize some of the properties of sequential convergence to arbitrary topological spaces. To do so in the absence of a metric, however, requires some topological assumptions.

3.51 Theorem. Let x be a point and let A be a subset of a first-countable space X. Then $x \in \operatorname{cls} A$ if and only if some sequence of points of A converges to x in X.

Proof. Assume there is a sequence $(x_n \mid n \in \mathsf{N})$ in A that converges to x in X. If V is any neighborhood of x, there is an $m \in \mathsf{N}$ with $x_m \in V$, so that $x_m \in V \cap A$ and $V \cap A \neq \varnothing$. Hence $x \in \operatorname{cls} A$. (First-countability was not used here.)

Conversely, assume $x \in \operatorname{cls} A$. Choose a sequence $(V_n \mid n \in \mathsf{N})$ of subsets of X such that $\{V_n \mid n \in \mathsf{N}\}$ is a local base at x. For each $n \in \mathsf{N}$ the set $V_0 \cap \cdots \cap V_n$ is a neighborhood of x, so we may choose some

$$x_n \in A \cap (V_0 \cap \cdots \cap V_n).$$

Then the sequence $(x_n \mid n \in \mathsf{N})$ of points of A converges to x in X. In fact, if V is any neighborhood of x, then

$$V_m \subset V$$

for some $m \in \mathsf{N}$, and hence

$$n \geq m \quad \Rightarrow \quad x_n \in V_0 \cap \cdots \cap V_n \subset V_m \subset V. \quad \square$$

The first part of the corollary below is our generalization of Theorem 1.55.

3.52 Corollary. Let A be a subset of a first-countable space X. Then:
(1) The set A is closed in X if and only if each point of X to which some sequence of points of A converges in X itself belongs to A.
(2) The set A is open in X if and only if each sequence $(x_n \mid n \in \mathsf{N})$ that converges in X to a point of A is "eventually in A" in the sense that for some $m \in \mathsf{N}$ we have $x_n \in A$ for all $n \geq m$.

Proof. (1) The set A is closed in X if and only if $\operatorname{cls} A \subset A$. Now use 3.51.
(2) The set A is open in X if and only if $X \setminus A$ is closed in X. Now apply (1) to $X \setminus A$ in place of A. $\square$

The knowledge of which sequences converge to which points thus completely determines the topology of a first-countable space. Hence sequential convergence should determine which functions between first-countable spaces are continuous (compare 1.56).

3.53 Theorem. Let X and Y be first-countable spaces. Then a necessary and sufficient condition for a map $f\colon X \to Y$ to be continuous at a point $x \in X$ is that for each sequence $(x_n \mid n \in \mathsf{N})$ converging to x in X, the sequence $(f(x_n) \mid n \in \mathsf{N})$ must converge to $f(x)$ in Y.

Proof. Necessity. Assume f is continuous at x. Suppose $(x_n \mid n \in \mathsf{N})$ converges to x in X. Let V be any neighborhood of $f(x)$ in Y. Then $f^{-1}(V)$

is a neighborhood of x in X, so there is an $m \in \mathsf{N}$ with $x_n \in f^{-1}(V)$ for all $n \geq m$. Hence $f(x_n) \in V$ for all $n \geq m$.

Sufficiency. Assume the condition holds. Suppose there is some neighborhood V of $f(x)$ in Y such that $f^{-1}(V)$ is not a neighborhood of x in X. Choose a decreasing sequence $(U_n \mid n \in \mathsf{N})$ of sets in X such that $\{U_n \mid n \in \mathsf{N}\}$ is a local base at x. Then for each $n \in \mathsf{N}$ we have $U_n \not\subseteq f^{-1}(V)$, so there is some $x_n \in U_n$ with $x_n \notin f^{-1}(V)$. Clearly $(x_n \mid n \in \mathsf{N})$ converges to x, so by assumption $(f(x_n) \mid n \in \mathsf{N})$ converges to $f(x)$. Then $f(x_m) \in V$ for some m. Hence $x_m \in f^{-1}(V)$, which is impossible.   ☐

Now that we have shown how easily the definition of sequential convergence generalizes to topological spaces and how nicely sequential convergence works in first-countable spaces, we have the sad duty to report that sequential convergence is inadequate for describing the topology and continuous maps of arbitrary topological spaces.

3.54 Example. Let X be an infinite set provided with its countable-complement topology (Exercise 2.6), so that a subset U of X is open if and only if either $U = \varnothing$ or $X \setminus U$ is countable.

It is easy to determine all convergent sequences in X. If $(x_n \mid n \in \mathsf{N})$ is a sequence that is "eventually constant"—that is, there is an $x \in X$ and an $m \in \mathsf{N}$ with

$$(*) \qquad\qquad\qquad x_n = x \qquad\qquad\qquad (n \geq m)$$

—then obviously $(x_n \mid n \in \mathsf{N})$ converges to x in X. Conversely, suppose $(x_n \mid n \in \mathsf{N})$ converges to a point x in X. Since the set $\{x_n \mid n \in \mathsf{N}, x_n \neq x\}$ is countable, its complement

$$V = X \setminus \{x_n \mid n \in \mathsf{N}, x_n \neq x\}$$

is an open neighborhood of x in X. Then there exists an $m \in \mathsf{N}$ with $x_n \in V$ for all $n \geq m$, so $(*)$ holds.

Now suppose X is uncountable. Choose any $y \in X$ and let

$$A = X \setminus \{y\}.$$

By the preceding paragraph, no sequence in A can converge to y. Nonetheless,

$$y \in \operatorname{cls} A.$$

In fact, if V is any neighborhood of y, then $X \setminus V$ is countable, the set $(X \setminus V) \cup \{y\}$ is also countable, the set

$$V \cap A = V \cap (X \setminus \{y\}) = X \setminus [(X \setminus V) \cup \{y\}]$$

is uncountable since X is uncountable, and hence $V \cap A$ is nonempty.

The set A is not closed in X because it is not countable (or because $y \in \operatorname{cls} A$ whereas $y \notin A$). Nonetheless, if a sequence $(x_n \mid n \in \mathsf{N})$ of points of A converges in X to a point x, then $x \in A$ because $x = x_n$ for all sufficiently large n.

The set $\{y\}$ is not open because its complement A is not countable (or because A is not closed). Nonetheless, if a sequence $(x_n \mid n \in \mathsf{N})$ converges to the unique point y of this set, then $x_n = y$ and so $x_n \in \{y\}$ for all sufficiently large n.

Let Y be the set X provided with its discrete topology and let $f: X \to Y$ be the identity map. Then f is not continuous because the inverse image $\{y\} = f^{-1}(\{y\})$ of the open subset $\{y\}$ of Y is not open in X. Nonetheless, if a sequence $(x_n \mid n \in \mathsf{N})$ converges to a point x in X, then there is an m with $x_n = x$ for all $n \geq m$, and $(f(x_n) \mid n \in \mathsf{N}) = (x_n \mid n \in \mathsf{N})$ converges to $f(x) = x$ in Y [see Example 1.50 (5)].

Thus in the absence of first-countability the properties 3.51–3.53 of sequential convergence cease to hold. The intuitive notion of sequential convergence —points getting closer and closer to a given point—is so appealing, however, that it is desirable to find a more general notion than sequential convergence for which analogs of 3.51–3.53 do hold in arbitrary topological spaces. Just such a notion was found by E. H. Moore and H. L. Smith (but with different motivation), who in 1922 initiated the theory of convergence of nets.

The reason sequential convergence is inadequate is simply that a sequence has only countably many values, whereas in an arbitrary topological space it may take uncountably many neighborhoods of a point to determine all neighborhoods of the point. Hence the thing to do is to replace the countable set N which indexes sequences by more general index sets. Like N, these sets will need some kind of ordering but, it turns out, not necessarily a total ordering.

3.55 Definition. A *directed set* is a set I together with a relation $\leq$ in I to I, called a *direction of I*, such that

$$i \in I \quad \Rightarrow \quad i \leq i,$$

$$i, j, k \in I \quad \text{and} \quad i \leq j \quad \text{and} \quad j \leq k \quad \Rightarrow \quad i \leq k,$$

$$i, j \in I \quad \Rightarrow \quad \text{there exists a } k \in I \text{ with } i \leq k \text{ and } j \leq k.$$

A relation with these three properties is also said to *direct I*, and I is said to be *directed by* the relation. As usual, we write $j \geq i$ to mean $i \leq j$.

The first two of these properties are the same reflexivity and transitivity possessed by a total ordering, and the third is a substitute for comparability (0.34).

3.56 Examples

(1) Any totally ordered set is directed by its total ordering. In particular, N is a directed set under its usual ordering.

(2) Let x be a point in a topological space X. The neighborhood system

$\mathfrak{N}_x$ at x is directed by the relation $\leq$ of "reverse inclusion":

$$U \leq V \quad \Leftrightarrow \quad U \supset V \qquad (U, V \in \mathfrak{N}_x).$$

In fact, this relation is clearly reflexive and transitive; if $U, V \in \mathfrak{N}_x$, then $W = U \cap V \in \mathfrak{N}_x$ with $U \leq W$ and $V \leq W$. This direction of $\mathfrak{N}_x$ is called the *natural direction*.

Notice that comparability does not hold for this direction.

Whenever the neighborhood system at a point in a space is considered a directed set, it will be with respect to its natural direction.

(3) Again let x be a point in a space X. Let

$$I_x = \{ (U, t) \mid U \in \mathfrak{N}_x, t \in U \}.$$

Then the relation $\leq$ defined by

$$(U, t) \leq (V, z) \quad \Leftrightarrow \quad U \supset V$$

is a direction of I_x, called the *natural direction*. Whenever such a set I_x is considered a directed set, it will be with respect to this natural direction.

3.57 Definition. A *net* is a family indexed by some directed set. In accordance with the terminology for arbitrary families, a net $(x_i \mid i \in I)$ is *in* a set A when

$$i \in I \quad \Rightarrow \quad x_i \in A.$$

A net $(x_i \mid i \in I)$ is said to be *eventually in* a set A when

there exists $i \in I$ such that $j \in I$ and $j \geq i \Rightarrow x_j \in A$.

Sequences are special kinds of nets. To define convergence for nets we just mimic the definition for sequences.

3.58 Definition. Let $(x_i \mid i \in I)$ be a net in a topological space X. If $x \in X$, we say that $(x_i \mid i \in I)$ *converges to* x *(in X)* and write

$$(x_i \mid i \in I) \to x$$

or even

$$x_i \to x$$

when $(x_i \mid i \in I)$ is eventually in each neighborhood of x. We say that $(x_i \mid i \in I)$ *converges (in X)* when it converges to some point x in X.

Obviously a sequence $(x_n \mid n \in \mathsf{N})$ converges in the sense of Definition 3.50 to a point x in a topological space X precisely when it converges in the sense of this definition to x in X.

3.59 Examples

(1) Let $(x_i \mid i \in I)$ be a net in a space X that is eventually constant, that

is, there is an $x \in X$ and an $i \in I$ with

$$x_j = x \qquad\qquad (j \in I, j \geq i).$$

Then $(x_i \mid i \in I) \to x$.

(2) Let $\mathfrak{N}_x$ be the neighborhood system at a point x of a topological space X. Suppose $(x_U \mid U \in \mathfrak{N}_x)$ is any net indexed by $\mathfrak{N}_x$ (with its natural direction) such that

$$x_U \in U \qquad\qquad (U \in \mathfrak{N}_x).$$

Then

$$x_U \to x.$$

In fact, if U is any neighborhood of x, then

$$V \in \mathfrak{N}_x \quad \text{and} \quad V \geq U \quad \Rightarrow \quad x_V \in V \subset U.$$

(3) Let x be a point in a topological space X. Let I_x be the directed set of Example 3.56 (3). Then $(t \mid (U, t) \in I_x)$ is a net that converges to x in X.

(4) Let a and b be real numbers with $a < b$ and let

$$f \colon [a, b] \to \mathsf{R}$$

be a real-valued function on the closed interval $[a, b]$ which is bounded, that is, whose range has both a lower and an upper bound in R. We shall describe the Riemann integral of f by means of convergence of nets.

For any $n \geq 1$, any $(n + 1)$-tuple $(x_0, \ldots, x_n)$ of real numbers such that

$$a = x_0 < x_1 < \cdots < x_{n-1} < x_n = b$$

will be called a *partition* of $[a, b]$ (this is not a partition of $[a, b]$ in the set-theoretic sense of 0.53). Form the collection $\mathcal{P}$ of all such partitions of $[a, b]$. For partitions

$$P = (x_0, \ldots, x_n), \qquad Q = (y_0, \ldots, y_m)$$

of $[a, b]$ define

$$P \leq Q$$

to mean that Q *refines* P in the sense that

$$\{x_0, \ldots, x_n\} \subset \{y_0, \ldots, y_m\},$$

in other words, Q is obtained from P by inserting additional points between points of P. If $P, Q \in \mathcal{P}$, then by arranging in increasing order all the points of both P and Q we obtain a partition $R \in \mathcal{P}$ with $P \leq R$ and $Q \leq R$. Thus $\mathcal{P}$ is directed by the relation $\leq$.

Let

$$P = (x_0, \ldots, x_n) \in \mathcal{P}.$$

For each $i = 1, \ldots, n$ the set

$$f([x_{i-1}, x_i]) = \{f(x) \mid x_{i-1} \leq x \leq x_i\}$$

is a nonempty set of real numbers which, by our assumption that f is bounded,

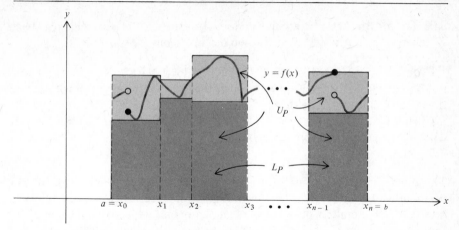

FIG. 3.29

has both a lower bound and an upper bound, so it is meaningful to form

$$m_i = \inf f([x_{i-1}, x_i]), \qquad M_i = \sup f([x_{i-1}, x_i]).$$

Then we may form the lower Riemann sum

$$L_P = \sum_{i=1}^{n} m_i(x_i - x_{i-1})$$

and the upper Riemann sum

$$U_P = \sum_{i=1}^{n} M_i(x_i - x_{i-1})$$

of f with respect to P. In geometric terms, if f is nonnegative, L_P is the area under the step function whose value is m_i on the ith interval $[x_{i-1}, x_i]$, and U_P is the area under the step function whose value is M_i on the ith interval (see Fig. 3.29).

We now have two nets, $(L_P \mid P \in \mathcal{P})$ and $(U_P \mid P \in \mathcal{P})$, in $\mathbf{R}$. If there is a number c with

$$L_P \to c, \qquad U_P \to c,$$

then f is *Riemann integrable*, and the *Riemann integral of f* is

$$\int_a^b f(x) \, dx = c.$$

When f is nonnegative, this number is interpreted as the area under the graph of f.

Other descriptions of the Riemann integral appear in the exercises.

By proving generalizations of 3.51–3.53 we now demonstrate that nets are indeed adequate for determining the topologies and continuous maps of arbitrary topological spaces.

3.60 Theorem. Let x be a point and let A be a subset of a topological space X. Then $x \in \operatorname{cls} A$ if and only if some net in A converges to x in X.

Proof. Assume there is a net $(x_i \mid i \in I)$ in A with $x_i \to x$. If V is any neighborhood of x, there is an $i \in I$ such that $x_j \in V$ for all $j \in I$ with $j \geq i$; then $x_i \in V$ since $i \geq i$, so that $x_i \in V \cap A$, and $V \cap A \neq \emptyset$. Hence $x \in \operatorname{cls} A$.

Conversely, assume $x \in \operatorname{cls} A$. Then each neighborhood U of x intersects A, and we may choose some point

$$x_U \in U \cap A.$$

The points so chosen constitute a net $(x_U \mid U \in \mathfrak{N}_x)$ in A indexed by the neighborhood system $\mathfrak{N}_x$ at x. According to 3.59 (2) this net converges to x in X. (The simultaneous choice of all the points x_U is an implicit application of the axiom of choice which may be avoided by a slightly more complicated construction—see Exercise 131.) ▯

In the same way we deduced 3.52 from 3.51 we can deduce from 3.60 the following criterion.

3.61 Corollary. Let A be a subset of a topological space X. Then:
(1) The set A is closed in X if and only if each point of X to which some net in A converges in X itself belongs to A.
(2) The set A is open in X if and only if each net that converges in X to a point of A is eventually in A.

3.62 Theorem. Let X and Y be topological spaces. Then a necessary and sufficient condition for a map $f: X \to Y$ to be continuous at a point $x \in X$ is that for each net $(x_i \mid i \in I)$ in X,

$$x_i \to x \quad \Rightarrow \quad f(x_i) \to f(x).$$

Proof. Necessity is proved as in 3.53. For sufficiency, assume the condition holds. Suppose there is a neighborhood V of $f(x)$ in Y such that $f^{-1}(V)$ is not a neighborhood of x in X. Then for each neighborhood U of x in X we may choose a point $x_U \in U$ with $x_U \notin f^{-1}(V)$. As in 3.59 (2) we obtain a net $(x_U \mid U \in \mathfrak{N}_x)$ that converges to x in X. By assumption, $f(x_U) \to f(x)$ in Y, so $f(x_U) \in V$ for some $U \in \mathfrak{N}_x$. Then $x_U \in f^{-1}(V)$, which contradicts the way x_U was chosen. ▯

In an arbitrary space there is nothing to prohibit a net, even a sequence, from converging to two different points.

3.63 Example. Let X be an infinite set provided with its finite-complement topology [2.3(7)]. Let $(x_i \mid i \in \mathsf{N})$ be any sequence in X with $x_i \neq x_j$ whenever $i \neq j$. Then

$$x \in X \quad \Rightarrow \quad x_i \to x.$$

In fact, if $x \in X$ and V is any neighborhood of x, then $X \setminus V$ is finite whereas $\{x_i \mid i \in \mathsf{N}\}$ is infinite, there exists $i \in \mathsf{N}$ with $x_j \notin X \setminus V$ for all $j \geq i$, and hence $x_j \in V$ for all $j \geq i$.

The net in the preceding example behaved in the worst possible way—it converged to *every* point of the space. In a wide class of spaces, however, a net cannot converge to more than one point.

3.64 Theorem. Let $(x_i \mid i \in I)$ be a net in a *Hausdorff* space X and let $x, y \in X$ with
$$x_i \to x, \qquad x_i \to y.$$
Then
$$x = y.$$

Proof. Suppose $x \neq y$. There are disjoint neighborhoods U and V of x and y, respectively. By hypothesis there exists an $i \in I$ with
$$k \in I \quad \text{and} \quad k \geq i \quad \Longrightarrow \quad x_k \in U$$
and a $j \in I$ with
$$k \in I \quad \text{and} \quad k \geq j \quad \Longrightarrow \quad x_k \in V.$$
Since I is directed, there is a $k \in I$ with $k \geq i$ and $k \geq j$. Then $x_k \in U \cap V$, which is impossible since $U \cap V = \varnothing$. $\qquad \Box$

This theorem guarantees, for example, that in Example 3.59 (3) there is only one number c such that $L_P \to c$ and $U_P \to c$, and hence that the Riemann integral of a Riemann integrable function is uniquely defined. The theorem also justifies the following definition.

3.65 Definition. Let $(x_i \mid i \in I)$ be a net that converges in a Hausdorff space X. Then the unique $x \in X$ for which
$$x_i \to x$$
is called the *limit of* $(x_i \mid i \in I)$ and is denoted by
$$\lim (x_i \mid i \in I)$$
or by
$$\lim_{i \in I} x_i$$
or even by
$$\lim x_i.$$

Recall from Example 1.53 that a sequence in the product space R^k converges exactly when it converges "coordinatewise". The same thing is true for a net in any product space. (Owing to our continued use of 'I' in this section for the directed set indexing a net, in the theorem below we depart from the convention of Section 3 and denote the index set of a family of spaces by 'A' and elements of A by lower-case Greek letters.)

3.66 Theorem. Let $(x_i \mid i \in I)$ be a net in the product X of a family $(X_\alpha \mid \alpha \in A)$ of topological spaces, so that

$$x_i = (x_{i\alpha} \mid \alpha \in A) \qquad\qquad (i \in I).$$

Let

$$y = (y_\alpha \mid \alpha \in A)$$

be a point of X. Then

$$(x_i \mid i \in I) \to y$$

in X if and only if

$$(x_{i\alpha} \mid i \in I) \to y_\alpha$$

in X_α for each $\alpha \in A$.

Proof. Assume $x_i \to y$ in X. Let $\alpha \in A$. The projection $p_\alpha: X \to X_\alpha$ is continuous, so by 3.62

$$(x_{i\alpha} \mid i \in I) = (p_\alpha(x_i) \mid i \in I) \to p_\alpha(y) = y_\alpha$$

in X_α.

Conversely, assume $x_{i\alpha} \to y_\alpha$ for each $\alpha \in A$. Let V be any neighborhood of y in X. Then V contains some basic open set

$$U = \underset{\alpha \in A}{\text{X}}\, U_\alpha$$

to which y belongs. Let

$$B = \{\beta \in A \mid U_\beta \neq X_\beta\}.$$

If $B = \varnothing$, then $V = U = X$ and surely $x_i \in V$ for all i. Suppose $B \neq \varnothing$. For each $\beta \in B$ the set U_β is a neighborhood of y_β in X_β, so by assumption there is an index $i_\beta \in I$ such that

$$j \in I \quad \text{and} \quad j \geq i_\beta \qquad \Rightarrow \qquad x_{j\beta} \in U_\beta.$$

Since B is finite and I is directed, there is an $i \in I$ with

$$i \geq i_\beta \qquad\qquad (\beta \in B).$$

Then

$$j \in I \quad \text{and} \quad j \geq i \qquad \Rightarrow \qquad x_{j\beta} \in U_\beta \quad \text{for each } \beta \in B$$

$$\qquad\qquad\qquad\qquad \Rightarrow \qquad x_j \in U.$$

Hence $x_j \in V$ for all $j \geq i$. This proves that $x_i \to y$. ☐

Although the product spaces treated in Section 3 were generally formed from families of spaces indexed by subsets of N, there is nothing in the statement or proof of 3.66 that needs to be altered in order to apply to products indexed by arbitrary sets (see 3.41). In particular, since the topology of pointwise convergence on the set of functions from one space to another is a product topology (see 3.42), Theorem 3.66 tells us which nets converge to which functions for this topology. It is nevertheless instructive to examine convergence for the topology of pointwise convergence directly, without reference to product spaces.

3.67 Examples

(1) Let $\mathfrak{F}$ be the set of all functions from a topological space X to a topological space Y. Recall from Exercise 2.71—and from Example 2.52 (3) in the case $X = Y = \mathsf{R}$—that the topology of pointwise convergence has as a base the collection of all sets of the form

$$B(x_1, \ldots, x_n; V_1, \ldots, V_n) = \{ f \in \mathfrak{F} \mid f(x_k) \in V_k \ (k = 1, \ldots, n) \},$$

where $x_1, \ldots, x_n \in X$ and $V_1, \ldots, V_n$ are open sets in Y.

Let $f \in \mathfrak{F}$ and let $(f_i \mid i \in I)$ be a net in $\mathfrak{F}$. We claim that

$$(f_i \mid i \in I) \to f \text{ in } \mathfrak{F}$$

for the topology of pointwise convergence if and only if $(f_i \mid i \in I)$ *converges pointwise to f* in the sense that

$$x \in X \quad \Rightarrow \quad (f_i(x) \mid i \in I) \to f(x) \text{ in } Y.$$

To verify our claim, assume first $f_i \to f$. Let $x \in X$. If V is any open neighborhood of $f(x)$ in Y, then $B(x; V)$ is an open neighborhood of f in $\mathfrak{F}$, by assumption there is an $i \in I$ with $f_j \in B(x, V)$ for all $j \geq i$, and hence $f_j(x) \in V$ for all $j \geq i$. Thus $f_i(x) \to f(x)$ in Y. Conversely, assume $f_i(x) \to f(x)$ in Y for each $x \in X$. Consider any open neighborhood W of f of the form

$$W = B(x_1, \ldots, x_n; V_1, \ldots, V_n)$$

as above. Then for each $k = 1, \ldots, n$

$$f(x_k) \in V_k,$$

and since $f_i(x_k) \to f(x_k)$ in Y, there is an $i_k \in I$ such that

$$j \geq i_k \quad \Rightarrow \quad f_j(x_k) \in V_k.$$

Since I is directed, there is an $i \in I$ with

$$i \geq i_k \qquad\qquad (k = 1, \ldots, n).$$

Then

$$j \geq i \quad \Rightarrow \quad f_j(x_k) \in V_k \ \text{ for } k = 1, \ldots, n$$

$$\Rightarrow \quad f_j \in W.$$

Hence $f_i \to f$ in $\mathfrak{F}$.

(2) Let X, Y, and $\mathfrak{F}$ be as in (1), and suppose the topology of Y is induced by a bounded metric d. Then for functions f, g belonging to $\mathfrak{F}$ the set $\{ d(f(x), g(x)) \mid x \in X \}$ of nonnegative real numbers has an upper bound,

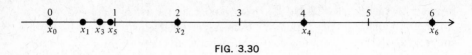

FIG. 3.30

so it is meaningful to form

$$d_\infty(f, g) = \sup_{x \in X} d(f(x), g(x)).$$

This formula defines a metric d_∞ on $\mathfrak{F}$ (compare Exercise 1.12 and the discussion preceding Theorem 1.69), and the topology on $\mathfrak{F}$ (and on any subset of $\mathfrak{F}$) induced by d_∞ is called the *topology of uniform convergence*. Note that

$$d_\infty(f, g) \leq \epsilon \quad \Leftrightarrow \quad d(f(x), g(x)) \leq \epsilon \quad \text{for all } x \in X.$$

A net $(f_i \mid i \in I)$ will converge to a function f in $\mathfrak{F}$ for this topology precisely when $(f_i \mid i \in I)$ is eventually in each d_∞-ball of f. Hence

$$(f_i \mid i \in I) \to f$$

for the topology of uniform convergence if and only if $(f_i \mid i \in I)$ *converges uniformly to f* in the sense that

for each $\epsilon > 0$ there exists an $i \in I$ such that $d(f_j(x), f(x)) < \epsilon$ for all $j \geq i$ and all $x \in X$.

Uniform convergence in the above sense is exactly the same uniform convergence used for sequences of functions in advanced calculus. Compare what it means for $(f_i \mid i \in I)$ to converge pointwise to f:

For each $\epsilon > 0$ and for each $x \in X$ there exists an $i \in I$ such that $d(f_j(x), f(x)) < \epsilon$ for all $j \geq i$.

Thus for pointwise convergence the value of i depends both on the given ϵ and the particular point x, whereas for uniform convergence the value of i depends solely on the given ϵ, so that the same value of i works for all points $x \in X$.

The sequence $(x_n \mid n \in \mathsf{N})$ in R given by

(*)
$$x_n = \begin{cases} n & \text{if } n \text{ is even,} \\ 1 - \dfrac{1}{n+1} & \text{if } n \text{ is odd} \end{cases}$$

does not converge in R, yet its values do "pile up" at the point 1 (see Fig. 3.30). In order to study nets behaving the same way as this sequence, we introduce next the notion of clustering, which is weaker than convergence. In the sequel, clustering will be applied only to sequences in metric spaces (in Section 2 of Chapter 4).

3.68 Definition. A net $(x_i \mid i \in I)$ is *frequently in* a set A when

for each $i \in I$ there exists a $j \in I$ such that $j \geq i$ and $x_j \in A$.

A net $(x_i \mid i \in I)$ in a topological space X *clusters at* a point x *(in X)* when it is frequently in each neighborhood of x, and $(x_i \mid i \in I)$ *clusters (in X)* when it clusters at some point of X.

Roughly speaking, a net $(x_i \mid i \in I)$ clusters at a point x when for each neighborhood V of x there are arbitrarily large indices $j \in I$ with $x_j \in I$. For a sequence in a metric space, clustering in the sense of this definition is the same as clustering in the sense of Exercise 1.65.

Although in a Hausdorff space a net cannot converge to more than one point (Theorem 3.64), it can cluster at more than one point: the sequence

$$x_n = \begin{cases} 0 & \text{if } n \text{ is even,} \\ 1 - \dfrac{1}{n} & \text{if } n \text{ is odd} \end{cases}$$

clusters at both 0 and 1 in R.

If a net converges to a point, then it clusters at that point. The converse is false: the sequence (*) above is a counterexample. Nevertheless, if a net $(x_i \mid i \in I)$ clusters at a point x, there will be some net which converges to x and whose values are among the values of $(x_i \mid i \in I)$. We shall demonstrate this fact first for sequences in a first-countable space.

3.69 Theorem. Let x be a point and $(x_n \mid n \in \mathsf{N})$ be a sequence in a first-countable space X. Then $(x_n \mid n \in \mathsf{N})$ clusters at x in X if and only if some subsequence of $(x_n \mid n \in \mathsf{N})$ converges to x in X.

By a *subsequence of* $(x_n \mid n \in \mathsf{N})$ we mean here a sequence of the form $(x_{n_j} \mid j \in \mathsf{N})$, where $(n_j \mid j \in \mathsf{N})$ is a sequence of natural numbers with

$$n_0 < n_1 < \cdots < n_j < n_{j+1} < \cdots .$$

Proof of 3.69. Assume first that $(x_n \mid n \in \mathsf{N})$ has a subsequence $(x_{n_j} \mid j \in \mathsf{N})$ converging to x. Let V be any neighborhood of x and let $i \in \mathsf{N}$. By assumption there is a $k \in \mathsf{N}$ such that

$$j \geq k \quad \Rightarrow \quad x_{n_j} \in V.$$

Since $\{ j \in \mathsf{N} \mid j \geq k \}$ is infinite whereas $\{ j \in \mathsf{N} \mid n_j < i \}$ is only finite, there is a $j \in \mathsf{N}$ with $j \geq k$ and $n_j \geq i$. Then $x_{n_j} \in V$.

Assume, conversely, that $(x_n \mid n \in \mathsf{N})$ clusters at x. Let $(V_n \mid n \in \mathsf{N})$ be a sequence of open sets in X such that $\{V_n \mid n \in \mathsf{N}\}$ is a local base at x. By replacing each V_n by $V_0 \cap \cdots \cap V_n$ if necessary, we may assume that

$$V_0 \supset V_1 \supset \cdots \supset V_n \supset V_{n+1} \supset \cdots .$$

Construct a subsequence $(x_{n_j} \,|\, j \in \mathbf{N})$ of $(x_n \,|\, n \in \mathbf{N})$ as follows. Choose $n_0 \in \mathbf{N}$ with

$$x_{n_0} \in V_0.$$

Next, choose $n_1 \in \mathbf{N}$ with

$$n_1 > n_0, \qquad x_{n_1} \in V_1.$$

In general, once $n_0, \ldots, n_{j-1}$ have been chosen, choose $n_j \in \mathbf{N}$ with

$$n_j > n_{j-1}, \qquad x_{n_j} \in V_j.$$

All these choices are possible because $(x_n \,|\, n \in \mathbf{N})$ clusters at x. The sequence $(x_{n_j} \,|\, j \in \mathbf{N})$ converges to x, for if V is any neighborhood of x, there is a $k \in \mathbf{N}$ such that $V_k \subset V$, and then

$$j \geq k \quad \Rightarrow \quad x_{n_j} \in V_j \subset V_k \subset V. \qquad \square$$

It is this theorem that will be used in Chapter 4; *nothing in the remainder of this section is needed for the sequel.*

Once the hypothesis that X be first-countable is removed from Theorem 3.69, it is no longer true that a sequence that clusters at some point must have a subsequence converging to that point—see Exercise 144. To see what *is* true we will have to generalize the notion of subsequence.

Let $(x_n \,|\, n \in \mathbf{N})$ be a sequence. A subsequence $(x_{n_j} \,|\, j \in \mathbf{N})$ of $(x_n \,|\, n \in \mathbf{N})$ can also be denoted by $(x_{\sigma(j)} \,|\, j \in \mathbf{N})$ by letting $\sigma(j) = n_j$ for each $j \in \mathbf{N}$. Then a subsequence of $(x_n \,|\, n \in \mathbf{N})$ is just a net $(y_j \,|\, j \in \mathbf{N})$ having the same index set $\mathbf{N}$ as the original sequence and having the form

$$y_j = x_{\sigma(j)} \qquad\qquad (j \in \mathbf{N})$$

where

$$\sigma : \mathbf{N} \to \mathbf{N}$$

is a strictly increasing map. By dropping the requirement that the index sets be the same and by modifying the requirement that σ be strictly increasing, we arrive at the proper generalization of subsequences.

3.70 Definition. Let $(x_i \,|\, i \in I)$ be a net. A *subnet of* $(x_i \,|\, i \in I)$ is a net $(y_j \,|\, j \in J)$ such that

$$y_j = x_{\sigma(j)} \qquad\qquad (j \in J)$$

for some map

$$\sigma : J \to I$$

having the properties:

(SN1) For each $j, j' \in J$, if $j \leq j'$, then $\sigma(j) \leq \sigma(j')$.
(SN2) For each $i \in I$, there exists a $j \in J$ with $\sigma(j) \geq i$.

The relationship between a net $(x_i \,|\, i \in I)$ in a set X and a subnet $(y_j \,|\, j \in J)$ of that net is indicated in Fig. 3.31. Although the directed set that indexes the subnet need not be the same as the one that indexes the net, the values of the subnet are among the values of the net. Property (SN2) says

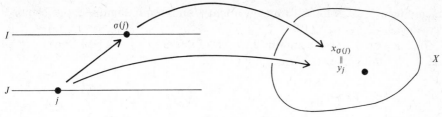

FIG. 3.31

that there are arbitrarily large indices $i \in I$ for which x_i is one of the values of the subnet.

A strictly increasing map $\sigma : \mathsf{N} \to \mathsf{N}$ certainly has property (SN1). It has property (SN2) as well, for if $i \in \mathsf{N}$, then the infinite set $\{\sigma(j) \mid j \in \mathsf{N}\}$ cannot be contained in the finite set $\{0, \ldots, i\}$. Hence a subsequence $(x_{\sigma(j)} \mid j \in \mathsf{N})$ of a sequence $(x_i \mid i \in \mathsf{N})$ is indeed a subnet of $(x_i \mid i \in \mathsf{N})$.

We can now establish the generalization of Theorem 3.69.

3.71 Theorem. Let x be a point and $(x_i \mid i \in I)$ be a net in a topological space X. Then $(x_i \mid i \in I)$ clusters at x in X if and only if some subnet of $(x_i \mid i \in I)$ converges to x in X.

Proof. Assume first that $(x_i \mid i \in I)$ has a subnet $(y_j \mid j \in J)$ converging to x. Just suppose $(x_i \mid i \in I)$ does not cluster at x. Then x has a neighborhood V such that $(x_i \mid i \in I)$ is not frequently in V, so $(x_i \mid i \in I)$ is eventually in $X \setminus V$. It follows that $(y_j \mid j \in J)$ is also eventually in $X \setminus V$, so $(y_j \mid j \in J)$ is not eventually in V. This contradicts the assumption that $(y_j \mid j \in J)$ converges to x.

Conversely, assume that $(x_i \mid i \in I)$ clusters at x. Define

$$J = \{(i, V) \mid i \in I, V \text{ is a neighborhood of } x, x_i \in V\}.$$

Then J is directed by

$$(i, V) \le (i', V') \qquad \Leftrightarrow \qquad i \le i' \text{ and } V \supset V'.$$

Define

$$\sigma : J \to I$$

by the formula

$$\sigma(i, V) = i.$$

The map σ clearly satisfies (SN1), and it satisfies (SN2) because if $i \in I$, then $(i, X) \in J$ and $\sigma(i, X) = i \ge i$. Thus $(x_i \mid (i, V) \in J)$ is a subnet of $(x_i \mid i \in I)$. It is a routine matter to verify that this subnet converges to x. $\quad \square$

The final topic involving convergence that we want to look at is the notion of limit of a function. Our aim here is not to prove any profound theorems, but merely to show that the diverse kinds of limits we meet in calculus—

including one-sided limits, limits at infinity, and infinite limits—are subsumed by one general notion of limit.

In calculus we often compute the limit of a function f at a point not belonging to the domain of f. For example,

$$\lim_{x \to 0} \frac{\sin x}{x} = 1,$$

but 0 does not belong to the domain of $\sin x/x$. Hence we will want to define the limit of a function $f: D \to Y$ at a point x of a space X containing the domain D of f even when $x \notin D$. The idea of the definition below is to call a point $y \in Y$ the limit of f at x if $f(t)$ can be made arbitrarily close to y by taking $t \in D$ sufficiently close to but different from x. Then it is reasonable to suppose there *are* points of D other than x itself that are arbitrarily close to x, that is, to suppose $x \in \text{cls}\,(D \backslash \{x\})$.

3.72 Definition. Let

$$f: D \to Y$$

be a map from a subset D of a topological space X into a topological space Y, and let $x \in X$ with

$$x \in \text{cls}\,(D \backslash \{x\}).$$

A point $y \in Y$ is said to be a *limit of f at x* when for each net $(x_i \mid i \in I)$ in $D \backslash \{x\}$

$$x_i \to x \quad \Rightarrow \quad f(x_i) \to y.$$

Suppose Y is a Hausdorff space. Since $x \in \text{cls}\,(D \backslash \{x\})$, there is at least one net $(x_i \mid i \in I)$ in $D \backslash \{x\}$ converging to x in X. If points y and z of Y are both limits of f at x, then $f(x_i) \to y$ and $f(x_i) \to z$, so $y = z$ by 3.64. Hence if some $y \in Y$ is a limit of f at x, it is unique, and then we call y *the* limit of f at x and denote it by

$$\lim_{t \to x} f(t).$$

To guarantee uniqueness of limits, *we assume below that Y is a Hausdorff space.*

When $x \in D$, the value of f at x has nothing to do with the limit of f at x: if $g: D \to Y$ is another map such that

$$t \in D \backslash \{x\} \quad \Rightarrow \quad f(t) = g(t),$$

then f has a limit at x precisely when g does, and in this case

$$\lim_{t \to x} f(t) = \lim_{t \to x} g(t).$$

When $x \notin D$, then

$$x \in \text{cls}\,(D \backslash \{x\}) \quad \Leftrightarrow \quad x \in \text{cls}\, D.$$

The following criterion, whose proof involves only minor modifications in the proof of 3.62, shows that Definition 3.72 does generalize the notion of limit familiar from calculus.

3.73 Proposition. Let $f: D \to Y$, x, and X be as in 3.72. Let $y \in Y$. Then a necessary and sufficient condition for

$$y = \lim_{t \to x} f(t)$$

to hold is that

> for each neighborhood V of y in Y, there exists some neighborhood U of x in X such that
> $$x \neq t \in U \cap D \quad \Rightarrow \quad f(t) \in V.$$

When $D = X$,

$$x \in \operatorname{cls}(D \setminus \{x\}) \quad \Leftrightarrow \quad \{x\} \text{ is not open in } X$$

(equivalently, in the language of Exercise 2.33, x is not an isolated point of X). Then just as in calculus we have the following corollary.

3.74 Corollary. Let $f: X \to Y$ be a map from a topological space X to a topological space Y. Let $x \in X$ such that $\{x\}$ is not open in X. Then f is continuous at x if and only if

$$\lim_{t \to x} f(t) = f(x).$$

To conclude our consideration of convergence we exhibit some limits from calculus as instances of our general definition.

3.75 Examples

(1) Let f be a real-valued function on a subset D of R, let $x \in \mathsf{R}$, and suppose

$$]a, x[\ \cup \ ]x, b[\ \subset D$$

for some $a < x < b$. Then $x \in \operatorname{cls}(D \setminus \{x\})$. By 3.73,

$$y = \lim_{t \to x} f(t)$$

if and only if

> for each $\epsilon > 0$ there exists a $\delta > 0$ such that
> $$t \in D \quad \text{and} \quad 0 < |x - t| < \delta \quad \Rightarrow \quad |f(t) - y| < \epsilon.$$

This δ-ϵ condition is the usual calculus definition of $y = \lim_{t \to x} f(t)$.

(2) Let g be a real-valued function on a subset E of R, let $x \in \mathsf{R}$, and

suppose

$$]a, x[\subseteq E$$

for some $a < x$. Take

$$D = E \cap]\leftarrow, x[, \qquad f = g \mid D.$$

Then $x \in \mathrm{cls}\,(D\backslash\{x\})$ since $]a, x[\subseteq D$. By 3.73,

$$y = \lim_{t \to x} f(t)$$

if and only if

for each $\epsilon > 0$ there exists a $\delta > 0$ such that
$$t \in E \quad \text{and} \quad x - \delta < t < x \quad \Rightarrow \quad |f(t) - y| < \epsilon.$$

Hence

$$\lim_{t \to x} f(t) = \lim_{t \to x-} g(t),$$

the left-hand limit of g at x.

A similar discussion holds for

$$\lim_{t \to x+} g(t),$$

the right-hand limit of g at x, when $]x, b[$ is contained in the domain of g for some $b > x$.

(3) Let f be a real-valued function on a subset D of R that has no upper bound in R. Take

$$X = \mathsf{R}^*,$$

the extended real line (1.37). According to Example 2.33,

$$+\infty \in \mathrm{cls}\,D = \mathrm{cls}\,(D\backslash\{+\infty\}).$$

By Lemma 1.39, a subset of X containing $+\infty$ is a neighborhood of $+\infty$ precisely when it contains some ray $]u, \to[$. Then by 3.73, a real number y satisfies

$$y = \lim_{t \to +\infty} f(t)$$

if and only if

for each $\epsilon > 0$ there exists a $u \in \mathsf{R}$ such that
$$t \in D \quad \text{and} \quad t > u \quad \Rightarrow \quad |f(t) - y| < \epsilon.$$

Hence $\lim_{t \to +\infty} f(t)$ in the sense given by Definition 3.72 is the usual limit of f at $+\infty$.

A similar discussion holds for

$$\lim_{t \to -\infty} f(t)$$

when the domain D of f has no lower bound in R.

(4) Let g be a real-valued function on a subset D of R, let $x \in \mathsf{R}$, and suppose as in (1) that

$$]a, x[\cup]x, b[\subseteq D$$

for some $a < x < b$. Form the map

$$f: D \to \mathsf{R}^*$$

having the same graph as the given g by enlarging the codomain of g to the extended real line R^*. By 3.73,

$$+\infty = \lim_{t \to x} f(t)$$

if and only if

for each $u \in \mathsf{R}$ there exists a $\delta > 0$ such that
$t \in D$ and $|t - x| < \delta \quad \Rightarrow \quad g(t) > u$.

Hence $+\infty$ is the limit of f at x precisely when $+\infty$ is the limit of g at x in the usual calculus sense.

A similar discussion holds for

$$\lim_{t \to x} g(t) = -\infty.$$

Additional examples appear in the exercises.

EXERCISES

124. Which sequences converge to which points in the following spaces?
   (**a**) The real line with its right-interval topology [2.18(1)].
   (**b**) The line with two origins [2.18(3)].
   (**c**) The half-disk space [2.20(3)].

125. Let $f: X \to Y$ be a map from a first-countable space X to a topological space Y. Using 3.52(2) but not the definition of convergence or Theorem 3.53, prove: f is continuous if and only if for each $x \in X$ and each sequence $(x_n \mid n \in \mathsf{N})$ converging to x in X, the sequence $(f(x_n) \mid n \in \mathsf{N})$ converges to $f(x)$.

126. Describe the interior of a set in a first-countable space by means of sequences and sequential convergence.

127. Let D be a subset of a topological space X and let $x \in X$ with $x \in \mathrm{cls}(D \backslash \{x\})$. Define

$$I = \{(U, t) \mid U \text{ is a neighborhood of } x \text{ in } X, x \neq t \in U \cap D\}.$$

Show that I is directed by

$$(U, t) \leq (V, z) \quad \Leftrightarrow \quad U \supset V$$

and that the net $(t \mid (U, t) \in I)$ converges to x in X.

128. Let $f: [a, b] \to \mathsf{R}$ be a bounded real-valued function on a closed interval $[a, b]$ in R. Let $\mathcal{P}$ be the directed set of all partitions of $[a, b]$ in the sense of 3.59(3).
   (**a**) If $P, Q \in \mathcal{P}$ with $P \leq Q$, show that $L_P \leq L_Q$ and $U_Q \leq U_P$.
   (**b**) If $P, Q \in \mathcal{P}$, show that $L_P \leq U_Q$.
   (**c**) Prove that f is Riemann integrable if and only if for each $\epsilon > 0$ there exist $P, Q \in \mathcal{P}$ with $|U_Q - L_P| < \epsilon$.
   (**d**) Show that $L = \sup\{L_P \mid P \in \mathcal{P}\}$ and $U = \inf\{U_P \mid P \in \mathcal{P}\}$ both exist. Deduce that f is Riemann integrable if and only if $L = U$, and in that case $L = U = \int_a^b f(x)\, dx$.

129. (Continuation of Exercise 128.) Suppose f is Riemann integrable. Associated with each partition

$$P = (x_0, \ldots, x_n)$$

of $[a, b]$ and each n-tuple

$$z = (z_1, \ldots, z_n)$$

of points with

$$x_0 \leq z_1 \leq x_1 \leq \cdots \leq x_{i-1} \leq z_i \leq x_i \leq \cdots \leq x_{n-1} \leq z_n \leq x_n$$

there is the *Riemann sum*

$$S_{P, z} = \sum_{i=1}^{n} f(z_i) \cdot (x_i - x_{i-1}).$$

By suitably directing the set I of all such pairs (P, z), obtain a net $(S_{P,z} \mid (P, z) \in I)$ that converges to $\int_a^b f(x)\ dx$.

130. Solve the analog of Exercise 125 for arbitrary topological spaces and nets.

131. Let A be a subset of a topological space X and let $x \in \operatorname{cls} A$. Prove anew the existence of a net in A converging to x in X as follows (thereby avoiding the use of the axiom of choice in 3.60). Let $\mathfrak{N}_x$ be the neighborhood system at x, suitably direct the set

$$I = \{(U, t) \mid U \in \mathfrak{N}_x, t \in U\},$$

and consider the net $(t \mid (U, t) \in I)$ [compare Exercise 127].

132. Theorem 3.64 says that a necessary condition for a topological space X to be a Hausdorff space is that no net in X converges to more than one point of X. Prove that this condition is sufficient as well. (*Hint:* Suppose $x, y \in X$ have no disjoint neighborhoods. Form a net indexed by $\mathfrak{N}_x \times \mathfrak{N}_y$, where $\mathfrak{N}_x$ and $\mathfrak{N}_y$ are the neighborhood systems at x and y.)

133. Let $\mathfrak{I}$ and $\mathfrak{S}$ be topologies on the same set X with $\mathfrak{I} \subset \mathfrak{S}$. What is the relationship between convergence of a net $(x_i \mid i \in I)$ to a point x in $(X, \mathfrak{I})$ and convergence of $(x_i \mid i \in I)$ to x in $(X, \mathfrak{S})$? Fill in the blank with 'fewer' or 'more': The finer the topology, the _____ the convergent nets.

134. Let $\mathfrak{I}$ and $\mathfrak{S}$ be topologies on the same set X. Suppose each net that converges to a point x in $(X, \mathfrak{S})$ converges to x in $(X, \mathfrak{I})$. Compare $\mathfrak{I}$ to $\mathfrak{S}$. Fill in the blank with 'finer' or 'coarser': The more convergent nets, the _____ the topology.

135. Provide the set $\mathfrak{F}$ of all functions from $\mathbb{R}$ to $\mathbb{R}$ with its topology of pointwise convergence. Let

$$\mathfrak{a} = \{f \in \mathfrak{F} \mid f(x) = 0 \text{ or } 1 \text{ for all } x, f(x) = 1 \text{ for almost all } x\}$$

(here 'almost all' means 'all but finitely many'). Construct a function $g \in \mathfrak{F}$ such that $g \in \operatorname{cls} \mathfrak{a}$, yet no sequence in $\mathfrak{a}$ converges to g in $\mathfrak{F}$.

136. Let $\mathfrak{F}$ be the set of all functions from a topological space X into X. Let $g: X \to X$ be one of these functions. Use nets to determine when the maps

$$R: \mathfrak{F} \to \mathfrak{F}, \qquad L: \mathfrak{F} \to \mathfrak{F}$$

$$f \mapsto f \circ g \qquad\qquad f \mapsto g \circ f$$

are continuous for the topology of pointwise convergence.

137. Let $\mathfrak{F}$ be the set of all functions from a set X to a bounded metric space (Y, d).
(**a**) Show that a net $(f_i \mid i \in I)$ in $\mathfrak{F}$ converges pointwise to a function $f \in \mathfrak{F}$ if it converges uniformly to f.
(**b**) Show that pointwise convergence does not necessarily imply uniform convergence by considering $Y = \mathsf{R}$ and taking d to be a bounded metric inducing the usual topology on R.

138. Let $\mathfrak{F}$ be the set of all functions from a set X to a metric space (Y, d), where d is not necessarily bounded. Let $\mathfrak{B}$ consist of those $f \in \mathfrak{F}$ that are bounded with respect to d—that is, for which $f(X)$ is d-bounded. Show that the topologies $\mathfrak{I}_1$ and $\mathfrak{I}_2$ described below are the same.
(i) Let d^* be the bounded metric on Y obtained from d as in 1.36. Form the topology of uniform convergence for $\mathfrak{F}$ using d^*. Then $\mathfrak{I}_1$ is the topology induced on the subset $\mathfrak{B}$ of $\mathfrak{F}$.
(ii) Let d_∞ be the sup metric on $\mathfrak{B}$ defined by

$$d_\infty (f, g) = \sup_{x \in X} d(f(x), g(x))$$

(compare the discussion preceding 1.69). Then $\mathfrak{I}_2$ is the topology induced by d_∞.

139. Do two equivalent bounded metrics on a set Y necessarily determine the same topology of uniform convergence for the collection of all functions from a set X to Y?

140. Let $(x_i \mid i \in I)$ be a net that converges to a point x in a Hausdorff space X. Then $(x_i \mid i \in I)$ clusters at x in X. Can $(x_i \mid i \in I)$ cluster at any point of X other than x?

141. Construct a net that clusters at every point of R.

142. Let $f : X \to Y$ be a map that is continuous at a point $x \in X$. If a net $(x_i \mid i \in I)$ clusters at x in X, must the net $(f(x_i) \mid i \in I)$ cluster at $f(x)$ in Y?

143. Construct nets $(x_i \mid i \in I)$ and $(y_i \mid i \in I)$ clustering at points x and y in R such that $((x_i, y_i) \mid i \in I)$ does not cluster at (x, y) in $\mathsf{R} \times \mathsf{R}$.

144. Let $X = \mathsf{N} \times \mathsf{N}$. By applying Theorem 2.17, show that there is a topology on X whose neighborhood systems are the following. If $(m, n) \neq (0, 0)$, then

$$V \in \mathfrak{N}_{(m,n)} \iff (m, n) \in V \subset X.$$

Also,

$$V \in \mathfrak{N}_{(0,0)} \iff (0, 0) \in V \subset X \quad \text{and for almost all } m \in \mathsf{N},$$
$$(m, n) \in V \quad \text{for almost all } n \in \mathsf{N}.$$

Construct a sequence in $X \setminus \{(0, 0)\}$ which clusters at $(0, 0)$ in X no subsequence of which converges to $(0, 0)$ in X.

145. Exhibit a sequence $(x_n \mid n \in \mathsf{N})$ in R and a subnet of $(x_n \mid n \in \mathsf{N})$ that is not a subsequence of $(x_n \mid n \in \mathsf{N})$. If possible, make $(x_n \mid n \in \mathsf{N})$ convergent.

146. Let I and J be directed sets. Prove that a map $\sigma \colon J \to I$ has properties (SN1) and (SN2) of 3.70 if and only if it has the single property

(SN) For each $i \in I$ there exists $j \in J$ such that for all $j' \in J$,
$$j \le j' \implies i \le \sigma(j').$$

147. Let $(y_j \mid j \in J)$ be a subnet of a net $(x_i \mid i \in I)$ in a topological space X.
(**a**) If $(x_i \mid i \in I)$ converges to x, must $(y_j \mid j \in J)$ converge to x? If not, must $(y_j \mid j \in J)$ converge in X?

(b) If $(x_i \mid i \in I)$ clusters at x, must $(y_j \mid j \in J)$ cluster at x? If not, must $(y_j \mid j \in J)$ cluster in X?

148. Let $f: D \to Y$, X, and x be as in Definition 3.72, and let $(t \mid (U, t) \in I)$ be the net defined in Exercise 127. Prove that a point $y \in Y$ is a limit of f at x precisely when the net $(f(t) \mid (U, t) \in I)$ converges to y in Y.

149. Explain each of the following kinds of limits from calculus as instances of Definition 3.72.

(a) $\lim\limits_{t \to +\infty} f(t) = +\infty$ **(e)** $\lim\limits_{t \to x-} f(t) = +\infty$

(b) $\lim\limits_{t \to +\infty} f(t) = -\infty$ **(f)** $\lim\limits_{t \to x-} f(t) = -\infty$

(c) $\lim\limits_{t \to -\infty} f(t) = +\infty$ **(g)** $\lim\limits_{t \to x+} f(t) = +\infty$

(d) $\lim\limits_{t \to -\infty} f(t) = -\infty$ **(h)** $\lim\limits_{t \to x+} f(t) = -\infty$

150. Construct a map $f: \mathsf{R} \times \mathsf{R} \to \mathsf{R}$ that is discontinuous at $(0, 0)$ and yet for which

$$\lim_{t \to 0} f(t, s) = f(0, s) \qquad (s \in \mathsf{R}),$$

$$\lim_{s \to 0} f(t, s) = f(t, 0) \qquad (t \in \mathsf{R}).$$

151. Let $f: X \to Y$ and $g: Y \to Z$ be maps between Hausdorff spaces. Let $x \in X$ with $x \in \mathrm{cls}\,(X \setminus \{x\})$. Assume f has a limit at x and

$$y = \lim_{t \to x} f(t) \in \mathrm{cls}\,(Y \setminus \{y\}).$$

(a) If g is continuous at y, show that

$$\lim_{t \to x} g(f(t)) = g\left(\lim_{t \to x} f(t)\right).$$

(b) Give an example where $g \circ f$ has a limit at x and yet

$$\lim_{t \to x} g(f(t)) \neq g\left(\lim_{t \to x} f(t)\right).$$

152. Let $(x_n \mid n \in \mathsf{N})$ be a convergent sequence in a Hausdorff space. Explain why $\lim_{n \to \infty} x_n$ is the limit of a function in the sense of Definition 3.72. (*Hint:* $\mathsf{N} \subset \mathsf{R}^*$.)

CHAPTER

Compactness

Several topological properties—metrizable, separated, second-countable, first-countable, separable—have already been introduced. Among the various properties a topological space may have, one of the most significant in its consequences, particularly those concerning the continuous functions on the space, is that of being compact. Just as a real-valued function on a finite set attains a maximum value, so a real-valued continuous function on a compact space attains a maximum value. Indeed, compactness is a topological analog of finiteness and can often be exploited to reduce consideration of all points of a space to the consideration of only finitely many points. That is why so many theorems in calculus involve closed bounded intervals in R, which are all compact spaces.

The first section of this chapter concerns compact spaces in general—their separation properties, continuous functions on them, and their subspaces, quotient spaces, and product spaces. The special features of compact spaces that are metrizable are examined in the second section, where compactness is related to completeness. Many important spaces, although not necessarily compact, have enough compact subspaces to allow compactness arguments to be used for them; these spaces are studied in the third section.

1. Compact Spaces

To motivate the definition of a compact space, we begin with a very concrete example.

4.1 Theorem. Let $\mathfrak{U}$ be an arbitrary collection of open subsets of $[0, 1]$ whose union is $[0, 1]$. Then there exist finitely many sets belonging to $\mathfrak{U}$ whose union is also $[0, 1]$.

Proof. Define A to be the set of those points $x \in [0, 1]$ having the property that $[0, x]$ is contained in the union of finitely many sets belonging to $\mathfrak{U}$. We must show that $1 \in A$. Since 0 belongs to some member of $\mathfrak{U}$, $0 \in A$, so A is nonempty. Since $A \subset [0, 1]$, 1 is an upper bound of A in R. In view of the order-completeness of R, it is meaningful to define

$$b = \sup A.$$

Our plan is to show that $b \in A$ and then that $b = 1$.

We show $b \in A$. Just suppose $b \notin A$. There is a $U_0 \in \mathfrak{U}$ with $b \in U_0$ since $b \in [0, 1]$. Now $b > 0$ since $0 \in A$ and $b \notin A$. Because U_0 is open, there is an $\epsilon > 0$ with

$$]b - \epsilon, b] \subset U_0.$$

Since $b - \epsilon$ is not an upper bound of A, there is some $a \in A$ with

$$b - \epsilon < a < b.$$

By definition of A,

$$[0, a] \subset U_1 \cup \cdots \cup U_n$$

for finitely many sets $U_1, \ldots, U_n \in \mathfrak{U}$. Then

$$[0, b] = [0, a] \cup [a, b] \subset U_1 \cup \cdots \cup U_n \cup U_0.$$

Hence $b \in A$, a contradiction.

Knowing now that $b \in A$, we complete the proof by showing that $b = 1$. Just suppose $b < 1$. By definition of A there are sets $U_1, \ldots, U_n \in \mathfrak{U}$ with

$$[0, b] \subset U_1 \cup \cdots \cup U_n.$$

Let $1 \leq j \leq n$ with $b \in U_j$. Since $b < 1$ and U_j is open, there is an $\epsilon > 0$ with

$$[b, b + \epsilon] \subset U_j.$$

Then

$$[0, b + \epsilon] = [0, b] \cup [b, b + \epsilon] \subset U_1 \cup \cdots \cup U_n \cup U_j.$$

Hence $b + \epsilon \in A$. This contradicts the definition of b as the least upper bound of A. $\square$

This theorem, first proved at the end of the last century by E. Heine and E. Borel, is profound in its consequences. To give just one, we prove the fact already assumed in Chapter 1 that *a continuous function $f: [0, 1] \rightarrow \mathsf{R}$ is bounded.*

To prove that f is bounded, it will suffice to find finitely many points $x_1, \ldots, x_n \in [0, 1]$ with the property that for each $x \in [0, 1]$ there is an

i with

$$|f(x) - f(x_i)| < 1.$$

Indeed, if $x \in [0, 1]$ and if i is such an index, then

$$|f(x)| \leq |f(x) - f(x_i)| + |f(x_i)|$$

$$< 1 + \max_{1 \leq i \leq n} |f(x_i)|.$$

Thus the boundedness of f reduces to the fact that the finitely many numbers $|f(x_1)|, \ldots, |f(x_n)|$ have a maximum.

To show that points $x_1, \ldots, x_n$ with the above property actually exist, we use the continuity of f. Denote the Euclidean metric on $[0, 1]$ by d. For each $x \in [0, 1]$ there is a $\delta(x) > 0$ such that

$$(*) \qquad u \in B_{\delta(x)}(x; d) \qquad \Rightarrow \qquad |f(x) - f(u)| < 1.$$

Now apply Theorem 4.1 to the collection

$$\mathcal{U} = \{B_{\delta(x)}(x; d) \mid x \in [0, 1]\}.$$

There exist points $x_1, \ldots, x_n \in [0, 1]$ with

$$[0, 1] = \bigcup_{i=1}^{n} B_{\delta(x_i)}(x_i; d).$$

If $x \in [0, 1]$, then $x \in B_{\delta(x_i)}(x_i; d)$ for some i, so

$$|f(x) - f(x_i)| < 1$$

by $(*)$.

The argument just concluded will still be valid for any metrizable space that shares with $[0, 1]$ the property enunciated in 4.1. This should be sufficient justification for embarking on a systematic study of spaces having this property. First, we introduce some simplifying terminology.

4.2 Definition. Let A be a set. A collection $\mathcal{U}$ of sets is called a *cover of A* if $A \subset \bigcup \mathcal{U}$, that is, each point of A belongs to some member of $\mathcal{U}$. ('Covering' is a synonym for 'cover'.)

Suppose now that A is a subset of a topological space X. Then a cover $\mathcal{U}$ of A is called an *open cover of A in X* if each $U \in \mathcal{U}$ is an open subset of X. An open cover of X in X is called simply an *open cover of X*.

To illustrate this terminology, let

$$0 < \epsilon_0 < \tfrac{1}{2}, \qquad 0 < \epsilon_1 < \tfrac{1}{2},$$

and for each $x \in \,]0, 1[$ let

$$0 < \epsilon_x < \min\{x, 1 - x\}.$$

Then

$$\{[0, \epsilon_0[,\ ]1 - \epsilon_1, 1]\}$$

is not a cover of $[0, 1]$. The collection

$$\{]x - \epsilon_x, x + \epsilon_x[\mid 0 \le x \le 1 \}$$

is an open cover of $[0, 1]$ in R, but it is not an open cover of $[0, 1]$ since its member $]-\epsilon_0, \epsilon_0[$ is not a subset of $[0, 1]$. The collection

$$\{ [0, \epsilon_0[, \,]1 - \epsilon_1, 1] \} \cup \{]x - \epsilon_x, x + \epsilon_x[\mid 0 < x < 1 \}$$

is an open cover of $[0, 1]$. Finally, the collection

$$\{ [0, \epsilon_0], \,]1 - \epsilon_1] \} \cup \{]x - \epsilon_x, x + \epsilon_x[\mid 0 < x < 1 \}$$

is a cover of $[0, 1]$ consisting of subsets of $[0, 1]$, but it is not an open cover of $[0, 1]$ since its member $[0, \epsilon_0]$ is not open in $[0, 1]$.

We now name the property that 4.1 says the closed interval $[0, 1]$ possesses.

4.3 Definition. A topological space X is said to be *compact* when each open cover of X contains a finite cover of X.

This definition was first applied to arbitrary topological spaces by P. Alexandroff and P. Urysohn in 1924 (however, they used the term 'bicompact' to distinguish it from the term 'compact' which had been introduced by Fréchet 18 years earlier for metric spaces having a special property equivalent to 4.3—see Section 2).

4.4 Examples

(1) With only minor modifications the proof of 4.1 also shows that any closed interval $[a, b]$ in R is a compact space.

An open interval $]a, b[$ in R is never compact. To see this, let $l = b - a$ and consider the collection

$$\mathfrak{U} = \{ U_n \mid n = 1, 2, \ldots \}$$

of open subsets of $]a, b[$ given by

$$U_n = \left] a + \frac{l}{n}, b \right[\qquad (n = 1, 2, \ldots).$$

Then $\mathfrak{U}$ is a cover of $]a, b[$; in fact, $a < x < b$ implies $x - a > l/n$ for some integer $n \ge 1$, and then $x \in U_n$. However, if $\mathfrak{F}$ is any finite subcollection of $\mathfrak{U}$, then

$$a + \frac{l}{m + 1} \notin U_n \qquad (U_n \in \mathfrak{F})$$

where

$$m = \max \{ n \mid U_n \in \mathfrak{F} \},$$

so $\mathfrak{F}$ is not a cover of $]a, b[$.

A similar argument shows that no half-open interval $]a, b]$ or $[a, b[$ is compact.

(2) The real line R is not compact, for

$$\{\,]-n, n[\mid n = 1, 2, \ldots\}$$

is an open cover of R that contains no finite cover of R.

(3) Any finite space is compact. Indeed, let $\mathfrak{U}$ be an open cover of a finite space $X = \{x_1, \ldots, x_n\}$. For each $1 \leq i \leq n$ choose a set $U_i \in \mathfrak{U}$ with $x_i \in U_i$. Then the finite collection $\{U_1, \ldots, U_n\} \subset \mathfrak{U}$ is also a cover of X.

(4) For a discrete space X to be compact it is both necessary and sufficient that X be a finite set. Sufficiency follows from (3). To prove necessity, suppose X is discrete and infinite. Then the infinite collection

$$\mathfrak{U} = \{\{x\} \mid x \in X\}$$

is an open cover of X, but no subcollection of $\mathfrak{U}$ is a cover of X except $\mathfrak{U}$ itself.

This example and the preceding one show that compactness may be regarded as a generalization of finiteness.

(5) Most of the interesting compact spaces we shall encounter are Hausdorff spaces (indeed, contemporary authors often stipulate as part of the definition of a compact space that it be a Hausdorff space, reserving the term 'quasicompact' for what we have called 'compact'). It is easy to give examples of compact spaces that are not separated, however. Any topological space X whose topology is indiscrete $[2.3\,(5)]$ is compact, for it has the single open cover $\{X\}$, but X will not be separated if it has at least two points.

A more substantial example is an infinite set X provided with its finite-complement topology $[2.3\ (7)]$. To see that X is not separated, let U and V be any two nonempty open sets in X, so that $X \setminus U$ and $X \setminus V$ are finite; then U cannot be disjoint from V, for otherwise

$$X = X \setminus \varnothing = X \setminus (U \cap V) = (X \setminus U) \cup (X \setminus V)$$

would be finite. To see that X is compact, let $\mathfrak{U}$ be an open cover of X. Arbitrarily choose some nonempty $U_0 \in \mathfrak{U}$. If $U_0 = X$, then $\{U_0\}$ is already a finite cover of X. If $U_0 \neq X$, then $X \setminus U_0$ is a finite set $\{x_1, \ldots, x_n\}$. For each $1 \leq i \leq n$ there is a $U_i \in \mathfrak{U}$ with $x_i \in U_i$. Then $\{U_0, \ldots, U_n\}$ is a finite cover of X.

The few examples above are but a meager sample of the totality of compact spaces. We will be able to present many more examples by using the general theory of compact spaces that we shall now begin to develop.

The definition of compactness of a topological space X may be reformulated as: Given any collection $\mathfrak{U}$ of open subsets of X, if each finite subcollection $\mathfrak{F}$ of $\mathfrak{U}$ fails to be a cover of X, then $\mathfrak{U}$ is not a cover of X. Now a collection $\mathfrak{a}$ of subsets of X is *not* a cover of X when $X \neq \bigcup \mathfrak{a}$, that is,

$$\bigcap_{A \in \mathfrak{a}} (X \setminus A) = X \setminus \bigcup_{A \in \mathfrak{a}} A \neq \varnothing.$$

Hence we are led to the following definition and the next theorem.

4.5 Definition. A collection $\mathcal{E}$ of sets is said to have the *finite intersection property* if $\mathcal{E}$ is nonempty and if each nonempty finite subcollection of $\mathcal{E}$ has nonempty intersection.

If $\mathcal{E}$ has the finite intersection property, then in particular for each $E \in \mathcal{E}$ the collection $\{E\}$ has nonempty intersection, that is, $E \neq \varnothing$.

4.6 Theorem. A topological space X is compact if and only if each collection of closed subsets of X having the finite intersection property itself has nonempty intersection.

Proof. Assume X is compact. Let $\mathcal{E}$ be a collection of closed subsets of X that has the finite intersection property. Define

$$\mathcal{U} = \{X \backslash E \mid E \in \mathcal{E}\},$$

so that $\mathcal{U}$ is a collection of open subsets of X. If $\mathcal{F}$ is any nonempty finite subcollection of $\mathcal{U}$, then $\{X \backslash U \mid U \in \mathcal{F}\}$ is a nonempty finite subcollection of $\mathcal{E}$,

$$X \backslash \bigcup_{U \in \mathcal{F}} U = \bigcap_{U \in \mathcal{F}} (X \backslash U) \neq \varnothing,$$

and $\mathcal{F}$ is not a cover of X. The empty collection is also not a cover of X, since the fact that $\mathcal{E}$ has a nonempty member tells us X is nonempty. Thus no finite subcollection of $\mathcal{U}$ is a cover of X. Since X is compact, $\mathcal{U}$ cannot be a cover of X. Hence

$$\bigcap_{E \in \mathcal{E}} E = \bigcap_{E \in \mathcal{E}} [X \backslash (X \backslash E)] = X \backslash \bigcup_{E \in \mathcal{E}} (X \backslash E) = X \backslash \bigcup_{U \in \mathcal{U}} U \neq \varnothing.$$

The proof of the converse is left as an exercise. $\square$

For a simple application of 4.6, observe that if $(E_n \mid n \in \mathbb{N})$ is a decreasing sequence of nonempty sets, then $\{E_n \mid n \in \mathbb{N}\}$ has the finite intersection property, for a nonempty finite collection $\mathcal{F} \subset \{E_n \mid n \in \mathbb{N}\}$ has as its intersection the nonempty set E_m, where $m = \max \{n \in \mathbb{N} \mid E_n \in \mathcal{F}\}$. Hence *a decreasing sequence of nonempty closed subsets of a compact space has nonempty intersection*. For $[0, 1]$ and other compact subspaces of $\mathbb{R}$ this result is the classic *Cantor product theorem* ('product' is an obsolete word for 'intersection'). To tell whether a subspace of a given topological space is compact we have the following criterion.

4.7 Lemma. A subspace K of a topological space X is compact if and only if each open cover of K in X contains some finite cover of K.

Proof. Assume K is compact. Let $\mathcal{U}$ be an open cover of K in X. Then $\{U \cap K \mid U \in \mathcal{U}\}$ is an open cover of K (in K), so there is a finite $\mathcal{F} \subset \mathcal{U}$ for which $\{U \cap K \mid U \in \mathcal{F}\}$ is a cover of K. Hence $\mathcal{F}$ is a cover of K.

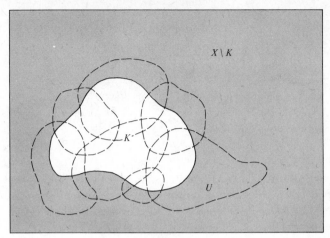

FIG. 4.1

Conversely, assume each open cover of K in X contains some finite cover of K. Let $\mathcal{V}$ be an open cover of K. Then $\mathcal{V} = \{U \cap K \mid U \in \mathcal{U}\}$ for a collection $\mathcal{U}$ of open subsets of X, $\mathcal{U}$ is an open cover of K in X, and by assumption some finite $\mathcal{F} \subset \mathcal{U}$ is a cover of K. Hence the finite subcollection $\{U \cap K \mid U \in \mathcal{F}\}$ of $\mathcal{V}$ is a cover of K.   ∏

Lemma 4.7 says that we do not have to inspect directly the relative topology of K to see whether it is compact, only the topology of the space X containing K. For this reason, we shall in the future refer to a *compact subset* of a topological space X to mean a subset of X that is compact when provided with its relative topology as a subspace of X.

Not every subset of a compact space is compact. For example, $[0, 1]$ is compact by 4.1, but by 4.4 (1) its subset $]0, 1[$ is not. There is one case, however, in which a subset of a compact space is necessarily compact.

4.8 Theorem. A closed subset K of a compact space X is itself compact.

Proof. We use 4.7. Let $\mathcal{U}$ be an open cover of K in X. The only points of X that could not belong to any member of $\mathcal{U}$ must belong to the open subset $X \setminus K$ of X (see Fig. 4.1). Hence $\mathcal{U} \cup \{X \setminus K\}$ is an open cover of X. Since X is compact, $\mathcal{U} \cup \{X \setminus K\}$ contains a finite cover $\mathcal{E}$ of X. Then $\mathcal{F} = \mathcal{E} \setminus \{X \setminus K\}$ is a finite subcollection of the original $\mathcal{U}$. Moreverover, $\mathcal{F}$ is a cover of K, for no point of K belongs to the member of $X \setminus K$ that was deleted in obtaining $\mathcal{F}$ from $\mathcal{E}$.   ∏

It is not true in general that a compact subset K of a topological space X must be closed in X. For example, let X be an infinite set with its finite-complement topology $[2.3\ (7)]$ and let K be a subset of X such that both K

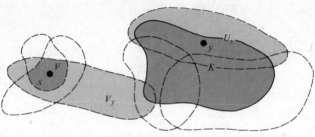

FIG. 4.2

and $X \setminus K$ are infinite. Then K is not closed in X, but K is compact by 4.4 (5) since the relative topology on K is the finite-complement topology on K. In this example the space X is not separated. For separated spaces the converse of 4.8 is true.

4.9 Theorem. Each compact subset of a Hausdorff space X is closed in X.

Proof. Let K be a compact subset of X. If K is empty, there is nothing to prove, so we assume K is nonempty. Let $x \in X \setminus K$. Since X is separated, for each $y \in K$ there are open subsets U_y and V_y of X with

$$y \in U_y, \qquad x \in V_y, \qquad U_y \cap V_y = \varnothing$$

(see Fig. 4.2). Then $\{U_y \mid y \in K\}$ is an open cover of the compact set K in X, so there exists a finite set $F \subset K$ with

$$K \subset \bigcup_{y \in F} U_y.$$

Set

$$V = \bigcap_{y \in F} V_y.$$

Then V, being the intersection of finitely many open sets each containing x, is an open subset of X containing x.

To complete the proof that K is closed in X, we show that V is disjoint from K. Just suppose there is some $z \in V \cap K$. Since $z \in K$, there is a y with

$$y \in F, \qquad z \in U_y.$$

Since $z \in V$, $z \in V_y$. This is impossible since U_y is disjoint from V_y.   □

4.10 Corollary. Let X be a compact Hausdorff space. Then a subset of X is compact if and only if it is closed in X.

With the aid of the following special result we will be able to determine all compact subsets of R.

4.11 Lemma. Let K be a compact set in a metric space (X, d). Then K is d-bounded.

Proof. We may suppose K is nonempty. Since $\{B_1(x; d) \mid x \in K\}$ is an open cover of K in X, there are points $x_1, \ldots, x_n \in K$ with

$$K \subset B_1(x_1; d) \cup \cdots \cup B_1(x_n; d).$$

Set

$$c = \max \{d(x_i, x_j) \mid 1 \leq i \leq n, 1 \leq j \leq n\}.$$

Then for any two points $x, y \in K$ there are indices i, j with

$$x \in B_1(x_i; d), \qquad y \in B_1(x_j; d),$$

and hence

$$d(x, y) \leq d(x, x_i) + d(x_i, x_j) + d(x_j, y) < 1 + c + 1. \qquad \square$$

4.12 Theorem (Heine-Borel-Lebesgue theorem). Let d be the Euclidean metric on the real line R. Then a subset K of R is compact if and only if it is both closed in R and d-bounded.

Proof. If K is compact, then it is closed in R by 4.9 and d-bounded by 4.11. Conversely, assume K is closed in R and d-bounded. Since K is d-bounded, K is contained in some closed interval $[a, b]$. Now $[a, b]$ is compact by 4.4 (1). Since K is closed in $[a, b]$, it follows from 4.8 that K is compact. $\qquad \square$

This theorem is extended to R^n below (see 4.25). However, *a d-bounded closed subset of a metric space (X, d) need not be compact, even when $X = \mathsf{R}$ and d induces the Euclidean topology.* In fact, take d to be a bounded metric on R which is equivalent to the Euclidean metric (such exists by 1.36); then R itself is d-bounded and closed in R, but by 4.12 the space R is not compact. An analog of 4.12 involving a strong form of boundedness is true and will be proved in Section 2.

To illustrate how strange compact subsets of the real line can be, we examine in some detail a famous example of G. Cantor.

4.13 Example. For any closed interval $[a, b]$ in R, let

$$[a, b]^* = [a, a + \tfrac{1}{3}(b - a)] \cup [a + \tfrac{2}{3}(b - a), b]$$

be the set obtained from $[a, b]$ by deleting its open middle-third interval $]a + (b - a)/3, a + 2(b - a)/3[$. For a union

$$E = \bigcup_{i=1}^{k} [a_i, b_i]$$

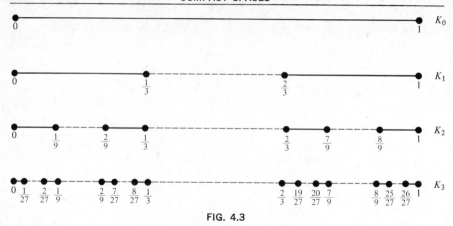

FIG. 4.3

of finitely many disjoint closed intervals, let

$$E^* = \bigcup_{i=1}^{k} [a_i, b_i]^*,$$

so that E^* is a closed set consisting of $2k$ disjoint closed intervals.

Starting with the closed unit interval

$$K_0 = [0, 1]$$

successively construct sets

$$K_1 = K_0{}^* = [0, \tfrac{1}{3}] \cup [\tfrac{2}{3}, 1],$$

$$K_2 = K_1{}^* = [0, \tfrac{1}{9}] \cup [\tfrac{2}{9}, \tfrac{1}{3}] \cup ([\tfrac{2}{3}, \tfrac{7}{9}] \cup [\tfrac{8}{9}, 1],$$

and so on, and in general

$$K_{n+1} = K_n{}^* \qquad\qquad (n \in \mathsf{N})$$

(see Fig. 4.3). This produces a decreasing sequence $(K_n \mid n \in \mathsf{N})$ of closed subsets of $[0, 1]$, with K_n being the union of 2^n pairwise disjoint closed intervals each of length 3^{-n}. Define

$$\mathsf{K} = \bigcap_{n=0}^{\infty} K_n.$$

The set K provided with its relative topology in R is called the *Cantor discontinuum* (or the *Cantor ternary set*, or the *Cantor middle-third set*).

Since K is closed in R and is contained in $[0, 1]$, it follows from 4.1 and 4.8 (or from 4.12) that *the Cantor discontinuum K is compact*. Of course, K is metrizable.

The set K definitely does contain some points—lots of them. Call a point $x \in [0, 1]$ an *endpoint* of K if for some $n \geq 0$ it is an endpoint of one of the open intervals deleted from K_n to obtain K_{n+1}. Then each endpoint of K

actually belongs to **K**. Among these endpoints are

$$0, 1, \frac{1}{3}, \frac{2}{3}, \frac{1}{9}, \frac{2}{9}, \frac{7}{9}, \frac{8}{9}, \frac{1}{27}, \frac{2}{27}, \frac{7}{27}, \frac{8}{27}, \frac{19}{27}, \frac{20}{27}, \frac{25}{27}, \frac{26}{27}.$$

We shall see that there are also many *nonendpoints* of **K**, points of **K** that are not endpoints of **K**.

The Cantor discontinuum **K** is *"self-dense"* in the sense that none of its open sets is a singleton [equivalently, each point of **K** is a limit point of **K** (Exercise 2.24)]. Just suppose, to the contrary, that $\{x\}$ is open in **K** for some x. Then

(*) $$]x - \epsilon, x + \epsilon[\cap \mathbf{K} = \{x\}$$

for some $\epsilon > 0$. Choose a positive integer n with $3^{-n} < \epsilon$. Since $x \in K_n$, $x \in [a, b]$, one of the 2^n intervals whose union is K_n, and then

$$[a, b] \subset]x - \epsilon, x + \epsilon[.$$

At least one endpoint y of $[a, b]$ is different from x. Since y is an endpoint of **K**, we have $y \in \mathbf{K}$, so we obtain a contradiction to (*).

In the next chapter we shall see that *the Cantor discontinuum **K** is "totally disconnected"*. This means, roughly speaking, that the only "connected pieces" of **K** are singletons.

The properties of **K** already mentioned characterize this space topologically: it is a theorem of Hausdorff that every totally disconnected, self-dense, compact, metrizable space containing at least two points is homeomorphic to the Cantor discontinuum. Moreover, a theorem of Alexandroff and Urysohn asserts that every nonempty compact metrizable space is a continuous image of the Cantor discontinuum. (Some concrete instances of the latter theorem appear in Exercise 14. Proofs of both theorems are given in Willard [34, pp. 216–218].)

To see precisely which points belong to **K**, we use the ternary (base 3) expansion of real numbers in $[0, 1]$. Given a sequence $(x_i \mid i = 1, 2, \ldots)$ with $x_i = 0$, 1, or 2 for each i, we see by comparing the series $\sum_{i=1}^{\infty} x_i 3^{-i}$ with the convergent geometric series $\sum_{i=1}^{\infty} 2 \cdot 3^{-i}$ that $\sum_{i=1}^{\infty} x_i 3^{-i}$ converges to a number $x \in [0, 1]$, and then we write

$$x = .x_1 x_2 \ldots x_i \ldots .$$

Of course, some numbers in $[0, 1]$ have two such distinct expansions; for example,

$$\tfrac{1}{9} = .010 \ldots 0 \ldots = .002 \ldots 2 \ldots ,$$

$$\tfrac{2}{3} = .200 \ldots 0 \ldots = .122 \ldots 2 \ldots .$$

Let $x \in [0, 1]$. We construct a certain sequence

$$s(x) = (x_i \mid i = 1, 2, \ldots)$$

in $\{0, 1, 2\}$ with

$$x = .x_1 x_2 \ldots x_i \ldots$$

as follows: If $x = 0$, take $x_i = 0$ for all i. Now let $x > 0$. Take x_1 to be the unique $k \in \{0, 1, 2\}$ such that

$$k3^{-1} < x \le (k+1)3^{-1}.$$

Once $x_1, \ldots, x_{i-1}$ have been constructed, take x_i to be the unique $k \in \{0, 1, 2\}$ such that

$$k3^{-i} < x - \sum_{j=1}^{i-1} x_j 3^{-j} \le (k+1)3^{-i}.$$

Since $0 < x - \sum_{j=1}^{i} x_j 3^{-j} \le 3 \cdot 3^{-(i+1)}$ for all $i \ge 1$, it follows that $x = \sum_{i=1}^{\infty} x_i 3^{-i}$. This construction yields, for example, the expansions

$$\tfrac{1}{3} = .022\ldots2\ldots, \qquad \tfrac{2}{3} = .122\ldots2\ldots, \qquad 1 = .222\ldots2\ldots,$$

$$\tfrac{1}{9} = .002\ldots2\ldots, \qquad \tfrac{2}{9} = .012\ldots2\ldots, \qquad \tfrac{4}{9} = .102\ldots2\ldots.$$

The sequence $s(x) = (x_i \mid i = 1, 2, \ldots)$ associated with an $x \in [0, 1]$ was so constructed that it is *not* the case that

$$x_n \ne 0, \qquad x_i = 0 \qquad (i > n)$$

for some n. In fact, if $x_n = 1$ and $x_i = 0$ for all $i > n$, then the definition of x_n gives

$$1 \cdot 3^{-n} < x - \sum_{j=1}^{n-1} x_j 3^{-j} = 3^{-n} \le 2 \cdot 3^{-n},$$

which is absurd; the case $x_n = 2$ and $x_i = 0$ for all $i > n$ is likewise absurd.

We must make one adjustment in the definition of $s(x) = (x_i \mid i = 1, 2, \cdots)$: If for some n we have $x_n = 1$ and $x_i = 2$ for all $i > n$, redefine $x_n, x_{n+1}, \ldots$ by

$$x_n = 2, \qquad x_i = 0 \qquad (i > n).$$

It is still true that

$$x = .x_1 x_2 \ldots x_i \ldots,$$

but now both the cases

$$x_n = 1, \qquad x_i = 0 \qquad (i > n),$$

$$x_n = 1, \qquad x_i = 2 \qquad (i > n)$$

are excluded. Thus our modified construction yields the expansions

$$\tfrac{1}{3} = .022\ldots2\ldots, \qquad \tfrac{2}{3} = .200\ldots0\ldots, \qquad 1 = .222\ldots2\ldots,$$

$$\tfrac{1}{9} = .002\ldots2\ldots, \qquad \tfrac{2}{9} = .020\ldots0\ldots, \qquad \tfrac{4}{9} = .102\ldots2\ldots.$$

Let $x \in [0, 1]$ and $s(x) = (x_i \mid i = 1, 2, \ldots)$. Induction on n shows that for all $n \ge 1$,

$$(**) \qquad x \in K_n \quad \Leftrightarrow \quad x_1 \ne 1, \ldots, x_n \ne 1.$$

To start the induction, observe that each $x \in [0, \frac{1}{3}] \cup [\frac{2}{3}, 1]$ has the expansion $x = .0x_2x_3\ldots$ or $x = .2x_2x_3\ldots$, whereas each $x \in]\frac{1}{3}, \frac{2}{3}[$ has the expansion $x = .1x_2x_3\ldots$. From (**) we conclude

$$x \in \mathsf{K} \quad \Leftrightarrow \quad x_n \in \{0, 2\} \qquad (n = 1, 2, \ldots).$$

Letting

$$P = \{1, 2, \ldots\},$$

we obtain a bijection

$$s: \mathsf{K} \to \{0, 2\}^P,$$

$$x \mapsto s(x).$$

A diagonal argument (compare the proof of 0.33) shows that $\{0, 2\}^P$ is uncountable. Hence *the Cantor discontinuum K is uncountable.*

Provide $\{0, 2\}$ with its discrete topology and $\{0, 2\}^P$ with its product topology. Then

$$\mathsf{K} \cong \{0, 2\}^P.$$

To prove this, we show that the inverse

$$f: \{0, 2\}^P \to \mathsf{K}$$

$$(x_i \mid i = 1, 2, \ldots) \mapsto \sum_{i=1}^{\infty} x_i 3^{-i}$$

of s is a homeomorphism. Let $x = (x_i \mid i = 1, 2, \ldots) \in \{0, 2\}^P$. Now $\{0, 2\}^P$ has as a local base at x the collection of sets of the form

$$V_n = \{ (y_i \mid i = 1, 2, \ldots) \mid y_1 = x_1, \ldots, y_n = x_n \}$$

for $n \geq 1$ [see the paragraph preceding 3.29]. Given $\epsilon > 0$, choose $n \geq 1$ with

$$3^{-n} < \epsilon;$$

then $y = (y_i \mid i = 1, 2, \ldots) \in V_n$ implies

$$|f(y) - f(x)| = \left| \sum_{i=n+1}^{\infty} (y_i - x_i) 3^{-i} \right| \leq \sum_{i=n+1}^{\infty} 2 \cdot 3^{-i} = 3^{-n} < \epsilon.$$

Hence f is continuous at x. A similar argument shows that f is open at x.

Let us return to general properties of compact spaces. The proof of Theorem 4.9 actually yields a stronger conclusion than the asserted theorem. Recall that for the compact subset K of the Hausdorff space X and a point $x \in X \setminus K$ we obtained open sets $(U_y \mid y \in F)$ and $(V_y \mid y \in F)$ with

$$x \in V = \bigcap_{y \in F} V_y, \qquad K \subset \bigcup_{y \in F} U_y,$$

$$U_y \cap V_y = \varnothing \qquad\qquad (y \in F).$$

Then the set

$$U = \bigcup_{y \in F} U_y$$

is disjoint from V. We have proved the following lemma.

4.14 Lemma. Let K be a compact subset of a Hausdorff space X, and let $x \in X \setminus K$. Then there exist disjoint open subsets U and V of X with

$$x \in V, \qquad K \subset U.$$

4.15 Theorem. Let K and L be disjoint compact subsets of a Hausdorff space X. Then there exist disjoint open subsets U and V of X with

$$K \subset U, \qquad L \subset V.$$

Proof. We may suppose neither K nor L is empty. By Lemma 4.14, for each $x \in L$ there are open subsets U_x and V_x of X with

$$x \in V_x, \qquad K \subset U_x, \qquad U_x \cap V_x = \varnothing.$$

Since $\{V_x \mid x \in L\}$ is an open cover of L in X, there exists a nonempty finite set $E \subset L$ with

$$L \subset \bigcup_{x \in E} V_x.$$

The desired sets U and V are now obtained by taking

$$U = \bigcap_{x \in E} U_x, \qquad V = \bigcup_{x \in E} V_x. \qquad \square$$

The reader should be sure he thoroughly understands the entire line of reasoning used to prove 4.9, 4.14, and 4.15, for it is the prototype of many compactness arguments.

By combining 4.8 and 4.15 we obtain the following theorem.

4.16 Theorem. Let X be a compact Hausdorff space. Then for any two disjoint closed subsets K and L of X there exist disjoint open subsets U and V of X with $K \subset U$ and $L \subset V$.

A topological space for which this same conclusion holds is said to be *normal*. Along with compact Hausdorff spaces, all metrizable spaces are normal (see Exercise 1.27). Among the many interesting properties of arbitrary normal spaces the foremost is expressed by *Urysohn's lemma*: If A and B are disjoint closed subsets of a normal space X, then there is a continuous function $f: X \to \mathsf{R}$ with $f(a) = 0$ for all $a \in A$, $f(b) = 1$ for all $b \in B$, and $0 \leq f(x) \leq 1$ for all $x \in X$ (see Willard [34, p. 102]). This says that if any two disjoint closed sets in a topological space can be separated by open sets, then any two disjoint closed sets in the space can be separated by a continuous real-valued function on the space.

As with any property meaningful for arbitrary topological spaces, we want to know whether compactness is preserved under the formation of subspaces, quotient spaces, and product spaces. We have already discussed subspaces of compact spaces. Concerning quotient spaces we have the key result that any continuous image of a compact space is compact.

4.17 Theorem. Let $f: X \to Y$ be a continuous map from a compact space X into a topological space Y. Then $f(X)$ is compact.

Proof. Let $\mathcal{U}$ be an open cover of $f(X)$ in Y. Now $f^{-1}(V)$ is open in X for each $V \in \mathcal{U}$, and $\{f^{-1}(V) \mid V \in \mathcal{U}\}$ is a cover of X since

$$X = f^{-1}(f(X)) \subset f^{-1}\left(\bigcup_{V \in \mathcal{U}} V\right) = \bigcup_{V \in \mathcal{U}} f^{-1}(V).$$

Then there exists a finite $\mathcal{F} \subset \mathcal{U}$ such that $\{f^{-1}(V) \mid V \in \mathcal{F}\}$ is a cover of X. Hence

$$f(X) = f\left(\bigcup_{V \in \mathcal{F}} f^{-1}(V)\right) = \bigcup_{V \in \mathcal{F}} f(f^{-1}(V)) = \bigcup_{V \in \mathcal{F}} V,$$

so $\mathcal{F}$ is a cover of $f(X)$.   ☐

4.18 Corollary. Any quotient space of a compact space is itself compact.

4.19 Corollary. Compactness is a topological property: If Y is a topological space homeomorphic to a compact space, then Y is compact.

For an application, recall that

$$\mathsf{R}^* \cong [0, 1], \qquad \mathsf{R} \cong \,]0, 1[.$$

From the compactness of $[0, 1]$ and the noncompactness of $]0, 1[$, it follows that *the extended real line* R^* *is compact, but the real line* R *is noncompact.*

It was emphasized earlier that continuity is not in general a sufficient condition for a bijection to be a homeomorphism. However, continuity is sufficient under special circumstances.

4.20 Corollary. Let $f: X \to Y$ be a continuous map from a compact space X into a Hausdorff space Y. Then f is a closed map. If f is bijective, then it is a homeomorphism.

Proof. If E is any closed subset of X, then E is compact by 4.8, $f(E)$ is compact by 4.17, and so $f(E)$ is closed in Y by 4.9. Hence f is closed. If in addition f is bijective, then from the discussion after 3.14 it follows that f is a homeomorphism.   ☐

Use of Corollary 4.20 will often shorten the work of showing that a given bijection is a homeomorphism.

At the beginning of this section we proved that any continuous real-valued function on $[0, 1]$ is bounded. We can now both generalize and strengthen that result.

4.21 Corollary (extreme value theorem). Let $f: X \to \mathsf{R}$ be a continuous real-valued function on a nonempty compact space X. Then f is bounded. Moreover, f attains both a minimum and a maximum value on X, that is, there exist $x_1, x_2 \in X$ with

$$f(x_1) \leq f(x) \leq f(x_2) \qquad\qquad (x \in X).$$

Proof. The boundedness of f is a consequence of 4.17 and 4.11. Set

$$m_1 = \inf f(X), \qquad m_2 = \sup f(X).$$

Now $f(X)$ is closed in R by 4.20, so $m_1 \in f(X)$ and $m_2 \in f(X)$. Take x_1, x_2 to be any points of X with $f(x_1) = m_1, f(x_2) = m_2$. $\square$

In the special case that X is a closed interval $[a, b]$ in R, this is a familiar fact having many uses in calculus. For an application to arbitrary metric spaces, consider a nonempty subset A of a metric space (X, d) and a point $x \in X$. We know that in general $d(x, A)$, the distance from x to A (1.26), need not equal $d(x, a)$ for any $a \in A$. However, if A is compact, then

$$d(x, A) = d(x, a)$$

for some $a \in A$; this follows from 4.21 and the continuity of the map

$$A \to \mathsf{R}$$

$$a \mapsto d(x, a)$$

(see Exercise 1.56).

The proof that compactness is preserved under the formation of products is considerably more difficult than the proof that it is preserved under the formation of quotients.

4.22 Lemma. Let X be a topological space, let Y be a compact space, let $x \in X$, and let $\mathscr{U}$ be an open cover of $\{x\} \times Y$ in $X \times Y$. Then there is an open neighborhood U of x in X and a finite subcollection of $\mathscr{U}$ that is a cover of $U \times Y$.

Proof. Since $\{x\} \times Y \cong Y$, the set $\{x\} \times Y$ is compact. Then there exists a finite $\mathscr{F} \subset \mathscr{U}$ which is a cover of $\{x\} \times Y$. For each $y \in Y$ there is then some $W_y \in \mathscr{F}$ with $(x, y) \in W_y$, and hence there are open neighborhoods U_y of x in X and V_y of y in Y with

$$U_y \times V_y \subset W_y.$$

The collection $\{V_y \mid y \in Y\}$ is an open cover of the compact space Y, so there exists a finite set $F \subset Y$ such that $\{V_y \mid y \in F\}$ is a cover of Y. Let

$$U = \bigcap_{y \in F} U_y$$

whence U is an open neighborhood of x in X (see Fig. 4.4).

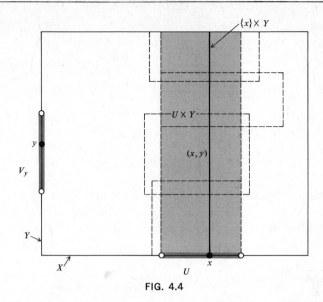

FIG. 4.4

The collection $\{W_y \mid y \in F\}$ is finite and is contained in $\mathcal{U}$. We complete the proof by showing that it is a cover of $U \times Y$. Let $(x', y') \in U \times Y$. By the definition of F, there is a $y \in F$ with

$$y' \in V_y,$$

and then

$$x' \in U_y$$

by definition of U. Hence

$$(x', y') \in U_y \times V_y \subset W_y. \qquad \square$$

4.23 Theorem. Let X and Y be compact spaces. Then $X \times Y$ is compact.

Proof. Let $\mathcal{U}$ be an open cover of $X \times Y$. For each $x \in X$, $\mathcal{U}$ is an open cover of $\{x\} \times Y$ in $X \times Y$ so by Lemma 4.22 there is an open neighborhood U_x of x in X and a finite collection $\mathcal{F}_x \subset \mathcal{U}$ with

$$U_x \times Y \subset \bigcup \mathcal{F}_x.$$

Now $\{U_x \mid x \in X\}$ is an open cover of the compact space X, so there is a finite set $E \subset X$ with

$$X \subset \bigcup_{x \in E} U_x.$$

Define

$$\mathcal{F} = \bigcup_{x \in E} \mathcal{F}_x.$$

Since $\mathcal{F}$ is the union of finitely many finite subcollections of $\mathcal{U}$, $\mathcal{F}$ is a finite subcollection of $\mathcal{U}$. To see that $\mathcal{F}$ is a cover of $X \times Y$, let $(x', y') \in X \times Y$.

There is an $x \in E$ with

$$x' \in U_x.$$

Then

$$(x', y') \in U_x \times Y \subset \bigcup \mathfrak{F}_x,$$

so

$$(x', y') \in V$$

for some $V \in \mathfrak{F}_x$. But then $V \in \mathfrak{F}$.   ◻

At last we reach our main result concerning compactness of products.

4.24 Theorem (Tychonoff product theorem—the finite case).

Let $(X_i \mid i = 1, \ldots, n)$ be a finite family of nonempty topological spaces. Then the product space $\mathsf{X}_{i=1}^n X_i$ is compact if and only if X_i is compact for each $i = 1, \ldots, n$.

Proof. Assume $\mathsf{X}_{i=1}^n X_i$ is compact. For each $i = 1, \ldots, n$, the ith projection

$$p_i: \mathsf{X}_{i=1}^n X_i \to X_i$$

is a continuous surjection, so by 4.17 the space X_i is compact.

The converse is proved by induction on n using 4.23 and the relation

$$\mathsf{X}_{i=1}^n X_i \cong \left(\mathsf{X}_{i=1}^{n-1} X_i \right) \times X_n$$

(see 3.37).   ◻

In the next section the same result will be proved for a sequence of spaces under the additional assumption that each of the spaces is metrizable. It so happens that the product of any family of compact spaces is compact; this is the *Tychonoff product theorem*, the single most important theorem in general topology. However, the proof of this theorem in its full generality uses methods that are beyond the elementary scope of this book (see Willard [34, p. 120], Kelley [21, p. 143], and Bourbaki [6, p. 88] for three different proofs).

Using 4.24 we can generalize the Heine-Borel-Lebesgue theorem (4.12) to n-dimensional Euclidean space.

4.25 Theorem.

Let d be the Euclidean metric on R^n. Then a subset K of R^n is compact if and only if it is both closed in R^n and d-bounded.

Proof. If $K \subset \mathsf{R}^n$ is compact, then K is closed in R^n by 4.9 and is d-bounded by 4.11.

Conversely, assume K is closed in R^n and d-bounded. Since K is d-bounded, it is contained in some d-disk and hence in some cube

$$E = [a_1, b_1] \times \cdots \times [a_n, b_n].$$

By 4.4 (1) the closed intervals $[a_1, b_1], \ldots, [a_n, b_n]$ in R are compact, so by 4.24 their product E is a compact subset of R^n. However, K is a closed subset of E, so K is compact by 4.8. $\Box$

Using 4.25 we can now list a number of standard compact spaces.

4.26 Examples

(1) The n-cube

$$I^n = [0, 1]^n$$

is compact. More generally, any product $[a_1, b_1] \times \cdots \times [a_n, b_n]$ of closed intervals is compact.

(2) The n-disk

$$D_n = \{x \in R^n : ||x|| \leq 1\}$$

is compact. However, neither the n-ball

$$B_n = \{x \in R^n : ||x|| < 1\}$$

nor R^n itself is compact.

(3) The n-sphere

$$S_n = \{x \in R^{n+1} : ||x|| = 1\}$$

is compact. In particular, the circle

$$S_1 = \{x \in R : |x| = 1\}$$

is compact.

(4) The n-torus

$$(S_1)^n$$

is compact.

(5) The n-dimensional real projective space RP_n is compact, for RP_n is a quotient space of S_n [3.49 (7)].

(6) The Klein bottle [3.49 (6)] and the Möbius strip [3.49 (5)] are compact, being quotient spaces of I^2.

To conclude our discussion of compact spaces in general, we give a characterization of compactness involving nets.

4.27 Theorem. A topological space X is compact if and only if each net in X clusters in X.

Proof. We shall use the characterization 4.6 of compactness in terms of closed sets. Assume X is compact. Let $(x_i \mid i \in I)$ be a net in X. We apply

4.6 to the collection $\{E_i \mid i \in I\}$, where

$$E_i = \text{cls } A_i,$$
$$A_i = \{x_j \mid j \in I, i \leq j\}$$

for each $i \in I$. If J is any nonempty finite subset of I, then since I is directed, there is a $k \in I$ with $j \leq k$ for all $j \in J$, so

$$\varnothing \neq A_k \subset \bigcap_{j \in J} A_j \subset \bigcap_{j \in J} E_j.$$

Thus $\{E_i \mid i \in I\}$ has the finite intersection property. By 4.6 there exists some

$$x \in \bigcap_{i \in I} E_i.$$

We show that $(x_i \mid i \in I)$ clusters at x. Let V be a neighborhood of x and let $i \in I$. Since $x \in E_i$, then $V \cap A_i \neq \varnothing$. Hence $x_j \in V$ for some $j \geq i$.

Conversely, assume each net in X clusters in X. Let $\mathcal{E}$ be a collection of closed subsets of X that has the finite intersection property. Construct a net $(x_i \mid i \in I)$ in X as follows. The set I is defined to be the set of all ordered pairs $(\mathcal{F}, x)$ such that $\mathcal{F}$ is a nonempty finite subset of $\mathcal{E}$ and $x \in \bigcap \mathcal{F}$. The direction $\leq$ of I is defined by

$$(\mathcal{F}_1, x_1) \leq (\mathcal{F}_2, x_2) \qquad \Leftrightarrow \qquad \mathcal{F}_1 \subset \mathcal{F}_2.$$

Clearly $\leq$ is reflexive on I and transitive; to see that it directs I note that if $(\mathcal{F}_1, x_1), (\mathcal{F}_2, x_2) \in I$, then $\mathcal{F}_1 \cup \mathcal{F}_2$ is a nonempty finite subset of $\mathcal{E}$, there is some $x \in \bigcap (\mathcal{F}_1 \cup \mathcal{F}_2)$, and then $(\mathcal{F}_1 \cup \mathcal{F}_2, x) \in I$ with

$$(\mathcal{F}_1, x_1) \leq (\mathcal{F}_1 \cup \mathcal{F}_2, x), \qquad (\mathcal{F}_2, x_2) \leq (\mathcal{F}_1 \cup \mathcal{F}_2, x).$$

Finally, for

$$i = (\mathcal{F}, x) \in I$$

the element $x_i \in X$ is defined by

$$x_i = x.$$

By assumption the net $(x_i \mid i \in I)$ clusters at some $x \in X$. We show $x \in \bigcap \mathcal{E}$. Let $E \in \mathcal{E}$. To show that x belongs to the closed set E it is enough to show that each neighborhood of x intersects E. Let V be any neighborhood of x. Since $E \in \mathcal{E}$, there is some $y \in E$. Then $(\{E\}, y) \in I$, so there exists some $(\mathcal{F}, z) \in I$ with

$$(\{E\}, y) \leq (\mathcal{F}, z), \qquad z \in V.$$

But $E \in \mathcal{F}$, so

$$z \in \bigcap \mathcal{F} \subset E.$$

Thus $z \in V \cap E$. $\quad \square$

In view of 3.71, Theorem 4.27 has the equivalent formulation that a topological space X is compact if and only if each net in X has some convergent subnet.

EXERCISES

1. Show that the set A constructed in the proof of 4.1 is both open and closed in $[0, 1]$. (*Note*: In Chapter 5 we will prove that $[0, 1]$ is "connected" in the sense that the only nonempty subset of $[0, 1]$ that is both open and closed in this space is $[0, 1]$ itself. Hence the compactness of $[0, 1]$ is a consequence of its connectedness. Conversely, the connectedness of $[0, 1]$ can be deduced from its compactness—see Exercise 5.29.)

2. Working directly with the definition of compactness, show that the extended real line R^* is compact.

3. Give examples of:
   (**a**) A compact space that is not pseudometrizable (Exercise 2.12).
   (**b**) A compact Hausdorff space that is not metrizable.
   (**c**) A compact T_1-space (Exercise 2.10) that is not a T_2-space.

4. Determine which of the following spaces are compact:
   (**a**) An uncountable set provided with its countable-complement topology (Exercise 2.6).
   (**b**) The real line provided with its right-interval topology [2.18(1)].
   (**c**) The subspace $[0, 1]$ of the space in (b).
   (**d**) The line with two origins [2.18(3)].
   (**e**) The half-disk space [2.20(3)].
   (**f**) The tangent disk space (Exercise 2.27).

5. (**a**) Let X be a totally ordered set provided with its order topology [2.52(1)]. Find a necessary and sufficient condition on the total ordering for X to be compact. (*Hint*: Which properties of the usual ordering on $[0, 1]$ were used in the proof of 4.1?)
   (**b**) Apply this condition to the set $[0, 1] \times [0, 1]$ with the total ordering given by
$$(x_1, x_2) \leq (y_1, y_2) \quad \Leftrightarrow \quad (x_1 < y_1) \quad \text{or} \quad (x_1 = y_1 \text{ and } x_2 \leq y_2)$$
(compare Exercise 2.29).

6. (**a**) Let $(x_i \mid i \in I)$ be a net in a topological space that converges to a point $x \in X$. Is $\{x\} \cup \{x_i \mid i \in I\}$ compact?
   (**b**) Is $\{x\} \cup \{x_i \mid i \in I\}$ compact if $(x_i \mid i \in I)$ only clusters at x?

7. A nonempty collection $\mathcal{B}$ of nonempty subsets of a topological space X is called a *filter base on* X if for any $B_1, B_2 \in \mathcal{B}$ there is some $B \in \mathcal{B}$ with $B \subset B_1 \cap B_2$. A filter base $\mathcal{B}$ on X is said to *cluster at* $x \in X$ if $x \in \text{cls } B$ for every $B \in \mathcal{B}$.
   (**a**) Prove: If X is compact, then each filter base on X clusters at some point of X.
   (**b**) Does the converse of (a) hold?

8. A collection $\mathcal{U}$ of subsets of a topological space X is said to be *locally finite* if each point of X has some neighborhood that intersects only finitely many members of $\mathcal{U}$ (compare Exercise 2.49).
   (**a**) Construct a locally finite open cover of R having infinitely many members.
   (**b**) Show that each locally finite open cover of a compact space is necessarily finite.

9. Let $\mathcal{B}$ be a base of a topological space X. Suppose each cover of X consisting of members of $\mathcal{B}$ contains a finite collection that is a cover of X. Prove that X must be compact.

10. (**a**) Prove that the union of finitely many compact subsets of a topological space is itself compact.

(**b**) What can be said about the intersection of two compact subsets of a topological space?

11. A subset A of a topological space X is said to be *relatively compact* if it is contained in some compact subset of X. Clearly A will be relatively compact if cls A is compact. When does the converse hold?

12. Characterize the endpoints of K in terms of their ternary expansions. How many endpoints does K have? How many nonendpoints?

13. Show that K is nowhere dense in $[0, 1]$ (see Definition 2.41).

14. (**a**) Define $g: \mathsf{K} \to [0, 1]$ by

$$g\left(\sum_{n=1}^{\infty} x_n 3^{-n} \right) = \sum_{n=1}^{\infty} \frac{x_n}{2} 2^{-n}.$$

Show that g maps K continuously onto $[0, 1]$.

(**b**) Given a positive integer k, construct a continuous map from K onto $[0, 1]^k$. [*Hint*: Show first that $(\{0, 2\}^P)^k \cong \{0, 2\}^P$, where $P = \{1, 2, \ldots\}$.]

15. Let $(K_n \mid n \in \mathsf{N})$ be a decreasing sequence of nonempty compact subsets of a Hausdorff space X. If U is an open set in X with $\bigcap_{n=0}^{\infty} K_n \subset U$, show that $K_n \subset U$ for some n.

16. Let X be a regular space (Exercise 2.60). If A is a compact subset and B is a closed subset of X, show that there are disjoint open subsets U and V of X with $A \subset U$ and $B \subset V$.

17. Prove the following special case of Urysohn's lemma: If A and B are disjoint closed sets in a metrizable space X, then there is a continuous function $f: X \to \mathsf{R}$ with $f(a) = 0$ for all $a \in A$, $f(b) = 1$ for all $b \in B$, and $0 \le f(x) \le 1$ for all $x \in X$. [*Hint*: If d induces the topology of X, consider the function defined by $f(x) = d(x, A)/[d(x, A) + d(x, B)]$.]

18. (**a**) Given topologies $\mathfrak{I}$ and $\mathcal{S}$ on the same set X with $\mathcal{S} \subset \mathfrak{I}$, prove: If $(X, \mathfrak{I})$ is compact, then $(X, \mathcal{S})$ is compact. Moreover, if $(X, \mathfrak{I})$ is compact and $(X, \mathcal{S})$ is a Hausdorff space, then $\mathfrak{I} = \mathcal{S}$.

(**b**) Can a set X have two different topologies $\mathfrak{I}$ and $\mathcal{S}$ such that $(X, \mathfrak{I})$ and $(X, \mathcal{S})$ are both compact Hausdorff spaces?

19. Must a set $K \subset \mathsf{R}$ be compact if every continuous function $f: K \to \mathsf{R}$ is bounded and attains a maximum value on K?

20. Let K be a nonempty subset of the plane R^2. If $x \in \mathsf{R}^2$, then a point $y \in K$ is called a "nearest point in K to x" when $d(x, K) = d(x, y)$. Prove that a necessary and sufficient condition for K to be both closed and convex is that for each $x \in \mathsf{R}^2$ there exists a *unique* nearest point in K to x. [*Hint*: To establish necessity you may want to use the continuity of the function $y \mapsto d(x, y)$ on K.]

21. (**a**) If X is a nonempty compact space, show that every function $f: X \to \mathsf{R}$ that is upper semicontinuous (Exercise 3.28) attains a maximum value on X.

(**b**) Prove the analog of (a) for lower semicontinuous functions.

22. Let K be a nonempty compact subset of a metric space (X, d).

(**a**) Show that diam $K = d(x, y)$ for some points $x, y \in K$.

(b) If L is another nonempty compact subset of X, show that there are points $x \in K$ and $y \in L$ with $d(x, y) = d(K, L)$, the distance from K to L (Exercise 1.28).

(c) Does the conclusion of (b) still hold if L is only assumed to be a nonempty closed subset of X?

23. Let K and L be compact subsets of topological spaces X and Y, respectively. If W is an open set in $X \times Y$ with $K \times L \subset W$, show that there are open sets U in X and V in Y with $K \times L \subset U \times V \subset W$.

24. When is the sum (Exercise 3.57) of a family of topological spaces compact?

25. In the second part of the proof of 4.27, suppose we had constructed a net $(x_i \mid i \in I)$ from $\mathcal{E}$ by taking I to be the set of all nonempty finite subsets of $\mathcal{E}$, I being directed by inclusion, and by taking $x_i \in \bigcap_{E \in i} E$ for each $i \in I$.

(a) If x is a point in X to which $(x_i \mid i \in I)$ clusters, show that $x \in \bigcap \mathcal{E}$.

(b) Discuss why such a net need exist.

26. Prove the *closed-graph theorem*: If $f \colon X \to Y$ is a map from a topological space X into a compact space Y whose graph $\{(x, f(x)) \mid x \in X\}$ is closed in $X \times Y$, then f is continuous. [*Hint:* Show that if $(x_i \mid i \in I)$ is a net converging to x in X, then $(f(x_i) \mid i \in I)$ converges to $f(x)$ in Y by applying 4.27 to $(f(x_i) \mid i \in I)$.]

27. A continuous surjection $f \colon X \to Y$ is said to be *proper* (or *perfect*) if f is closed and $f^{-1}(y)$ is compact for each $y \in Y$.

(a) Show that a continuous map from a compact space onto a Hausdorff space is proper.

(b) Must a continuous closed map from a Hausdorff space onto a compact Hausdorff space be proper?

(c) If $f \colon X \to Y$ is proper and X is a Hausdorff space, prove that Y is also a Hausdorff space.

28. Let $\sim$ be a closed equivalence relation on a compact Hausdorff space X. Show that the quotient space $X/\sim$ is a Hausdorff space.

29. Let $f \colon X \to Y$ be a closed surjection such that $f^{-1}(y)$ is compact for each $y \in Y$. Prove that $f^{-1}(K)$ is compact for each compact subset K of Y. (It is *not* assumed that f is continuous).

30. Let X be a topological space such that for any two distinct points $a, b \in X$ there is a continuous function $h \colon X \to [0, 1]$ with $h(a) = 0$, $h(b) = 1$.

(a) Let B be a compact subset of X. If $a \in X \setminus B$, show that there is a continuous function $g \colon X \to [0, 1]$ with $g(a) = 0$ and $g(b) = 1$ for all $b \in B$. [*Hint:* First obtain a continuous $g' \colon X \to [0, 1]$ with $g'(a) = 0$ and $g'(b) > \frac{1}{2}$ for all $b \in B$. To do this note that for $b \in B$ and for a continuous $h \colon X \to [0, 1]$ with $h(b) = 1$, the set $h^{-1}(]\frac{1}{2}, 1])$ is a neighborhood of b. You will probably want to use 3.16(3).]

(b) Now let A be another compact subset of X that is disjoint from B. Prove that there is a continuous function $f \colon X \to [0, 1]$ with $f(a) = 0$ for all $a \in A$ and $f(b) = 1$ for all $b \in B$.

31. Let G be a topological group (Exercise 3.84).

(a) If K and L are compact subsets of G, show that the set KL is also compact. (*Hint:* Use 4.23.)

(b) If K and L are compact subsets of G and if W is an open set in G

with $KL \subset W$, show there exist open sets U and V in G with $K \subset U$, $L \subset V$, and $UV \subset W$. (*Hint:* Use Exercise 23.)

(**c**) If A is a closed subset of G and B is a compact subset of G, prove that AB is closed in G. [*Hint:* If $x \in G \setminus AB$, then $xB^{-1} \subset X \setminus A$; use (b).]

(**d**) By considering the subsets $A = \mathsf{Z}$ and $B = \{\sqrt{2}n \mid n \in \mathsf{Z}\}$ of R, show that AB need not be closed in G when A and B are both closed subsets of G (for $G = \mathsf{R}$, of course, $AB = A + B$).

2. Compact Metric Spaces

Many of the most important compact spaces are subspaces of Euclidean spaces and hence are metrizable. For this reason we now explore the special properties that result from the presence of a metric on a compact space.

A compact subset of R^n, like any subspace of R^n, is second-countable. The same is true of any compact metrizable space.

4.28 Theorem. Let X be a compact space. If X is metrizable, then X is second-countable.

Proof. Let d be a metric inducing the topology of X. For each positive integer n the collection $\{B_{1/n}(x; d) \mid x \in X\}$ of balls of radius $1/n$ is an open cover of X, so there is a finite set $F_n \subset X$ with

$$X = \bigcup_{x \in F_n} B_{1/n}(x; d).$$

Set

$$\mathscr{B} = \{B_{1/n}(x; d) \mid n = 1, 2, \ldots, x \in F_n\},$$

whence $\mathscr{B}$ is a countable collection of open sets in X.

To see that $\mathscr{B}$ is actually a base of X, let U be an open set in X and let $x \in U$. Choose n large enough so that

$$B_{2/n}(x; d) \subset U.$$

For this n there is some $y \in F_n$ with

$$x \in B_{1/n}(y; d).$$

Then $B_{1/n}(y; d) \in \mathscr{B}$ with $x \in B_{1/n}(y; d) \subset U$. □

We mention without proof that 4.28 has a converse: *Any second-countable compact Hausdorff space is metrizable.* This is a special case of the Urysohn metrization theorem (see Willard [34, p. 166] or Kelley [21, p. 125]).

Recall that a metrizable space is second-countable precisely when it is separable (2.57). Recall also that any separable metrizable space can be homeomorphically embedded in the Hilbert cube (3.23). Now the Hilbert cube is a Hausdorff space, so from 4.28 we obtain the following "concrete" representation of compact metrizable spaces.

4.29 Corollary. Any compact metrizable space is homeomorphic to a closed subspace of the Hilbert cube.

Conversely, each closed subspace of the Hilbert cube is a compact metrizable space, for the Hilbert cube itself is metrizable and, as will follow from 4.40, compact as well.

Suppose X is a compact metrizable space, and let d be a metric inducing the topology of X. Then X is d-bounded (4.11). We are going to prove that X must actually have a very strong form of d-boundedness—namely, that X can be written as the union of finitely many sets whose diameters are as small as we wish.

4.30 Definition. A metric space (X, d) is said to be *totally bounded* if for each real number $\epsilon > 0$ there is a finite cover of X consisting of d-bounded sets each of d-diameter at most ϵ.

The following criterion provides an equivalent formulation of total boundedness that is often simpler to work with than the definition.

4.31 Lemma. A metric space (X, d) is totally bounded if and only if for each $\epsilon > 0$ there is a finite set $F \subset X$ that is "ϵ-dense in X" in the sense that each point of X is at a distance less than ϵ from some point of F.

Proof. If $\epsilon > 0$ and $\mathcal{Q}$ is a finite cover of X consisting of d-bounded sets each of d-diameter at most $\epsilon/2$, then a set consisting of exactly one point from each member of $\mathcal{Q}$ will be finite and ϵ-dense in X. Conversely, if $F \subset X$ is a finite set which is $\epsilon/2$-dense in X, then $\{B_{\epsilon/2}(y; d) \mid y \in F\}$ will be a finite cover of X with diam $B_{\epsilon/2}(y; d) \leq \epsilon$ for each $y \in F$.   ☐

4.32 Proposition. Let (X, d) be a compact metric space. Then (X, d) is totally bounded.

Proof. Let $\epsilon > 0$. The collection $\{B_\epsilon(x; d) \mid x \in X\}$ of all d-balls of radius ϵ is an open cover of X, so X has a finite subset F with

$$X = \bigcup_{y \in F} B_\epsilon(y; d).$$

Then F is ϵ-dense in X.   ☐

Notice that from the topological property of compactness we have deduced what appears to be a nontopological conclusion concerning the metric d. This point is clarified by the following examples.

4.33 Examples

(1) If a metric space (X, d) is isometric to a metric space (Y, d'), then (X, d) is totally bounded if and only if (Y, d') is. Hence total boundedness is a metric property.

(2) If Y is a subset of a totally bounded metric space (X, d) and if d' is the metric on Y induced by d, then (Y, d') is totally bounded.

(3) Let d' be the Euclidean metric on the open interval $]0, 1[$. Since the closed interval $[0, 1]$ is compact, it follows from 4.32 and (2) that the metric space $(]0, 1[, d')$ is totally bounded. However, $]0, 1[$ is not compact.

(4) Let d be the Euclidean metric on R. Then (R, d) is not totally bounded. In fact, if (R, d) were totally bounded we could write R as the union of finitely many sets each of diameter at most 1, so R would then be d-bounded.

(5) Let d' and d be as in (3) and (4). Then $(]0, 1[, d')$ is topologically equivalent to (R, d) since $]0, 1[$ is homeomorphic to R, $(]0, 1[, d')$ is totally bounded, but (R, d) is not totally bounded.

Although the class of all totally bounded metric spaces strictly contains the class of all compact metric spaces, the two classes share the property of being second-countable.

4.34 Proposition. Let (X, d) be a totally bounded metric space. Then X is second-countable.

Proof. By 2.57 it suffices to show that X is separable. For each positive integer n there is a finite subset F_n of X which is $(1/n)$-dense in X in the sense of 4.31. Then $\bigcup_{n=1}^{\infty} F_n$ is a countable set that is easily seen to be dense in X. $\quad \square$

A totally bounded metric space need not be compact. However, from 4.34 and the next theorem it will follow that such a space is close to being compact: each open cover of the space contains a *countable* cover of the space.

4.35 Theorem (Lindelöf theorem). Let $\mathfrak{u}$ be an open cover of a second-countable topological space X. Then $\mathfrak{u}$ contains some countable cover of X.

Proof. Let $\mathfrak{B}$ be a countable base of X. Define

$$\mathfrak{a} = \{B \in \mathfrak{B} \mid B \subset U \text{ for some } U \in \mathfrak{u}\}.$$

Since $\mathfrak{a} \subset \mathfrak{B}$, the collection $\mathfrak{a}$ is countable. For each $B \in \mathfrak{a}$ choose some one $U_B \in \mathfrak{u}$ with

$$B \subset U_B.$$

Define

$$\mathfrak{v} = \{U_B \mid B \in \mathfrak{a}\}.$$

Then $\mathfrak{v} \subset \mathfrak{u}$, and $\mathfrak{v}$ is countable since $\mathfrak{a}$ is.

We show that $\mathcal{V}$ is a cover of X. Let $x \in X$. By hypothesis, $x \in U$ for some $U \in \mathcal{U}$. Since $\mathcal{B}$ is a base of X, then $x \in B \subset U$ for some $B \in \mathcal{B}$. Then $B \in \mathcal{C}$, $U_B \in \mathcal{V}$, and $x \in U_B$.   ☐

This theorem is named after E. Lindelöf, who in 1903 proved it for Euclidean spaces. Although of considerable interest in itself, in the present context it is purely a technical result needed to prove the next theorem.

Theorem 4.27 said that an arbitrary topological space is compact precisely when each net in the space clusters. We know that sequences suffice to describe the topology of a metrizable space. Then it is to be expected that for a metrizable space compactness can be characterized in terms of sequences. This is happily the case and provides the principal tool for dealing with compactness in metrizable spaces.

4.36 Theorem. For a metrizable space X the following conditions are equivalent:

(1) X is compact.

(2) X is "countably compact" in the sense that each countable open cover of X contains some finite cover of X.

(3) Each sequence in X clusters in X.

(4) X is "sequentially compact" in the sense that each sequence in X has a subsequence that converges in X.

Proof. Clearly (1) implies (2).

Assume (2). We show (3). Let $(x_n \mid n \in \mathsf{N})$ be a sequence in X, and just suppose this sequence does not cluster in X. For each $n \in \mathsf{N}$ define an open subset U_n of X by

$$U_n = \bigcup \{V \mid V \text{ is open in } X, x_i \notin V \text{ for all } i \geq n\}.$$

The countable collection $\{U_n \mid n \in \mathsf{N}\}$ is a cover of X; in fact, if $x \in X$, then $(x_n \mid n \in \mathsf{N})$ does not cluster at x, there is a neighborhood V of x and an $n \in \mathsf{N}$ with $x_i \notin V$ for all $i \geq n$, and hence $x \in U_n$.

By assumption, $\{U_n \mid n \in \mathsf{N}\}$ contains a finite cover of X. Now $(U_n \mid n \in \mathsf{N})$ is an increasing sequence, so $X = U_n$ for some $n \in \mathsf{N}$. Since $x_n \in X$, we have $x_n \in U_n$, but this contradicts the definition of U_n.

Conditions (3) and (4) are equivalent, because in a first-countable space a sequence clusters at a point x if and only if some subsequence converges to x (see 3.69).

Assume (3). We show (2). Let $\{U_n \mid n \in \mathsf{N}\}$ be a countable open cover of X that contains no finite cover of X. For each $n \in \mathsf{N}$ we may choose some point

$$x_n \in X \setminus \bigcup_{i=0}^{n} U_i.$$

By assumption, the sequence $(x_n \mid n \in \mathsf{N})$ clusters at some $x \in X$. Let $i \in \mathsf{N}$

with $x \in U_i$. Since $(x_n \mid n \in \mathsf{N})$ clusters at x, there is an $n \geq i$ with $x_n \in U_i$, but this contradicts the choice of x_n.

Finally, assume (2). We show (1). Let d be a metric inducing the topology of X [this is the first place we use the full strength of the hypothesis that X is metrizable]. In view of 4.34 and 4.35, it suffices to show that (X, d) is totally bounded.

Suppose (X, d) is not totally bounded. Then there is an $\epsilon > 0$ such that no finite subset of X is ϵ-dense in X (see 4.31). Starting with any point $x_0 \in X$, we may successively choose points $x_1, x_2, \ldots, x_n, \ldots$ with

$$d(x_1, x_0) \geq \epsilon,$$

$$d(x_2, x_0) \geq \epsilon, d(x_2, x_1) \geq \epsilon,$$

$$\cdots\cdots\cdots$$

$$d(x_n, x_0) \geq \epsilon, d(x_n, x_1) \geq \epsilon, \ldots, d(x_n, x_{n-1}) \geq \epsilon.$$

The sequence $(x_n \mid n \in \mathsf{N})$ certainly cannot cluster in X. By the consequence (3) of our assumption (2), however, this sequence must cluster in X. $\square$

Sequential compactness, the form of compactness originally defined by Fréchet for arbitrary metric spaces, is not in general the same as compactness. There are compact Hausdorff spaces that are not sequentially compact, and first-countable sequentially compact Hausdorff spaces that are not compact (see Greever [14, p. 54]). Although countable and sequential compactness are topological properties that can be studied in their own right, in contrast to compactness they suffer the limitation of not being preserved under the formation of products of uncountably many spaces.

To relate Theorem 4.36 to a result in classical analysis, we recall a notion introduced in Exercise 2.24. Given a subset A of a topological space X, a point $x \in X$ is called a *limit point of A in X* if each neighborhood of x contains at least one element of A different from x.

4.37 Corollary. A metrizable space X is compact if and only if each infinite subset of X has a limit point in X.

Proof. Assume X is compact. Let A be an infinite subset of X. Choose a sequence $(x_n \mid n \in \mathsf{N})$ of distinct points of A. By 4.36, this sequence clusters at some $x \in X$. Then x is a limit point of A in X. In fact, let V be any neighborhood of x. There is an $n \in \mathsf{N}$ with $x \neq x_i$ for all $i \geq n$ (possibly $n = 0$). Since $(x_n \mid n \in \mathsf{N})$ clusters at x, there is an $i \geq n$ with $x_i \in V$. Then $x \neq x_i \in V \cap A$.

Conversely, assume each infinite subset of X has a limit point in X. To show that X is compact, by 4.36 it suffices to show that each sequence in X clusters in X. Let $(x_n \mid n \in \mathsf{N})$ be any sequence in X. If the range $\{x_n \mid n \in \mathsf{N}\}$ of this sequence is finite, the sequence certainly clusters in X. Now suppose $\{x_n \mid n \in \mathsf{N}\}$ is an infinite set. By assumption, this set has a limit point x in X.

We show that $(x_n \mid n \in \mathbb{N})$ clusters at x. Let V be any neighborhood of x and let $n \in \mathbb{N}$. The neighborhood

$$V \setminus \{x_i \mid 1 \leq i < n, x_i \neq x\}$$

of x contains some point x_i of $\{x_n \mid n \in \mathbb{N}\}$ different from x. Then $i \geq n$ and $x_i \in V$.   ☐

From this corollary and the Heine-Borel-Lebesgue theorem (4.12) we obtain the classical *Bolzano-Weierstrass theorem*: Each bounded infinite set of real numbers has a limit point in $\mathbb{R}$.

For our first substantial application of Theorem 4.36, we at last establish the exact relationship between compactness and boundedness.

4.38 Theorem. Let d be any metric inducing the topology of a metrizable space X. Then a necessary and sufficient condition for the topological space X to be compact is that the metric space (X, d) be both complete and totally bounded.

Proof. Necessity. Assume X is compact. That (X, d) is totally bounded is just 4.32. To see that (X, d) is also complete, let $(x_n \mid n \in \mathbb{N})$ be a Cauchy sequence in (X, d). By 4.36 (or by 4.27), this sequence clusters at some $x \in X$.

We show that $(x_n \mid n \in \mathbb{N})$ converges to x. Let $\epsilon > 0$. Since $(x_n \mid n \in \mathbb{N})$ is a Cauchy sequence, there is an $n \in \mathbb{N}$ with

$$d(x_i, x_j) < \frac{\epsilon}{2} \qquad (i \geq n, j \geq n),$$

and since it clusters at x, there is an $m \geq n$ with

$$d(x_m, x) < \frac{\epsilon}{2}.$$

Then $d(x_i, x) < \epsilon$ for all $i \geq n$. (Observe that this part of the proof did not require 4.36, only 4.27.)

Sufficiency. Assume (X, d) is complete and totally bounded. We are going to show that each sequence in X clusters in X, and from 4.36 it will then follow that X is compact. Let $(x_n \mid n \in \mathbb{N})$ be any sequence in X. If the range

$$R = \{x_n \mid n \in \mathbb{N}\}$$

of $(x_n \mid n \in \mathbb{N})$ is finite, then already this sequence will cluster in X, so we now suppose that R is infinite. To obtain a point at which $(x_n \mid n \in \mathbb{N})$ clusters we shall use the total boundedness of (X, d) to construct a certain sequence of sets and then apply the nested set theorem (1.71) to this sequence of sets.

Since (X, d) is totally bounded, for each $n \in \mathbb{N}$ there is a finite cover $\mathcal{E}_n$

of X consisting of d-bounded sets with

$$\text{diam } E \leq \frac{1}{n+1} \qquad\qquad (E \in \mathcal{E}_n).$$

Since the diameter of the closure of a d-bounded set is the same as the diameter of the set (Exercise 2.40), we may assume that each member of each $\mathcal{E}_n$ is closed in X.

We construct now a decreasing sequence $(E_n \mid n \in \mathsf{N})$ of closed subsets of X such that

$$\text{diam } E_n \leq \frac{1}{n+1}, \qquad E_n \cap R \text{ is infinite} \qquad (n \in \mathsf{N}).$$

Since the cover $\mathcal{E}_0$ of X is finite and R is infinite, we may choose $E_0 \in \mathcal{E}_0$ with $E_0 \cap R$ infinite. Now suppose $n > 0$ and we have already constructed $E_0, \ldots, E_{n-1}$. Since the cover $\mathcal{E}_n$ of X is finite and $E_{n-1} \cap R$ is infinite, we may choose $E_n' \in \mathcal{E}_n$ with $E_n' \cap (E_{n-1} \cap R)$ infinite, and then we set $E_n = E_n' \cap E_{n-1}$.

By 1.71, there is some point $x \in \bigcap_{n=0}^{\infty} E_n$. To complete the proof we show that $(x_n \mid n \in \mathsf{N})$ clusters at x. Let $\epsilon > 0$, and let $m \in \mathsf{N}$. Choose $n \in \mathsf{N}$ with $1/(n+1) < \epsilon$. Since $E_n \cap R$ is infinite, there is an $i \geq m$ with $x_i \in E_n$; but $x \in E_n$ and diam $E_n \leq 1/(n+1) < \epsilon$, so $d(x, x_i) < \epsilon$. $\quad\square$

The remarkable thing about this theorem is its assertion that in a metrizable space the topological property of compactness is equivalent to the metric property of completeness and total boundedness. Note that neither completeness nor total boundedness of (X, d) alone suffices to ensure the compactness of X; consider R and $]0, 1[$ with their Euclidean metrics.

We shall say that a subset K of a metric space (X, d) is d-*totally bounded* to mean, of course, that the metric space (K, d') is totally bounded, where d' is the metric on K induced by d. By combining 4.38 with 1.68, we obtain the following corollary.

4.39 Corollary. Let (X, d) be a complete metric space. Then a subset K of X is compact if and only if K is both closed in X and d-totally bounded.

This corollary is an analog of the Heine-Borel-Lebesgue theorem (4.12) and at the same time is an explanation of why that theorem does not generalize intact to arbitrary complete metric spaces.

For a second application of 4.36, we now prove, as promised in the preceding section, that the product of a sequence of compact metrizable spaces is itself compact. (Actually, our proof uses Theorem 4.38, but the proof of that result rested on 4.36).

4.40 Theorem (Tychonoff product theorem—the denumerable case). Let $(X_i \mid i = 1, 2, \ldots)$ be a sequence of nonempty metrizable spaces. Then the product space $\mathsf{X}_{i=1}^{\infty} X_i$ is compact if and only if X_i is compact for all $i = 1, 2, \ldots$.

Proof. Set $X = \bigtimes_{i=1}^{\infty} X_i$. The proof that each X_i is compact if X is compact is the same as in the finite case (4.24) and requires no metrizability assumption. To prove the converse we do use metrics. By 1.36, for each i there is a bounded metric d_i inducing the topology of X_i with diam $X_i \leq 1$. Then

$$d(x, y) = \sum_{i=1}^{\infty} \frac{d_i(x_i, y_i)}{2^i}$$

defines a metric on X that induces the product topology on X (3.25).

Assume now that X_i is compact for all $i = 1, 2, \ldots$. By 4.38, (X_i, d_i) is complete and totally bounded for each i. By Exercise 1.73, (X, d) is complete. If we can show that (X, d) is also totally bounded, then from 4.38 it will follow that X is compact.

Let $\epsilon > 0$. We shall construct a finite subset of X that is ϵ-dense in X in the sense of Lemma 4.31. Choose n so large that

$$\sum_{i=n+1}^{\infty} \frac{1}{2^i} < \frac{\epsilon}{2}.$$

Fix any point

$$(z_{n+1}, z_{n+2}, \ldots) \in X_{n+1} \times X_{n+2} \times \cdots.$$

For $i = 1, \ldots, n$ the totally bounded metric space (X_i, d_i) has a finite subset F_i that is $(\epsilon/2)$-dense in X_i. Set

$$F = F_1 \times \cdots \times F_n \times \{z_{n+1}\} \times \{z_{n+2}\} \times \cdots,$$

so that F is a finite subset of X.

To complete the proof, we show that F is ϵ-dense in X. Consider any point $x = (x_i \mid i = 1, 2, \ldots) \in X$. For each $i = 1, \ldots, n$ there is a point $y_i \in F_i$ with

$$d_i(x_i, y_i) < \frac{\epsilon}{2}.$$

Set $y_{n+1} = z_{n+1}, y_{n+2} = z_{n+2}, \ldots$. Then the point $y = (y_i \mid i = 1, 2, \ldots)$ of F satisfies

$$d(x, y) = \sum_{i=1}^{n} \frac{d_i(x_i, y_i)}{2^i} + \sum_{i=n+1}^{\infty} \frac{d_i(x_i, y_i)}{2^i}$$

$$< \sum_{i=1}^{n} \frac{\epsilon}{2^{i+1}} + \sum_{i=n+1}^{\infty} \frac{1}{2^i}$$

$$< \frac{\epsilon}{2} \sum_{i=1}^{\infty} \frac{1}{2^i} + \frac{\epsilon}{2}$$

$$= \epsilon. \qquad \square$$

An immediate application of this theorem is that *the Hilbert cube is compact.*

The main thrust of our discussion of compact metrizable spaces up until now has been the use of metrics to provide sufficient criteria for compactness. In the remainder of the discussion we proceed in the opposite direction, deducing metric properties from the topological assumption of compactness.

Suppose $\mathcal{U}$ is an open cover of a compact space X. Although some (possibly very large) number of members of $\mathcal{U}$ will also form a cover of X, a given subset of X may require many of these members to cover all its points. When the topology of X is induced by a metric d, however, any subset of X of sufficiently small d-diameter will lie completely within a single one of these members.

4.41 Theorem (Lebesgue covering lemma). Let $\mathcal{U}$ be an open cover of a compact metric space (X, d). Then there exists a real number $\delta > 0$ such that any two points $x, t \in X$ with $d(x, t) \leq \delta$ belong to the same member of $\mathcal{U}$.

Proof. For each $x \in X$ we may choose a set $U_x \in \mathcal{U}$ with $x \in U_x$ and then choose a number $\delta(x) > 0$ with

$$B_{\delta(x)}(x; d) \subset U_x.$$

The collection $\{B_{\delta(x)/2}(x; d) \mid x \in X\}$ is an open cover of X, so there is a finite set $F \subset X$ with

$$X = \bigcup_{z \in F} B_{\delta(z)/2}(z; d)$$

(we may assume $F \neq \varnothing$, for otherwise $X = \varnothing$ and there is nothing to prove). Set

$$\delta = \min_{z \in F} \frac{\delta(z)}{2}$$

Let $x, t \in X$ with $d(x, t) \leq \delta$. Choose $z \in F$ with $x \in B_{\delta(z)/2}(z; d)$. Then

$$d(t, z) \leq d(t, x) + d(x, z) < \delta + \frac{\delta(z)}{2} \leq \frac{\delta(z)}{2} + \frac{\delta(z)}{2} = \delta(z).$$

Hence x and t both belong to $B_{\delta(z)}(z; d)$ and consequently to the member U_z of $\mathcal{U}$. $\square$

An evidently equivalent way of stating the property of δ is that if $A \subset X$ with diam $A \leq \delta$, then $A \subset U$ for some $U \in \mathcal{U}$. Any positive number δ having this property is called a *Lebesgue number* of the open cover $\mathcal{U}$ of X.

We are going to apply the Lebesgue covering lemma to prove an important result concerning uniform continuity, a notion introduced in Exercise 1.66 and which we now recall.

4.42 Definition. Given metric spaces (X, d) and (Y, d'), a map $f: X \to Y$ is said to be (d, d')-*uniformly continuous* if for each $\epsilon > 0$ there is some $\delta > 0$

such that

$$d(x, t) < \delta \quad \Rightarrow \quad d'(f(x), f(t)) < \epsilon$$

for all $x, t \in X$.

Contrast this definition with what it means to say that f is continuous. The map f is continuous when it is (d, d')-continuous at each $x \in X$—in other words, when for each $\epsilon > 0$ and each $x \in X$, there is some $\delta(x) > 0$ such that

$$d(x, t) < \delta(x) \quad \Rightarrow \quad d'(f(x), f(t)) < \epsilon.$$

Corresponding to a given ϵ, then, ordinary continuity of f says there is a separate $\delta(x)$ for each $x \in X$ which "works at x", whereas uniform continuity says there is a single δ which "works at every $x \in X$".

Clearly f will be continuous if it is (d, d')-uniformly continuous, but the converse fails (Exercise 1.66). A partial converse, however, does hold.

4.43 Theorem. Let $f: X \to Y$ be a continuous map from a compact metric space (X, d) into a metric space (Y, d'). Then f is (d, d')-uniformly continuous.

Proof. Let $\epsilon > 0$. The collection

$$\{f^{-1}(B_{\epsilon/2}(y; d')) \mid y \in Y\}$$

is an open cover of X, so by 4.41 it has a Lebesgue number $\delta > 0$. If $x, t \in X$ with $d(x, t) < \delta$, there is a $y \in Y$ with $x, t \in f^{-1}(B_{\epsilon/2}(y; d'))$, so $f(x), f(t) \in B_{\epsilon/2}(y; d')$, and hence $d'(f(x), f(t)) < \epsilon$. $\square$

This theorem was first proved by E. Heine for the case of a continuous real-valued function on a closed interval $[a, b]$ in R. In that case, especially, it has many significant applications to the theory of integration and other areas of analysis. We give here just one simple application.

4.44 Example. Consider a closed interval $[a, b]$ in R. A function $g: [a, b] \to$ R which, for suitable points

$$a = x_0 < x_1 < \cdots < x_{n-1} < x_n = b,$$

is linear on each of the open intervals $]x_0, x_1[, \ldots,]x_{n-1}, x_n[$ is said to be *piecewise linear*. Such a function will be continuous precisely when it is linear on each of the closed intervals $[x_0, x_1], \ldots, [x_{n-1}, x_n]$.

Let $f: [a, b] \to$ R be a given continuous function, and let $\epsilon > 0$ be a given number. We claim there is a continuous piecewise linear function $g: [a, b] \to$ R with

$$|f(x) - g(x)| < \epsilon \qquad\qquad (a \le x \le b),$$

in other words, with g approximating f to within ϵ uniformly over $[a, b]$.

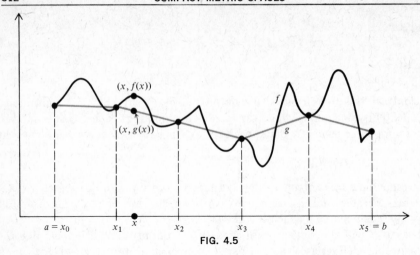

FIG. 4.5

To justify our claim we first invoke the uniform continuity of f to obtain a $\delta > 0$ such that $x, t \in [a, b]$ and $|x - t| < \delta$ implies $|f(x) - f(t)| < \epsilon/2$. Next we choose a positive integer n with $(b - a)/n < \delta$ and define points $x_0, x_1, \ldots, x_n$ in $[a, b]$ by

$$x_i = a + i\frac{b - a}{n} \qquad (i = 0, 1, \ldots, n).$$

Then for each $i = 1, \ldots, n$ we have $|x_i - x_{i-1}| = (b - a)/n < \delta$ so that

$$x, t \in [x_{i-1}, x_i] \quad \Rightarrow \quad |f(x) - f(t)| < \frac{\epsilon}{2}.$$

Define $g: [a, b] \to \mathbf{R}$ to be the function that is linear on each of the closed intervals $[x_{i-1}, x_i]$ and takes the values

$$g(x_i) = f(x_i) \qquad (i = 0, 1, \ldots, n)$$

at the endpoints of these intervals (see Fig. 4.5).

Suppose $x \in [x_{i-1}, x_i]$, where $1 \leq i \leq n$. Then $g(x)$ lies between the numbers $g(x_{i-1}) = f(x_{i-1})$ and $g(x_i) = f(x_i)$, so

$$|f(x_i) - g(x)| \leq |f(x_i) - f(x_{i-1})| < \frac{\epsilon}{2}.$$

Since also

$$|f(x) - f(x_i)| < \frac{\epsilon}{2},$$

we conclude that $|f(x) - g(x)| < \epsilon$.

EXERCISES

32. Show that a metric space (X, d) is totally bounded precisely when each infinite subset of X contains distinct points that are an arbitrarily small distance apart.

33. If d is the Euclidean metric on R^n, show that a subset of R^n is d-totally bounded if and only if it is d-bounded. Is the same thing true if d is instead the max metric on R^n? the taxicab metric?

34. Given two equivalent metrics d and d' on the same set X with (X, d) totally bounded, must (X, d') also be totally bounded?

35. Suppose a metric space (X, d) has a dense d-totally bounded subset. Will (X, d) necessarily be totally bounded?

36. Does the Lindelöf theorem (4.35) remain valid if the hypothesis that X be second-countable is weakened to X being first-countable? If in addition X is metrizable?

37. A topological space X is said to be *σ-compact* if $X = \bigcup_{n=0}^{\infty} K_n$ for some sequence $(K_n \mid n \in \mathsf{N})$ of compact subsets of X. For example, any compact space is σ-compact; the noncompact space R is σ-compact since $\mathsf{R} = \bigcup_{n=0}^{\infty} [-n, n]$.

(**a**) Show that the Lindelöf theorem (4.35) remains valid if the hypothesis that X be second-countable is changed to X being σ-compact (of course, it is trivially valid if X is actually compact).

(**b**) Let $\Im$ be the usual topology on R and let $S = \{1/n \mid n = 1, 2, \ldots\}$. Verify that the collection

$$\mathcal{S} = \{U \backslash A \mid U \in \Im, A \subset S\}$$

is a topology on R and that the topological space $(\mathsf{R}, \mathcal{S})$ is a noncompact σ-compact Hausdorff space that is not first-countable (and hence is neither second-countable nor metrizable).

(**c**) Prove that any σ-compact metrizable space is necessarily second-countable.

38. Prove the following converse of the Lindelöf theorem: If each open cover of a metrizable space X contains a countable cover of X, then X is second-countable.

39. Let S be the set of all sequences in $\{0, 1\}$. Let X be the set of all functions $x: S \to \{0, 1\}$, and provide X with its topology of pointwise convergence, where $\{0, 1\}$ has the discrete topology. Form the sequence $(x_n \mid n \in \mathsf{N})$ in X defined by

$$x_n(s) = s_n \qquad\qquad (s = (s_i \mid i \in \mathsf{N}) \in S).$$

Show that $(x_n \mid n \in \mathsf{N})$ has no convergent subsequence, thereby demonstrating that X is not sequentially compact. (The general Tychonoff product theorem says that X is, nonetheless, compact.)

40. Does Theorem 4.8 generalize to countably compact topological spaces?

41. (**a**) Prove that any continuous image of any countably compact topological space is itself countably compact.

(**b**) Deduce from (a) that a continuous function $f: X \to \mathsf{R}$ on a countably compact topological space is bounded and attains both a maximum and a minimum value on X.

42. (**a**) Show that a subset A of a metrizable space X is relatively compact (Exercise 11) if and only if each sequence in A clusters in X.

(**b**) Give an analog of (a) concerning limit points.

43. (**a**) If X is a compact space, show that each infinite subset of X has a limit point in X.

(**b**) Verify that the countable collection $\{\{2n, 2n + 1\} \mid n \in \mathsf{Z}\}$ of subsets of Z is a base for a topology $\Im$ on Z such that $(\mathsf{Z}, \Im)$ is noncompact but each infinite subset of Z has a limit point in this space.

44. Must an infinite d-totally bounded subset of a metric space (X, d) have a limit point in X? If (X, d) is complete?

45. A *compactification* of a topological space X is a compact space Y containing X as a dense subspace. For example, the extended real line is a compactification of the real line. Prove that a separable metrizable space always has a compactification. (*Hint*: Look at Theorem 3.23.)

46. Let X be a second-countable metrizable space. Show that there is a totally bounded metric on X that induces the given topology of X.

47. Prove that a metric space is totally bounded if and only if each sequence in X has a subsequence that is a Cauchy sequence in (X, d).

48. Is the Hilbert sequence space (1.9) compact?

49. Let K be a compact subset of the product X of a sequence of noncompact metrizable spaces. Prove that K has an empty interior in X.

50. Given a compact space X, let U be an open subset of $X \times X$ containing the diagonal $\{(x, x) \mid x \in X\}$ of $X \times X$.
  (**a**) Show there is a finite open cover $\{V_1, \ldots, V_n\}$ of X with $\bigcup_{i=1}^{n} (V_i \times V_i) \subset U$.
  (**b**) If a metric d induces the topology of X, show there is an $\epsilon > 0$ such that $\{(x, y) \in X \times X \mid d(x, y) < \epsilon\} \subset U$.

51. Let d' be the Euclidean metric on $\mathbf{R}$. Prove that the following is a necessary and sufficient condition for a real-valued function f on a metric space (X, d) to be (d, d')-uniformly continuous: If $(x_n \mid n \in \mathbf{N})$ and $(y_n \mid n \in \mathbf{N})$ are any sequences in X with $\lim_{n \to \infty} d(x_n, y_n) = 0$, then $\lim_{n \to \infty} |f(x_n) - f(y_n)| = 0$.

52. (**a**) Let $f: [0, 1] \to \mathbf{R}$ be continuous and let $\epsilon > 0$. Show there is a positive integer n and a continuous function $g: [0, 1] \to \mathbf{R}$ that is linear on each of the intervals $[0, 1/n]$, $[1/n, 2/n]$, $\ldots$, $[(n-1)/n, 1]$, takes rational values at the $n + 1$ endpoints $0, 1/n, \ldots, 1$ of these intervals, and approximates f to within ϵ uniformly over $[0, 1]$.
  (**b**) Deduce that the space $\mathcal{C}([0, 1])$ of all continuous real-valued functions on $[0, 1]$ is separable when given the topology induced by its max metric d_∞ (1.7).

53. Again let $f: [0, 1] \to \mathbf{R}$ be continuous and $\epsilon > 0$. Establish the existence of a positive integer n and functions g and h on $[0, 1]$ that are constant on each of the open intervals $]0, 1/n[$, $]1/n, 2/n[$, $\ldots$, $](n-1)/n, 1[$ such that $g(x) \leq f(x) \leq h(x)$ and $|h(x) - g(x)| < \epsilon$ for all $x \in [0, 1]$. (The "step functions" g and h need not be continuous, of course.) (*Note*: This result is crucial for proving that f is integrable.)

54. Let d' be the Euclidean metric on $\mathbf{R}$, and let (X, d) be a metric space with the property that every continuous function $f: X \to \mathbf{R}$ is (d, d')-uniformly continuous.
  (**a**) If A and B are any two disjoint nonempty closed subsets of X, show that $0 < d(A, B)$, the distance from A to B (Exercise 1.28). (*Hint*: Use Exercise 17.)
  (**b**) Prove that (X, d) must be complete. [*Hint*: If $(x_n \mid n \in \mathbf{N})$ is a Cauchy sequence of distinct points, apply (a) taking $A = \{x_{2n} \mid n \in \mathbf{N}\}$ and $B = \{x_{2n+1} \mid n \in \mathbf{N}\}$.]

55. Let (X, d) be a metric space, let d_1 be the metric induced by d on a dense subset D of X, and let (Y, d') be a complete metric space. Let $f: D \to Y$ be a map.
  (**a**) If f is (d_1, d')-uniformly continuous, prove that f has a unique continuous

extension $F: X \to Y$ and that F is (d, d')-uniformly continuous. [*Hint*: If $x \in X$, there is a sequence $(x_n \mid n \in \mathsf{N})$ in D converging to x in X; if such a continuous extension exists, then $(f(x_n) \mid n \in \mathsf{N})$ must converge to $F(x)$.]

(**b**) Must f have a continuous extension to X if f is (d_1, d')-uniformly continuous but (Y, d') is not complete? If f is only continuous but (X, d) and (Y, d') are both complete?

56. Let (X, d) and (Y, d') be metric spaces. A surjection $f: X \to Y$ is said to be "(d, d')-uniformly open" if for each $\delta > 0$ there is some $\epsilon > 0$ such that $y, z \in Y$ with $d'(y, z) < \epsilon$ implies $y = f(x)$ and $z = f(t)$ for some $x, t \in X$ with $d(x, t) < \delta$.

(**a**) Show that a (d, d')-uniformly open surjection $f: X \to Y$ must be an open map, but that the converse need not hold.

(**b**) If X is compact, prove that each continuous open surjection $f: X \to Y$ is (d, d')-uniformly open.

57. Let $(E_n \mid n \in \mathsf{N})$ be a decreasing sequence of nonempty closed subsets of a compact metric space. Compare diam $\bigcap_{n=0}^{\infty} E_n$ with inf $\{\text{diam } E_n \mid n \in \mathsf{N}\}$.

58. Let $\mathfrak{F}$ be the collection of all nonempty closed subsets of a compact metric space (X, d), and provide $\mathfrak{F}$ with its topology induced by the Hausdorff metric (Exercise 1.37). Show that the function $E \mapsto \text{diam } E$ from $\mathfrak{F}$ to R is continuous.

59. (**a**) Prove: A Hausdorff space that is the continuous image of a compact metrizable space is second-countable. [*Hint*: You may want to use Exercise 27(a).]

(**b**) Deduce from (a) and the Urysohn metrization theorem (see the paragraph following 4.28) that the quotient space of a compact metrizable space under a closed equivalence relation is itself metrizable.

60. (**a**) Prove that an isometric embedding (Exercise 1.38) of a compact metric space (X, d) into itself actually maps X onto X and hence is an isometry from (X, d) to (X, d).

(**b**) If each of two compact metric spaces can be isometrically embedded in the other, prove that the two metric spaces are isometric to one another. [Compare Exercise 1.40.]

3. Locally Compact Spaces

It is a fact of life that numerous topological spaces, including Euclidean spaces, fail to be compact. The pleasant properties of compact spaces can still be exploited in many of these noncompact spaces provided they are "locally" compact in the sense of having arbitrarily small compact neighborhoods at each point.

4.45 Definition. A topological space X is said to be *locally compact* if at each point $x \in X$ there is a local base consisting of compact sets—in other words, for each neighborhood U of x there is a compact neighborhood V of x with $V \subset U$.

Most of the locally compact spaces encountered are separated as well, and for such spaces the following proposition provides an especially simple criterion for local compactness.

4.46 Proposition. For a Hausdorff space X the following conditions are equivalent:

(1) X is locally compact.

(2) For each $x \in X$ and each neighborhood U of x there is an open neighborhood V of x such that cls V is compact and cls $V \subset U$.

(3) Each $x \in X$ has some compact neighborhood in X.

Proof. Assume (1). We show (2). Let $x \in X$ and let U be a neighborhood of x. By (1) there is a compact neighborhood W of x with $W \subset U$. Take

$$V = \text{int } W,$$

so that V is an open neighborhood of x. Now W, being a compact subset of the Hausdorff space X, is closed in X. Hence cls $V \subset$ cls $W = W \subset U$, and cls V is compact since it is a closed subset of the compact set W.

Clearly (2) implies (3).

Assume (3). We show (1). Let $x \in X$ and let U be any neighborhood of x. By assumption there exists some compact neighborhood K of x in X. Set

$$W = \text{int } (U \cap K),$$

so that W is an open neighborhood of x in X. Since K is a compact Hausdorff space, the same is true of its subspace

$$Y = \text{cls } W.$$

In the compact Hausdorff space Y the set $Y \backslash W$ is closed and $x \notin Y \backslash W$, so by 4.14 there are open sets M and N in Y with $x \in M$, $Y \backslash W \subset N$, and $M \cap N = \varnothing$. Take

$$V = Y \backslash N.$$

Since $x \in M \subset V \subset W$, then V is a neighborhood of x in Y with $V \subset U$. The set V is compact since it is closed in the compact space Y. Finally, V is a neighborhood of x in X because $V \subset W$ and W is open in X. □

The main import of this proposition is that a Hausdorff space will be locally compact as soon as each of its points has at least *one* compact neighborhood. In particular, since a topological space is itself a neighborhood of each of its points, we obtain the following corollary.

4.47 Corollary. Any compact Hausdorff space is locally compact.

4.48 Examples

(1) Any discrete space X is locally compact, since for $x \in X$ the set $\{x\}$ is a neighborhood of X. Hence any infinite discrete space (Z, for example) is a locally compact Hausdorff space that is not compact.

(2) Euclidean space R^n, which is a noncompact Hausdorff space, is locally compact. In fact, if d is the Euclidean metric on R^n, then for each $x \in \mathsf{R}^n$ and each $\epsilon > 0$ the d-disk $D_\epsilon(x; d)$ is a neighborhood of x that is compact by 4.25.

(3) The Hilbert sequence space H (1.9) is not locally compact, for no neighborhood of a point $x \in \mathsf{H}$ is compact. To see this, note first that if x had a compact neighborhood W, then the closed neighborhood $D_\epsilon(x; d_2) \subset W$ would also be compact for sufficiently small ϵ.

We show that $D_\epsilon(x; d_2)$ is not compact. To prove this, by 4.36 it suffices to construct a sequence $(x^n \mid n \in \mathsf{N})$ in $D_\epsilon(x; d_2)$ that does not cluster in H (we use superscripts to denote the values of the sequence because subscripts are used to denote the coordinates of points in H). Define

$$x^0 = x^1 = \left(x_1 + \sqrt{\frac{\epsilon}{2}}, \, x_2, \, x_3, \, . \, . \, . \right),$$

$$x^2 = \left(x_1, \, x_2 + \sqrt{\frac{\epsilon}{2}}, \, x_3, \, . \, . \, . \right),$$

$$\cdot$$
$$\cdot$$
$$\cdot$$

$$x^n = \left(x_1, \, . \, . \, . \, , \, x_{n-1}, \, x_n + \sqrt{\frac{\epsilon}{2}}, \, x_{n+1}, \, . \, . \, . \right)$$

$$\cdot$$
$$\cdot$$
$$\cdot$$

Then $d_2(x^n, x^m) = \sqrt{\epsilon}$ for all $n, m \geq 1$ with $n \neq m$, so $(x^n \mid n \in \mathsf{N})$ cannot cluster in H.

(4) Let X be an infinite set provided with its finite-complement topology [2.3 (7)]. By 4.4 (5), X is a compact space that is not a Hausdorff space. Nonetheless, X is locally compact. In fact, every subset A of X is compact since the relative topology on A is the finite-complement topology on A.

(5) The space Q of rational numbers is not locally compact, for by the Heine-Borel-Lebesgue theorem (4.12) no closed neighborhood in Q of the form $\mathsf{Q} \cap [x - \epsilon, x + \epsilon]$, $\epsilon > 0$, can be compact. Likewise, the space $\mathsf{Q} \cap [0, 1]$ is not locally compact.

(6) Let z be an object with $z \notin \mathsf{Q}$, and let $X = \mathsf{Q} \cup \{z\}$. Define

$$\mathfrak{N}_z = \{X \backslash F \mid F \subset \mathsf{Q}, \, F \text{ finite}\}$$

and for each $x \in \mathsf{Q}$ define

$$\mathfrak{N}_x = \{V \mid V \text{ a neighborhood of } x \text{ in } \mathsf{Q}\}$$

$$\cup \{V \cup \{z\} \mid V \text{ a neighborhood of } x \text{ in } \mathsf{Q}\}.$$

It is easy to see that the conditions (N1)–(N5) listed in 2.16 are satisfied, and then Theorem 2.17 furnishes a topology on X such that for each $y \in X$ the collection $\mathfrak{N}_y$ is the collection of all neighborhoods of y. Provided with this topology, the space X is compact [the argument is like that in 4.4 (5)], so each point of X has a compact neighborhood. However, X is not locally compact, for no point $x \in$ Q has arbitrarily small compact neighborhoods in X.

Additional examples are provided by the following proposition.

4.49 Proposition. Let A be a subspace of a locally compact space X that is either open in X or closed in X. Then A is locally compact.

Proof. Let $x \in A$ and let U be any neighborhood of x in A.

Case (i). A is open in X. In this case U is also a neighborhood of x in X, so there is a compact neighborhood V of x in X with $V \subset U$. Since $V \subset A$, V is also a neighborhood of x in A.

Case (ii). A is closed in X. Choose a neighborhood U_0 of x in X such that $U = U_0 \cap A$. There is some compact neighborhood V_0 of x in X with $V_0 \subset U_0$. Set

$$V = V_0 \cap A.$$

Then V is a neighborhood of x in A with $V \subset U$. Now V_0 is compact, and V is closed in V_0 since A is closed in X. Hence V is compact by 4.8.   ⬚

This result is the analog of 4.8 for locally compact spaces. Unfortunately, 4.9, 4.17, and 4.20 do not admit direct analogs for locally compact spaces.

4.50 Examples

   (1) By 4.48 (2) and 4.49 the open interval $]0, 1[$ in R is a locally compact subspace of the Hausdorff space R, but $]0, 1[$ is not closed in R.
   (2) Let X be the set of rational numbers provided with its discrete topology and Y be the same set provided with its Euclidean topology. Then X is a locally compact Hausdorff space, Y is a Hausdorff space, and the identity map $f \colon X \to Y$ is continuous. By 4.48 (5) the image Y of X under f is not locally compact, and the map f is not closed.

Although the image of a locally compact space under an arbitrary continuous map thus need not be locally compact, we do have a limited result in this direction.

4.51 Proposition. Let $f \colon X \to Y$ be a continuous open surjection with X a locally compact space. Then Y is locally compact.

Proof. Let V be any neighborhood of a point $y \in Y$. Choose $x \in X$ with $f(x) = y$. Then $f^{-1}(V)$ is a neighborhood of x in X, so there is some compact neighborhood W of x in X with $W \subset f^{-1}(V)$. Since f is an open map, $f(W)$ is a neighborhood of y in Y, and clearly $f(W) \subset V$. Finally, $f(W)$ is compact by 4.17. $\square$

It follows from 4.51 that *local compactness is a topological property*.
Local compactness is preserved under the formation of finite products.

4.52 Theorem. Let $(X_i \mid i = 1, \ldots, n)$ be a finite family of nonempty topological spaces. Then the product space $\mathsf{X}_{i=1}^n X_i$ is locally compact if and only if X_i is locally compact for each $i = 1, \ldots, n$.

Proof. Let $X = \mathsf{X}_{i=1}^n X_i$. If X is locally compact, then for each i Proposition 4.51 may be applied to the ith projection $p_i \colon X \to X_i$ to show that X_i is locally compact.
Conversely, assume X_i is locally compact for each i. Let U be any neighborhood of a point $x = (x_1, \ldots, x_n) \in X$. There are neighborhoods V_1 of x_1 in $X_1, \ldots, V_n$ of x_n in X_n with $V_1 \times \cdots \times V_n \subset U$. By assumption, for each $i = 1, \ldots, n$ there is a compact neighborhood W_i of x_i in X_i with $W_i \subset V_i$. Then $W = W_1 \times \cdots \times W_n$ is a neighborhood of x in X with $W \subset U$, and W is compact by the Tychonoff product theorem 4.24. $\square$

Unlike compactness, local compactness is not preserved under the formation of the product of even denumerably many spaces. Instead, the following limited result holds.

4.53 Theorem. Let $(X_i \mid i = 1, 2, \ldots)$ be a sequence of nonempty metrizable spaces. Then the product space $\mathsf{X}_{i=1}^\infty X_i$ is locally compact if and only if X_i is locally compact for every i and X_i is compact for all but finitely many i.

Proof. Let $X = \mathsf{X}_{i=1}^\infty X_i$. Assume first that X is locally compact. Exactly as in the proof of 4.52 it follows that X_i is locally compact for all i. Now let V be a compact neighborhood of any point $x \in X$. By replacing V if necessary by a smaller closed neighborhood of x, we may assume V has the form

$$V = V_1 \times \cdots \times V_n \times X_{n+1} \times X_{n+2} \times \cdots$$

for some $n \geq 1$ and some neighborhoods V_1 of x_1 in $X_1, \ldots, V_n$ of x_n in X_n. Then for all $i > n$ the ith projection $p_i \colon X \to X_i$ maps the compact set V onto X_i, so X_i is compact.
Conversely, assume X_i is locally compact for all i and X_i is compact for all but finitely many i. Choose $n \geq 1$ such that X_i is compact for all $i > n$. Let $x \in X$ and let U be any neighborhood of x in X. There is an $m \geq n$ such

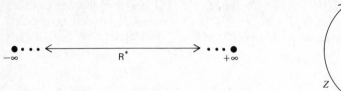
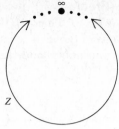

FIG. 4.6

that U contains a neighborhood V of x of the form

$$V = V_1 \times \cdots \times V_m \times X_{m+1} \times X_{m+2} \times \cdots,$$

where V_1 is a neighborhood of x_1 in $X_1, \ldots, V_m$ is a neighborhood of x_m in X_m. By assumption, for each $i = 1, \ldots, m$ there is a compact neighborhood W_i of x_i in X_i with $W_i \subset V_i$. Set

$$W = W_1 \times \cdots \times W_m \times X_{m+1} \times X_{m+2} \times \cdots.$$

Then W is a neighborhood of x in X with $W \subset V \subset U$. From the Tychonoff product theorem 4.40 it follows that W is compact.   ☐

The only place metrizability of the spaces X_i was used in this proof was in the citation of Theorem 4.40. Since the Tychonoff product theorem is true without any metrizability assumption (although we have not proved it in that generality), so is Theorem 4.53.

According to 4.53, the product of denumerably many copies of R is not locally compact [actually, this product is homeomorphic to the Hilbert sequence space H—see 3.35 (6)—and we have already proved directly that H is not locally compact].

The very definition of locally compact spaces refers to compact spaces, and the preceding theorems about locally compact spaces were proved by using corresponding theorems about compact ones. We now look at another way locally compact spaces are related to compact spaces.

Consider the locally compact space R. Although R itself is noncompact, it is a dense subspace of the compact space R* [see 1.38 and 2.36 (4)], and R*\R = $\{-\infty, +\infty\}$. Thus by adjoining two points to the locally compact space R we obtain a compact space. But we can do even better than this. We can obtain a compact space by adjoining a single point to R.

Form the quotient space

$$Z = R^*/\{-\infty, +\infty\}$$

obtained from R* by collapsing the closed subset $\{-\infty, +\infty\}$ of R* to a single point. (Imagine grabbing hold of the extended real line at its ends $-\infty$ and $+\infty$ and pulling them together—see Fig. 4.6.) Denote the point $\{-\infty, +\infty\}$

of Z by ∞. The space Z is compact since R^* is. By 3.49 (8), the quotient map

$$q: \mathsf{R}^* \to Z$$

maps R homeomorphically onto the open subspace

$$q(\mathsf{R}) = Z \setminus \{\infty\}$$

of Z. Moreover, the homeomorphic image $q(\mathsf{R})$ of R is dense in Z since R is dense in R^*.

As in 3.22 (2) we may now regard R as a subspace of Z by replacing $q(\mathsf{R})$ with its homeomorphic copy R. Thus R becomes a dense open subspace of Z whose complement in Z consists of the single point ∞. Since R is an open subspace of Z, the open sets in Z that do not contain ∞ are precisely the open subsets of the real line R.

Let us next determine the open subsets of Z that contain ∞. Let U be an open set in Z with $\infty \in U$. Then $V = q^{-1}(U)$ is an open set in R^* with $-\infty, +\infty \in V$. By 1.39, $[-\infty, a[\cup]b, +\infty] \subset V$ for suitable real $a < b$. Then the closed subset $\mathsf{R} \setminus V$ of R satisfies $\mathsf{R} \setminus V \subset [a, b]$. Hence $\mathsf{R} \setminus V$ is compact. Setting

$$K = q(\mathsf{R} \setminus V),$$

we see that K is a compact subset of $Z \setminus \{\infty\} = \mathsf{R}$ with

$$U = Z \setminus K.$$

Now Z is easily seen to be separated, so each set U of this form is, conversely, an open subset of Z containing ∞.

To summarize,

$$Z = \mathsf{R} \cup \{\infty\}, \qquad \infty \notin \mathsf{R},$$

and the topology of Z is just

$$\{U \mid U \text{ open in } \mathsf{R}\} \cup \{Z \setminus K \mid K \subset \mathsf{R}, K \text{ compact}\}.$$

With this topology, Z is a compact space containing R as a dense subspace.

This description of the compact space obtained by adjoining a single point to R can immediately be generalized if we start with an arbitrary locally compact Hausdorff space.

4.54 Theorem (Alexandroff). Let X be a noncompact, locally compact Hausdorff space whose topology is $\mathfrak{I}$. Set

$$X_\infty = X \cup \{\infty\},$$

where $\infty \notin X$. Then the collection

$$\mathfrak{I}_\infty = \mathfrak{I} \cup \{X_\infty \setminus K \mid K \subset X, K \text{ compact}\}$$

is a topology on X_∞ making X_∞ a compact Hausdorff space in which X (with its original topology $\mathfrak{I}$) is a dense subspace.

Proof. Note first that if $V \in \mathfrak{I}$ and if K is a compact subset of X, then $V \setminus K \in \mathfrak{I}$ since X is a Hausdorff space.

We show that $\mathfrak{I}_\infty$ is a topology on X_∞. Clearly $\varnothing \in \mathfrak{I}_\infty$ and $X_\infty \in \mathfrak{I}_\infty$. We leave to the reader the verification that the union of any collection of members of $\mathfrak{I}_\infty$ belongs to $\mathfrak{I}_\infty$. Now let $V_1, V_2 \in \mathfrak{I}_\infty$. To show that $V_1 \cap V_2 \in \mathfrak{I}_\infty$, we consider the three possibilities. If $V_1 \in \mathfrak{I}$ and $V_2 \in \mathfrak{I}$, then $V_1 \cap V_2 \in \mathfrak{I} \subset \mathfrak{I}_\infty$. If $V_1 \in \mathfrak{I}$ but $V_2 \notin \mathfrak{I}$, then $V_2 = X_\infty \setminus K$ for some compact $K \subset X$, so

$$V_1 \cap V_2 = V_1 \cap (X_\infty \setminus K) = V_1 \setminus K \in \mathfrak{I} \subset \mathfrak{I}_\infty.$$

Finally, if $V_1 \notin \mathfrak{I}$ and $V_2 \notin \mathfrak{I}$, then $V_1 = X_\infty \setminus K_1$, $V_2 = X_\infty \setminus K_2$ for some compact sets $K_1, K_2 \subset X$, and so

$$V_1 \cap V_2 = (X_\infty \setminus K_1) \cap (X_\infty \setminus K_2) = X_\infty \setminus (K_1 \cup K_2) \in \mathfrak{I}_\infty$$

since $K_1 \cup K_2$ is compact.

We show that $(X, \mathfrak{I})$ is a subspace of $(X_\infty, \mathfrak{I}_\infty)$. Since $\mathfrak{I} \subset \mathfrak{I}_\infty$, we need only show that $V \cap X \in \mathfrak{I}$ for each $V \in \mathfrak{I}_\infty$. Let $V \in \mathfrak{I}_\infty$. If $V \in \mathfrak{I}$, we are done. Otherwise, if $V \notin \mathfrak{I}$, write $V = X_\infty \setminus K$ for a compact $K \subset X$; then $V \cap X = X \setminus K \in \mathfrak{I}$.

We show that X_∞ is compact. Let $\mathfrak{U}$ be an open cover of X_∞. Choose $U_\infty \in \mathfrak{U}$ with $\infty \in U_\infty$. Then

$$U_\infty = X_\infty \setminus K$$

for some compact $K \subset X$. Since $\mathfrak{U}$ is also an open cover of K in X_∞, some finite $\mathcal{E} \subset \mathfrak{U}$ is a cover of K. Then $\mathcal{E} \cup \{U_\infty\}$ is a finite cover of X_∞ contained in $\mathfrak{U}$.

We show that X_∞ is a Hausdorff space. Since $\mathfrak{I} \subset \mathfrak{I}_\infty$, any two distinct points of X have disjoint neighborhoods in X_∞. Now let $x \in X_\infty$ with $x \neq \infty$, that is, $x \in X$. By hypothesis there is at least one compact neighborhood W of x in X. Then $X_\infty \setminus W$ and W are disjoint neighborhoods of ∞ and x in X_∞.

We complete the proof by showing that X is dense in X_∞. Let V be any nonempty open subset of X_∞. If $V \in \mathfrak{I}$, then $V \cap X = V \neq \varnothing$. If $V \notin \mathfrak{I}$, then

$$V = X_\infty \setminus K$$

for some compact subset K of X, $K \neq X$ since X is noncompact, and hence

$$V \cap X = X \setminus K \neq \varnothing. \qquad \square$$

The space X_∞ constructed in 4.54 is called the *one-point* (or Alexandroff) *compactification of* X, and ∞ is called the *point at infinity* in X_∞.

Is there another way to construct a compact Hausdorff space in which X is a dense subspace whose complement is a single point? No, the one-point compactification X_∞ as just constructed is, up to homeomorphism, the only such space.

4.55 Proposition. Let X be a noncompact, locally compact Hausdorff space. Suppose Y is a compact Hausdorff space and $p \in Y$ with

$$X \cong Y \setminus \{p\}.$$

Then there is a homeomorphism

$$g : X_\infty \cong Y$$

with

$$g(\infty) = p.$$

Proof. Let $f : X \cong Y \setminus \{p\}$. Define a bijection $g : X_\infty \to Y$ by

$$g(x) = f(x) \qquad\qquad\qquad (x \in X),$$

$$g(\infty) = p.$$

Since X_∞ is compact and Y is a Hausdorff space, it now remains by 4.20 only to show that g is continuous.

Let $x \in X$. We show that g is continuous at x. Let V be any open neighborhood of $g(x)$ in Y. Since $g(x) = f(x) \neq p$, there is an open neighborhood W of $g(x)$ in Y with $p \notin W \subset V$. Then $U = f^{-1}(W)$ is an open neighborhood of x in X, and hence in X_∞, with $g(U) = f(U) \subset V$.

We show that g is continuous at ∞. Let V be any open neighborhood of $p = g(\infty)$ in Y. Now $Y \setminus V$ is compact, so

$$K = f^{-1}(Y \setminus V)$$

is a compact subset of X. Then

$$U = X_\infty \setminus K$$

is an open neighborhood of ∞ in X_∞ with $g(U) \subset V$. $\square$

It is at least intuitively evident that the space Z constructed from R in the discussion preceding 4.54 is a circle, so the one-point compactification of the real line is homeomorphic to S_1. This example generalizes to higher dimensions.

4.56 Example. For each $n \geq 1$, the one-point compactification $(\mathsf{R}^n)_\infty$ of n-dimensional Euclidean space is homeomorphic to the n-sphere S_n. In fact, let p be the "north pole" $(0, \ldots, 0, 1) \in \mathsf{S}_n$. Our earlier discussion of the stereographic projection [3.16 (11)] shows that

$$\mathsf{R}^n \cong \mathsf{S}_n \setminus \{p\}.$$

It follows from 4.26 (3) and 4.55 that

$$(\mathsf{R}^n)_\infty \cong \mathsf{S}_n.$$

The case $n = 2$ is of particular importance in complex analysis, for R^2 is as a topological space the complex plane C. In this context the one-point compactification of R^2 is called the *Riemann sphere* (the Alexandroff compactification of an arbitrary locally compact space is the natural generalization of the Riemann sphere). We can often describe the behavior of a complex function $f: \mathbf{C} \to \mathbf{C}$ at complex numbers z of large modulus $|z|$ in terms of the behavior at the point $\infty \in \mathbf{C}_\infty$ of a suitable extension of f to the Riemann sphere $\mathbf{C}_\infty$ (compare Exercises 77–79).

The development in the preceding section indicates that among all compact spaces, the metrizable ones are especially nice. We mentioned there that the second-countable compact Hausdorff spaces are precisely the compact metrizable spaces. It is therefore of interest to know that the one-point compactification of a second-countable locally compact Hausdorff space is itself second-countable. To prove this we need a bit of preparation.

4.57 Lemma. Let K be a compact subset of a locally compact Hausdorff space X. Then $K \subset U$ for some open subset U of X with cls U compact.

Proof. For each $x \in X$ there is by 4.46 an open set U_x in X with $x \in U_x$ and cls U_x compact. Since K is compact, it contains a finite set F such that $K \subset \bigcup_{x \in F} U_x$. Now take

$$U = \bigcup_{x \in F} U_x \qquad \square$$

4.58 Proposition. Let X be a second-countable locally compact Hausdorff space. Then there is a sequence $(U_n \mid n \in \mathbf{N})$ of open sets in X such that $\{U_n \mid n \in \mathbf{N}\}$ is a cover of X, cls U_n is compact for every $n \in \mathbf{N}$, and

$$\text{cls } U_n \subset U_{n+1}$$

for every $n \in \mathbf{N}$.

Proof. For each $x \in X$ there is by 4.46 an open neighborhood V_x of x with cls V_x compact. By the Lindelöf theorem (4.35) the open cover $\{V_x \mid x \in X\}$ contains a countable cover of X. Replacing each member of this countable cover by its closure, we obtain a sequence $(K_n \mid n \in \mathbf{N})$ of compact subsets of X with

$$X = \bigcup_{n=0}^{\infty} K_n.$$

By 4.57 there is an open set U_0 in X with

$$K_0 \subset U_0, \qquad \text{cls } U_0 \text{ compact.}$$

Now cls $U_0 \cup K_1$ is compact, so by 4.57 again there is an open set U_1 in X with

$$\text{cls } U_0 \cup K_1 \subset U_1, \qquad \text{cls } U_1 \text{ compact.}$$

Continuing in this way, we construct open sets $U_1, U_2, \ldots$ in X with

$$\text{cls } U_{n-1} \cup K_n \subset U_n, \qquad \text{cls } U_n \text{ compact}$$

for each $n \geq 1$. Since $\{K_n \mid n \in \mathbf{N}\}$ is a cover of X, so is $\{U_n \mid n \in \mathbf{N}\}$. $\qquad \square$

When $X = \mathbf{R}^k$ we can construct the sequence $(U_n \mid n \in \mathbf{N})$ directly by taking $U_n = \mathbf{B}_{n+1}(0; d)$, d being the Euclidean metric.

4.59 Theorem. Let X be a noncompact, locally compact Hausdorff space. Then X is second-countable if and only if the one-point compactification X_∞ of X is second-countable.

Proof. If X_∞ is second-countable, then its subspace X is certainly second-countable.

Conversely, assume X is second-countable. Let $(U_n \mid n \in \mathbf{N})$ be as in 4.58. Let V be an open neighborhood of ∞ in X_∞. We show that

$$X_\infty \backslash \mathrm{cls}\, U_m \subset V$$

for some $m \in \mathbf{N}$. Write

$$V = X_\infty \backslash K$$

for some compact $K \subset X$. Since $(U_n \mid n \in \mathbf{N})$ is increasing and $\{U_n \mid n \in \mathbf{N}\}$ is an open cover of K in X, there is an $m \in \mathbf{N}$ with $K \subset U_m$. Then $K \subset \mathrm{cls}\, U_m$, so $X_\infty \backslash \mathrm{cls}\, U_m \subset X_\infty \backslash K = V$.

Now let $\mathfrak{U}$ be a countable base of X. It follows from what we just proved that

$$\mathfrak{U} \cup \{X_\infty \backslash \mathrm{cls}\, U_n \mid n \in \mathbf{N}\}$$

is a countable base of X_∞. $\square$

EXERCISES

61. Which of the following spaces are locally compact?
  (**a**) An uncountable set provided with its countable-complement topology (Exercise 2.6).
  (**b**) The set of real numbers provided with its right-interval topology [2.18(1)].
  (**c**) The line with two origins [2.18(3)].
  (**d**) The half-disk space [2.20(3)].
  (**e**) The tangent disk space (Exercise 2.27).

62. Let

$$S = \left\{ \left(x, \sin \frac{1}{x} \right) \,\middle|\, 0 < x \le 1 \right\}.$$

  (**a**) Is the subspace $S \cup \{(0,0)\}$ of the plane locally compact?
  (**b**) Is the subspace $S \cup (\{0\} \times [-1, 1])$ of the plane locally compact?

63. Prove that any topological manifold (3.17) is locally compact.

64. Verify that the collection

$$\left\{ \varnothing, \, \rbrack 0, 1\lbrack \right\} \cup \left\{ \left\rbrack \frac{1}{n}, 1 \right\lbrack \,\middle|\, n = 2, 3, \ldots \right\}$$

is a topology on the set $\rbrack 0, 1 \lbrack$. Show that $\rbrack 0, 1 \lbrack$, provided with this topology, is a locally compact space but that cls U is not compact for any neighborhood U of any point of this space. (*Hint*: Show that every open set in this space is compact, but no nonempty closed set is compact.)

65. Verify that the collection

$$\{\varnothing\} \cup \{U \mid U \subset \mathsf{R}, \mathsf{R}\backslash U \text{ compact}\}$$

is a topology on R. Is R provided with this topology compact? locally compact? Hausdorff?

66. Let Y be a topological space that is not locally compact but is a dense subspace of some compact Hausdorff space X. Can $X\backslash Y$ be a singleton? a finite set? a denumerable set?

67. Let X be a Hausdorff space. According to 4.49, if X is locally compact, then each subspace Y of X of the form $Y = U \cap E$ with U open in X and E closed in X is a locally compact space. Prove that, conversely, each locally compact subspace Y of X has this form. (*Hint*: Find an open subset of X containing Y in which Y is closed.)

68. By considering the quotient space of R obtained by collapsing Z to a point, show that a Hausdorff space need not be locally compact when it is the image of a locally compact Hausdorff space under a continuous closed map. (Compare Exericse 85.)

69. Suppose in 4.54 that X is actually a compact Hausdorff space. Show that $\mathfrak{I}_\infty$ is still a topology on X_∞ making X_∞ a compact Hausdorff space containing X as a subspace, but X is no longer dense in X_∞.

70. Let X be any topological space with topology $\mathfrak{I}$, let $X_\infty = X \cup \{\infty\}$ with $\infty \notin X$, and let

$$\mathfrak{I}_\infty = \mathfrak{I} \cup \{X_\infty\backslash K \mid K \text{ is a closed compact subset of } X\}.$$

Verify that $\mathfrak{I}_\infty$ is a topology on X_∞. Provide X_∞ with $\mathfrak{I}_\infty$. Prove:

(**a**) X is a subspace of X_∞.

(**b**) X_∞ is compact.

(**c**) X_∞ is a Hausdorff space if and only if X is a locally compact Hausdorff space.

71. (**a**) Embed the one-point compactifications of the discrete spaces N and Z in the real line.

(**b**) Embed the one-point compactification of $]0, 1[\cup]2, 3[$ in the plane.

72. Show that the n-sphere S_n is homeomorphic to the quotient space of the n-disk D_n obtained by collapsing the $(n-1)$-sphere S_{n-1} to a point. [*Hint*: Show first that $(\mathsf{D}_n/\mathsf{S}_{n-1})\backslash\{\mathsf{S}_{n-1}\} \cong \mathsf{B}_n.$]

73. Let X and Y be two noncompact, locally compact Hausdorff spaces. Prove that $X_\infty \cong Y_\infty$ if $X \cong Y$. Is the converse true?

74. Let X be a locally compact Hausdorff space and let K be a compact subset of X. Given any open subset U of X with $K \subset U$, show that there is an open subset V of X with $K \subset V \subset \text{cls } V \subset U$ and cls V compact.

75. In the notation of 4.58, show that if K is any compact subset of X, then $K \subset U_n$ for some n.

76. Let X be a noncompact, locally compact Hausdorff space. Prove that X is σ-compact (Exercise 37) if and only if there is a countable local base at ∞ in X_∞. (*Hint*: The conclusion of 4.58 still holds if X is σ-compact but not necessarily second-countable.)

77. Let X be a noncompact, locally compact Hausdorff space. A real-valued function f on X is said to *vanish at infinity* provided that for each $\epsilon > 0$ there is a compact subset K of X with $|f(x)| < \epsilon$ for all $x \in X \setminus K$.

(**a**) Exhibit a continuous function $f \colon \mathsf{R} \to \mathsf{R}$ that vanishes at infinity such that $f(x) \neq 0$ for all $x \in \mathsf{R}$.

(**b**) Prove that $f \colon X \to \mathsf{R}$ vanishes at infinity precisely when its extension $f_\infty \colon X_\infty \to \mathsf{R}$ with $f_\infty(\infty) = 0$ is continuous at the point ∞.

(**c**) Show that a continuous function $f \colon X \to \mathsf{R}$ that vanishes at infinity is necessarily bounded. Must such a function attain a maximum or a minimum value on X?

78. Let $f \colon X \to Y$ be a map from a Hausdorff space X to a Hausdorff space Y, and let $y \in Y$. We write '$\lim_{x \to \infty} f(x) = y$' to mean that for each neighborhood V of y in Y there is some compact subset K of X with $f(x) \in V$ for all $x \in X \setminus K$.

(**a**) When $Y = \mathsf{R}$, show that f vanishes at infinity (Exercise 77) precisely when $\lim_{x \to \infty} f(x) = 0$.

(**b**) Denoting the collection of all compact subsets of X by $\mathcal{K}$, verify that the set

$$I = \{ (x, K) \mid K \in \mathcal{K}, x \in X \setminus K \}$$

is directed by the relation $\leq$ defined by

$$(x, K) \leq (y, L) \quad \Leftrightarrow \quad L \subset K.$$

Then show that $\lim_{x \to \infty} f(x) = y$ if and only if the net $(f(x) \mid (x, K) \in I)$ converges to y in Y.

(**c**) If X is noncompact and locally compact, prove that $\lim_{x \to \infty} f(x) = y$ if and only if the extension $f_\infty \colon X_\infty \to Y$ of f with $f_\infty(\infty) = y$ is continuous at the point ∞ of X_∞.

(**d**) Explain convergence of sequences in Y in the context of this exercise. (*Hint*: Compare 1.57.)

79. Let X and Y be noncompact, locally compact Hausdorff spaces and let $f \colon X \to Y$ be a given map. We write '$\lim_{x \to \infty} f(x) = \infty$' to mean that for each compact $K \subset Y$ there is some compact $F \subset X$ with $f(x) \notin K$ for all $x \in X \setminus F$.

(**a**) If $Y = \mathsf{R}$, show that $\lim_{x \to \infty} f(x) = \infty$ if and only if for each $c > 0$ there is a compact set $F \subset X$ with $|f(x)| > c$ for all $x \in X \setminus F$. Generalize to $Y = \mathsf{R}^n$.

(**b**) If $X = Y = \mathsf{R}$, show that $\lim_{x \to \infty} f(x) = \infty$ if and only if $\lim_{x \to +\infty} f(x) = \infty = \lim_{x \to -\infty} f(x)$ in the usual calculus meaning of the latter two limits.

(**c**) Prove that, in general, $\lim_{x \to \infty} f(x) = \infty$ if and only if $f^{-1}(K)$ is compact for each compact subset K of Y.

(**d**) Prove also that $\lim_{x \to \infty} f(x) = \infty$ if and only if the extension $f_\infty \colon X_\infty \to Y_\infty$ of f with $f_\infty(\infty) = \infty$ is continuous at the point ∞ of X_∞. (We should really use different symbols to denote the points at infinity of the one-point compactifications of X and Y.)

(**e**) Take $Y = \mathsf{R}$. Suppose $f \colon X \to \mathsf{R}$ is continuous and $\lim_{x \to \infty} f(x) = \infty$. Show that f attains a minimum value on X.

80. Establish the following analog of the Baire category theorem (1.72): The intersection of any sequence of dense open subsets of a locally compact Hausdorff space X is itself dense in X. (*Hint*: Mimic the construction used in the proof of 1.72, replacing d-balls by open neighborhoods and d-disks by compact closures of open neighborhoods. What is a substitute for the nested set theorem in the absence of a metric?)

81. Show that any denumerable locally compact Hausdorff space has a subspace homeomorphic to the discrete space N. (*Hint*: Prove that the set E of all $x \in X$ for which $\{x\}$ is open in X is denumerable by applying Exercise 80 to the subspace $X \setminus E$ of X.)

82. Let $(f_i \mid i \in I)$ be a nonempty family of continuous real-valued functions on a locally compact Hausdorff space X. Suppose the family is "pointwise bounded" in the sense that the set $\{f_i(x) \mid i \in I\}$ of real numbers is bounded for each $x \in X$. Prove there exists a nonempty open set U in X and a constant $c > 0$ with $|f_i(x)| \leq c$ for all $x \in U$ and all $i \in I$. (*Hint*: Consider the sets

$$A_n = \{x \in X : |f_i(x)| \leq n \text{ for all } i \in I\}$$

for all $n \in \mathsf{N}$.)

83. It is known that for any two disjoint closed subsets A and B of a compact Hausdorff space Y there is a continuous function $g : Y \to [0, 1]$ with $g(a) = 0$ for all $a \in A$ and $g(b) = 1$ for all $b \in B$ (compare Exercise 30). Assuming this, prove that for any two disjoint subsets K and E of a locally compact Hausdorff space X with K compact and E closed in X, there is a continuous function $f : X \to [0, 1]$ with $f(x) = 0$ for all $x \in K$ and $f(x) = 1$ for all $x \in E$.

84. A *k-space* is a topological space X whose open sets are those subsets U of X for which $U \cap K$ is open in K for each compact subspace K of X. Prove:

(**a**) A space X is a *k*-space if and only if the closed sets in X are those subsets E of X for which $E \cap K$ is closed in K for each compact subspace K of X.

(**b**) Any locally compact Hausdorff space is a *k*-space.

(**c**) Any first-countable space is a *k*-space. [*Hint*: This has nothing to do with local compactness. Use sequences.)

85. Let $\sim$ be a closed equivalence relation (Exercise 3.117) on a locally compact Hausdorff space X such that each equivalence class under $\sim$ is compact. Prove that the quotient space $X/\sim$ is then a locally compact Hausdorff space. [*Hint*: Use Exercise 27(c).]

5

Connectedness

Of the many topological properties a space may have, the most geometrical are those involving the notion of connectedness. A connected space resembles an interval on the real line in consisting of a single piece. Actually, that a space X be connected may be defined in two conceptually different ways: the existence of just one "component" of X (ordinary connectedness), or the existence of a path joining any two points of X (path-connectedness). Fortunately these two definitions turn out to be equivalent for a large class of spaces, including all manifolds.

Connectedness provides a proof of the intermediate-value theorem used so often in calculus. It also allows us to distinguish topologically between a circle and a sphere, and between the real line and higher dimensional Euclidean spaces.

We shall be examining spaces such as Euclidean spaces that have arbitrarily small connected neighborhoods of each point (locally connected spaces) and spaces such as the Cantor discontinuum that have only points as connected subsets (totally disconnected spaces).

The final two sections deal with continuous deformations of one map into another. Here the focus is on simply connected spaces, such as convex subspaces of Euclidean spaces, in which each closed path can be continuously shrunk to a point. The machinery developed to prove the intuitively obvious fact that a circle is not simply connected provides a proof of the fundamental theorem of algebra as a bonus.

1. Connected Spaces

The idea of connectedness we wish to make precise is that a connected space is one which consists of just a single "piece". It is geometrically reasonable, for example, to call each of the spaces $\{0\}$, $[0, 1]$, and R connected, although we would not want to call any of the spaces $\{0, 1\}$, $[0, 1] \cup [2, 3]$, or $\mathsf{R} \backslash \{0\}$ connected. Each of the latter three spaces can be split up into more than one nonempty open subset, but it is at least plausible that this cannot be done with any of the former three spaces.

5.1 Definition. A pair $\{A, B\}$ of two disjoint nonempty open subsets A and B of a topological space X such that $X = A \cup B$ is called a *separation of* X.

A topological space X is said to be *connected* if there does *not* exist any separation of X. A topological space X is said to be *disconnected* if it is not connected—that is, if there does exist some separation of X.

Evidently neither the empty space nor any space consisting of just one point has a separation, so these spaces are connected. A discrete space X consisting of more than one point is disconnected, for the pair

$$\{\{x\}, X \backslash \{x\}\}$$

is a separation of X for any $x \in X$. The space $[0, 1] \cup [2, 3]$ is disconnected, for

$$\{[0, 1], [2, 3]\}$$

is a separation of this space. The interval $[0, 1]$ in R is connected, but this will require some proof; the evident fact that $[0, 1]$ cannot be written as the union of two disjoint nonempty subintervals open in $[0, 1]$ does not rule out the existence of some more complicated separation of $[0, 1]$.

A separation $\{A, B\}$ of a topological space X is just a partition of X consisting of two open subsets of X. If $\{A, B\}$ is any partition of X into two subsets, then $A = X \backslash B$ and $B = X \backslash A$. If A is a subset of X with $\varnothing \neq A \neq X$, then $\{A, X \backslash A\}$ is a partition of X. Now a subset of X is open in X precisely when its complement in X is closed in X. From these observations we see that *both members of a separation of X are closed* as well as open *in X*, and we obtain the following criterion.

5.2 Lemma. The following assertions about a topological space X are equivalent:

(1) X is disconnected; that is, there exist disjoint nonempty open subsets U and V of X with $X = U \cup V$.

(2) There exist disjoint nonempty closed subsets E and F of X with $X = E \cup F$.

(3) There exist disjoint nonempty open, closed subsets A and B of X with $X = A \cup B$.

(4) There exists an open and closed subset A of X with $\varnothing \neq A \neq X$.

By definition, a topological space X is connected when it cannot be partitioned into *two* open subsets. Then a connected space X cannot be partitioned into more than two open subsets, either; if $\mathcal{C}$ is a partition of X consisting of more than two open subsets of X, then we obtain a separation $\{A, B\}$ of X by taking any member A of $\mathcal{C}$ and letting B be the union of the remaining members.

With the aid of 5.2 we can discuss the connectedness of $[0, 1]$ and other subspaces of the real line. Recall that an interval in R is just a subset X of R having the property that $a, b \in X$ with $a < b$ implies $[a, b] \subset X$. According to 0.48, a subset X of R is an interval in R precisely when it has one of the following forms:

$$\varnothing, \quad]a, b[, \quad [a, b[, \quad]a, b], \quad [a, b],$$

$$]a, \to[, \quad [a, \to[, \quad]\leftarrow, b[, \quad]\leftarrow, b], \quad \mathsf{R}.$$

5.3 Theorem. A subspace X of R is connected if and only if X is an interval in R.

Proof. Assume X is a connected subspace of R. Let $a, b \in X$ with $a < b$. If $c \in [a, b]$ but $c \notin X$, then

$$\{]\leftarrow, c[\cap X, \ ]c, \to[\cap X\}$$

is a separation of X. Hence $[a, b] \subset X$.

Conversely, assume X is an interval in R. Just suppose X is disconnected, so that there exists a separation $\{A, B\}$ of X. Choose any $a \in A$, $b \in B$. Since A is disjoint from B, $a \neq b$. We may assume $a < b$, for otherwise we simply reverse the roles of A and B. The set $A \cap [a, b]$ is nonempty since a belongs to it, and b is an upper bound of this set in R. Invoking the order-completeness of R, we set

$$c = \sup \ (A \cap [a, b]).$$

We show $c \in A$. Clearly $c \in \operatorname{cls} A$ (this closure being taken in R), and $c \in X$ because $a \leq c \leq b$ and X is an interval in R. Thus $c \in X \cap \operatorname{cls} A$. Now this set is the closure of A in the space X, and A is closed in X, so $X \cap \operatorname{cls} A = A$. Hence $c \in A$.

Since A is also open in the subspace X of R, there exist $x_1, x_2 \in \mathsf{R}$ with

$$x_1 < x_2 < b, \qquad c \in \ ]x_1, x_2[\cap X \subset A.$$

Choose any $t \in \ ]c, x_2[$. Now $c, b \in X$ and X is an interval, so $t \in X$. Thus

$$t \in \ ]x_1, x_2[\cap X,$$

and hence $t \in A$. However, $t \notin A$ because $c < t$ and c is an upper bound of $A \cap [a, b]$ in R. We have reached a contradiction. $\square$

5.4 Corollary. The real line R is connected.

We shall refer to a *connected subset* of a topological space X to mean a subset of X which is connected when provided with its relative topology making it a subspace of X. In this terminology, 5.3 says that the connected subsets of R are precisely the intervals in R.

A subset of a connected space need not itself be connected—even if it is open or is closed in the space. For example, the space $[0, 3]$ is connected but its subsets $]0, 1[\cup]2, 3[$ and $\{0, 2\}$ are not. Nonetheless, many connected spaces occur naturally as subspaces of (not necessarily connected) topological spaces. For testing whether a subset of a space is connected, the following relativized version of 5.2 is frequently useful.

5.5 Lemma. Each of the two conditions below is both necessary and sufficient for a subset A of a topological space X to be connected:

(1) There do not exist two open subsets U and V of X with

$$U \cap A \neq \varnothing \neq V \cap A, \quad U \cap V \cap A = \varnothing, \quad A \subset U \cup V.$$

(2) There do not exist two closed subsets E and F of X with

$$E \cap A \neq \varnothing \neq F \cap A, \quad E \cap F \cap A = \varnothing, \quad A \subset E \cup F.$$

One application of this lemma follows.

5.6 Theorem. Let A be a connected subset of a topological space X. Then cls A is connected. More generally, each set B with

$$A \subset B \subset \text{cls } A$$

is connected.

Proof. Let $A \subset B \subset \text{cls } A$ and just suppose B is not connected. By 5.5 there exist open subsets U and V of X with

(*) $$U \cap B \neq \varnothing \neq V \cap B, \quad U \cap V \cap B = \varnothing, \quad B \subset U \cup V.$$

Then $U \cap \text{cls } A$ and $V \cap \text{cls } A$ are nonempty open subsets of the space cls A, so

$$U \cap A \neq \varnothing, \quad V \cap A \neq \varnothing.$$

Since $A \subset B$, from (*) we obtain

$$U \cap V \cap A = \varnothing, \quad A \subset U \cup V.$$

According to 5.5 the set A is therefore not connected, which contradicts the hypothesis on A.   ☐

5.7 Corollary. A topological space that has a connected dense subset is itself connected.

The real line R is dense in the extended real line R^*, so from 5.4 and 5.7 it follows that R^* is connected. Notice that a dense subset of a connected space need not be connected. For example, the real line R is connected; the dense subspace Q of R is disconnected, for from the irrationality of $\sqrt{2}$ we see that

$$\{\mathsf{Q} \cap]\leftarrow, \sqrt{2}[, \mathsf{Q} \cap]\sqrt{2}, \rightarrow[\}$$

is a separation of Q.

Although connectedness is not preserved under the formation of arbitrary subspaces, it is preserved under the formation of quotient spaces. This is a consequence of the following more general fact.

5.8 Theorem. Let $f\colon X \to Y$ be a continuous map from a connected space X into a topological space Y. Then $f(X)$ is connected.

Proof. By replacing Y by $f(X)$, we may assume that $f(X) = Y$. Just suppose Y is disconnected, and let $\{A, B\}$ be a separation of Y. Set

$$U = f^{-1}(A), \qquad V = f^{-1}(B).$$

We claim that $\{U, V\}$ is a separation of X, in contradiction to the hypothesis that X is connected. In fact, U and V are both open in X since f is continuous and the sets A and B are open in Y. Both U and V are nonempty since f maps X onto Y and the sets A and B are nonempty. The sets U and V are disjoint because

$$U \cap V = f^{-1}(A) \cap f^{-1}(B) = f^{-1}(A \cap B) = f^{-1}(\varnothing) = \varnothing.$$

Finally,

$$U \cup V = f^{-1}(A) \cup f^{-1}(B) = f^{-1}(A \cup B) = f^{-1}(Y) = X. \qquad \square$$

The 1-sphere

$$\mathsf{S}_1 = \{(x_1, x_2) \in \mathsf{R}^2 \mid x_1{}^2 + x_2{}^2 = 1\}$$

is by 3.49 (1) the image of the unit interval $\mathsf{I} = [0, 1]$ under the continuous map $t \mapsto (\cos 2\pi t, \sin 2\pi t)$, and by 5.3 the interval I is connected. It follows from 5.8 that the 1-*sphere* S_1 *is connected*. We shall prove below that the n-sphere S_n is connected for every $n \geq 1$.

5.9 Corollary. Any quotient space of a connected space is itself connected.

5.10 Corollary. Connectedness is a topological property: If Y is a topological space that is homeomorphic to a connected space, then Y is connected.

Corollary 5.10 provides some examples which will shortly aid us in obtaining still further, more interesting examples.

5.11 Examples

(1) Given any two distinct points $x, y \in \mathbb{R}^n$, the line

$$L = \{ (1 - t)x + ty \mid t \in \mathbb{R} \}$$

passing through x and y is connected, for by 3.16 (6) the map

$$\mathbb{R} \to L$$

$$t \mapsto (1 - t)x + ty$$

is a homeomorphism, and by 5.4 the real line $\mathbb{R}$ is connected.

(2) The restriction of the above map to $[0, 1]$ gives a homeomorphism from the connected space $[0, 1]$ onto the line segment

$$\{ (1 - t)x + ty \mid 0 \le t \le 1 \}$$

joining x to y, so by 5.10 this line segment is connected.

A particularly important case of Theorem 5.8 arises when the continuous function f on the connected space X is real-valued. In that case Theorem 5.8 tells us that if f assumes two distinct values, then it must assume as values all numbers lying between those two values.

5.12 Theorem (intermediate-value theorem). Let $f: X \to \mathbb{R}$ be a continuous function on a connected space X. Let $a, b \in X$ with $f(a) < f(b)$, and suppose c is a real number with

$$f(a) < c < f(b).$$

Then there exists some $x \in X$ with

$$f(x) = c.$$

Proof. By 5.8 the image $f(X)$ of X under f is a connected subset of $\mathbb{R}$, and by 5.3 this image is an interval in $\mathbb{R}$. Now $f(a), f(b) \in f(X)$, so $[f(a), f(b)] \subset f(X)$. In particular, $c \in f(X)$. ☐

A closed interval in $\mathbb{R}$ is connected. Hence when we take $X = [a, b] \subset \mathbb{R}$ in 5.12 we obtain the classic intermediate-value theorem of R. Bolzano so familiar from calculus: A continuous function $f: [a, b] \to \mathbb{R}$ takes all numbers between $f(a)$ and $f(b)$ as values.

Suppose, in the notation of 5.12, that

$$f(a) < 0, \qquad f(b) > 0.$$

Then Theorem 5.12 asserts the existence of at least one solution $x \in X$ to the equation

$$f(x) = 0.$$

5.13 Example. Any polynomial

$$p(x) = a_0 x^n + a_1 x^{n-1} + \cdots + a_{n-1} x + a_n$$

with real coefficients $a_0, a_1, \ldots, a_n$ and of *odd* degree n has a real root; that is, the equation

$$p(x) = 0$$

has some solution $x \in \mathbf{R}$. To prove this we shall show that

$$p(-b) < 0, \qquad p(b) > 0$$

for some real number $b > 0$, and the result will then follow from the continuity of p and the intermediate-value theorem.

To say p is of degree n means $a_0 \neq 0$. Now for a real number x, we have $p(x) = 0$ if and only if $a_0^{-1} p(x) = 0$. By replacing $p(x)$ by $a_0^{-1} p(x)$, we may therefore assume without loss of generality that

$$a_0 = 1.$$

Then for $x \neq 0$ we may write

$$p(x) = x^n q(x)$$

where

$$q(x) = 1 + \frac{a_1}{x} + \frac{a_2}{x^2} + \cdots + \frac{a_n}{x^n}.$$

We shall find a $b > 0$ with

$$q(b) > 0, \qquad q(-b) > 0.$$

It will then follow that $p(b) > 0$ and, since n is odd, $p(-b) < 0$.

For $|x| > 1$ we have

$$|q(x) - 1| \leq \frac{|a_1|}{|x|} + \frac{|a_2|}{|x|^2} + \cdots + \frac{|a_n|}{|x|^n}$$

$$< \frac{|a_1|}{|x|} + \frac{|a_2|}{|x|} + \cdots + \frac{|a_n|}{|x|}$$

$$= \frac{A}{|x|}$$

where

$$A = |a_1| + |a_2| + \cdots + |a_n|.$$

Then

$$|x| > \max \{1, 2A\}$$

implies

$$|q(x) - 1| < \tfrac{1}{2}$$

and therefore

$$q(x) > 0.$$

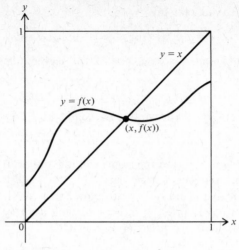

FIG. 5.1

Now choose any $b > \max\{1, 2A\}$. Since $|-b| = b$, $q(b)$ and $q(-b)$ are both positive, as desired.

The existence of roots of polynomials of even as well as odd degree is guaranteed by the fundamental theorem of algebra (5.78), which is proved later.

One sometimes wants to solve an equation not of the form $f(x) = 0$, but of the form $f(x) = x$—in other words, to find a *fixed-point* $x \in X$ of a map $f \colon X \to X$. This is the case, for example, in establishing the existence of solutions of a first-order differential equation, where the Banach contraction mapping principle (Exercise 1.85) may be used. Of course, not every continuous map $f \colon X \to X$ has a fixed-point, even when the space X is connected; consider a rotation of S_1 through an angle that is not an integral multiple of 2π. The existence of a fixed-point of a continuous map $f \colon X \to X$ can, however, be inferred from the connectedness of special spaces X.

5.14 Corollary. Let
$$f \colon [0, 1] \to [0, 1]$$
be a continuous map. Then f has a fixed-point; that is, there exists an $x \in [0, 1]$ with
$$f(x) = x.$$

Proof. To say that $f(x) = x$ is equivalent to saying that $x - f(x) = 0$. Hence we introduce the function
$$g \colon [0, 1] \to \mathsf{R}$$
defined by
$$g(x) = x - f(x).$$

If $f(0) = 0$ or $f(1) = 1$, we are done, so we now suppose $f(0) \neq 0$ and $f(1) \neq 1$. Since f maps into $[0, 1]$, we have

$$f(0) > 0, \qquad f(1) < 1,$$

so that

$$g(0) = -f(0) < 0, \qquad g(1) = 1 - f(1) > 0.$$

Since g is continuous and $[0, 1]$ is connected, it follows from 5.12 that $g(x) = 0$ for some $x \in [0, 1]$. Then $f(x) = x$. $\Box$

Geometrically, 5.14 says that the graph of a continuous function $f: [0, 1] \to [0, 1]$ must cross the line $y = x$ (see Fig. 5.1).

It is a fact, more general than 5.14, that for arbitrary dimension $n \geq 1$ any continuous map $f: I^n \to I^n$ from the n-cube into itself must have a fixed-point. This is the *Brouwer fixed-point theorem*. We shall give a proof for the case of dimension $n = 2$ in Section 5.

The intermediate-value theorem provides a surprising result concerning real-valued functions on the circle S_1. Observe that if $x = (x_1, x_2) \in S_1$, then $-x = (-x_1, -x_2) \in S_1$; in fact, $-x$ is that point on S_1 at which the line through x and the origin intersects S_1 (see Fig. 5.2).

5.15 Corollary. Let

$$f: S_1 \to R$$

be a continuous function. Then there exists $x \in S_1$ with

$$f(x) = f(-x).$$

Proof. Define the continuous map

$$g: S_1 \to R$$

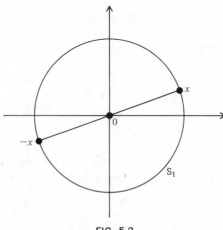

FIG. 5.2

by
$$g(x) = f(x) - f(-x).$$

We seek an $x \in S_1$ for which $g(x) = 0$. Clearly
$$g(-x) = -g(x) \qquad\qquad (x \in S_1),$$

so to find a solution of $g(x) = 0$ we need only look at the points in the upper semicircle $\{x \in S_1 \mid x_2 \geq 0\}$ of S_1. Therefore, we introduce the map
$$p: [0, 1] \to S_1$$
$$t \mapsto (\cos \pi t, \sin \pi t)$$

which is continuous and maps $[0, 1]$ onto this upper semicircle. To show that $g(x) = 0$ for some x, and hence to complete the proof, we need only show that $g(p(t)) = 0$ for some $t \in [0, 1]$.

Form the continuous map
$$F = g \circ p: [0, 1] \to R.$$

Direct computation gives
$$F(1) = -F(0).$$

If $F(0) \neq 0$, then one of the numbers $F(0)$, $F(1)$ is positive and the other is negative, so by 5.12 there is a $t \in \,]0, 1[$ with $F(t) = 0$. Thus
$$g(p(t)) = F(t) = 0$$

for some $t \in [0, 1]$, as needed.   ☐

Corollary 5.15 has an amusing physical interpretation. At a given moment of time the temperatures at all points on a given meridian of the earth may be described by a function $f: S_1 \to R$. It is a reasonable physical assumption that the temperature varies continuously around this meridian, in other words, that f is continuous. The conclusion of 5.15 then says that at a given moment there are two diametrically opposite points on the meridian at which the temperatures are the same. (Question: Could one experimentally verify— or refute—this conclusion or the physical assumption leading to it?)

Like the fixed-point theorem 5.14, Corollary 5.15 can be extended to arbitrary dimension $n \geq 1$. The *Borsuk-Ulam theorem* asserts that if $f: S_n \to R^n$ is a continuous map, then $f(x) = f(-x)$ for some $x \in S_n$.

The next theorem and its corollaries will allow us to enlarge our repertory of connected spaces.

5.16 Lemma. Let C be a connected subset of a topological space X, and let U be an open, closed subset of X which intersects C. Then $C \subset U$.

Proof. The set $C \cap U$ is a nonempty subset of C that is both open and closed in C. By 5.2, $U \cap C = C$. Hence $C \subset U$.   ☐

5.17 Theorem. Let X be a topological space. Suppose there is a covering $\mathcal{C}$ of X consisting of connected subsets of X such that for any two members A and B of $\mathcal{C}$ there are finitely many sets $A_1, \ldots, A_n \in \mathcal{C}$ with

$$A_1 = A, \qquad A_n = B,$$

and

$$A_i \cap A_{i+1} \neq \varnothing \qquad\qquad (i = 1, \ldots, n - 1).$$

Then X is connected.

Proof. Just suppose there exists some separation $\{U, V\}$ of X. Since U and V are nonempty and $\mathcal{C}$ is a covering of X, there are $A, B \in \mathcal{C}$ with

$$A \cap U \neq \varnothing, \qquad B \cap V \neq \varnothing.$$

By hypothesis there are sets $A_1, \ldots, A_n \in \mathcal{C}$ with $A_1 = A$, $A_n = B$, and

$$A_i \cap A_{i+1} \neq \varnothing \qquad\qquad (i = 1, \ldots, n - 1).$$

We are going to show that $B = A_n \subset U$; this will produce the desired contradiction, for then B will intersect both of the disjoint sets U and V.

To show that $A_n \subset U$ we show something more, namely, $A_i \subset U$ for each $i = 1, \ldots, n$. Since A_1 is connected and intersects the open, closed subset U of X, it follows from 5.16 that $A_1 \subset U$. Now let $1 \leq i < n$ and assume $A_i \subset U$. Then A_{i+1} is a connected set which, since

$$\varnothing \neq A_i \cap A_{i+1} \subset U \cap A_{i+1},$$

intersects U. By 5.16 again, $A_{i+1} \subset U$ also.   ☐

Suppose, in the notation of 5.17, that the covering $\mathcal{C}$ of X actually has nonempty intersection. Then any two members A and B of $\mathcal{C}$ will intersect, so the hypothesis of 5.17 will hold with $n = 2$. This proves the following corollary.

5.18 Corollary. A topological space X is connected if there is a covering $\mathcal{C}$ of X consisting of connected sets and having nonempty intersection.

5.19 Corollary. Suppose that for each two points x and y of a topological space X there is a connected subset $C_{x,y}$ of X with $x \in C_{x,y}$ and $y \in C_{x,y}$. Then X is connected.

Proof. If $X = \varnothing$, there is nothing to prove, so we assume $X \neq \varnothing$. Arbitrarily choose some $y \in X$. Then

$$X = \bigcup_{x \in X} C_{x,y}, \qquad y \in \bigcap_{x \in X} C_{x,y}.$$

Hence 5.18 applies with $\mathcal{C} = \{C_{x,y} \mid x \in X\}$.   ☐

5.20 Examples

(1) Any convex subset of R^n is connected. In fact, if X is a convex subset of R^n, then for each $x, y \in X$ the line segment

$$L_{x,y} = \{ (1 - t)x + ty \mid 0 \le t \le 1\}$$

is a subset of X containing both x and y which by 5.11 (2) is connected, so X is connected by 5.19.

In particular, the n-cube

$$\mathsf{I}^n = [0, 1]^n,$$

the n-disk

$$\mathsf{D}_n = \{x \in \mathsf{R}^n : ||x|| \le 1\},$$

the n-ball

$$\mathsf{B}_n = \{x \in \mathsf{R}^n : ||x|| < 1\},$$

and R^n itself are all connected spaces.

(2) The Klein bottle [3.49 (6)] and the Möbius strip [3.49 (5)] are connected, being quotient spaces of I^2.

(3) For every $n \ge 1$ the n-sphere

$$\mathsf{S}_n = \{x \in \mathsf{R}^n : ||x|| = 1\}$$

is connected. To see this, let

$$p = (0, 0, \ldots, 0, 1)$$

be the north pole of S_n and $q = -p$ be the south pole. Define

$$A = \mathsf{S}_n \backslash \{p\}, \qquad B = \mathsf{S}_n \backslash \{q\}.$$

Now

$$A \cong \mathsf{R}^n$$

by 3.16 (11), and likewise

$$B \cong \mathsf{R}^n.$$

In (1) we proved that R^n is connected, so by 5.10 both A and B are connected. Since

$$A \cup B = \mathsf{S}_n, \qquad A \cap B \ne \varnothing,$$

it follows from 5.18 that S_n is connected.

(4) The n-dimensional real projective space RP_n is connected, for RP_n is a quotient space of S_n [3.49 (7)].

(5) Let $n > 1$ and let E be a countable subset of R^n. Then $\mathsf{R}^n \backslash E$ is connected. To prove this we shall show that any two points $x, y \in \mathsf{R}^n \backslash E$ belong to some connected subset of $\mathsf{R}^n \backslash E$. Fix $x, y \in \mathsf{R}^n \backslash E$ with $x \ne y$. For any $u, v \in \mathsf{R}^n$ let

$$L_{u,v} = \{ (1 - t)u + tv \mid 0 \le t \le 1\},$$

the line segment joining u to v.

Choose any

$$z \in L_{x,y}, \qquad x \ne z \ne y$$

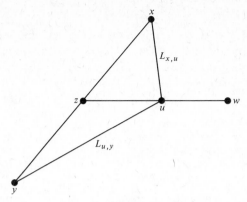

FIG. 5.3

(for example, $z = \frac{1}{2}u + \frac{1}{2}v$). Since $n > 1$, there exists $w \in \mathbf{R}^n \backslash L_{x,y}$ with

$$L_{z,w} \cap L_{x,y} = \{z\}$$

(see Fig. 5.3) [this is where we use the hypothesis $n > 1$]. For each $u \in L_{z,w}$ let

$$C_u = L_{x,u} \cup L_{u,y}$$

so that $x \in C_u$ and $y \in C_u$. For each $u \in L_{z,w}$ the set C_u is connected, because both the line segments $L_{x,u}$ and $L_{u,y}$ are connected and contain the point u. Hence it suffices to show that $C_u \subset \mathbf{R}^n \backslash E$ for some $u \in L_{z,w}$.

Just suppose

$$C_u \cap E \neq \varnothing \qquad\qquad (u \in L_{z,w}).$$

We have

$$C_u \cap C_v = \{x, y\} \qquad (u, v \in L_{z,w}; u \neq v)$$

and $\{x, y\} \cap E = \varnothing$. Therefore for distinct $u, v \in L_{z,w}$ the points in $C_u \cap E$ are all different from the points in $C_v \cap E$. Since $L_{z,w}$ is an uncountable set, we conclude

$$E \cap \bigcup_{u \in L_{z,w}} C_u$$

is uncountable. This is impossible since E was supposed to be countable.

The fact that connectedness is a topological property tells us that a connected space cannot be homeomorphic to a disconnected space. For example, $[0, 1]$ is not homeomorphic to $[0, 1] \cup [2, 3]$. This fact can even be used to show that two connected spaces X and Y are not homeomorphic to one another by applying it to suitable subspaces of X and Y.

5.21 Definition. A subset A of a topological space is said to *disconnect* X when $X \backslash A$ is disconnected.

5.22 Examples

(1) An interval J in R containing at least two points is disconnected by $\{x\}$ for any nonendpoint $x \in J$, because $J \backslash \{x\}$ is not an interval.

In particular, R is disconnected by $\{x\}$ for each $x \in \mathsf{R}$.

(2) Let $p \in \mathsf{S}_1$. Then $\{p\}$ does not disconnect S_1, for by 3.16 (11)

$$\mathsf{S}_1 \backslash \{p\} \cong \mathsf{R}$$

and R is connected.

We can now prove that the circle S_1 is not homeomorphic to the real line R, or to any interval in R. Just suppose there exists some

$$f \colon \mathsf{S}_1 \cong J$$

with J an interval in R (possibly R itself). Certainly S_1 contains at least two points, so the same thing is true of J. Choose any $p \in \mathsf{S}_1$ with $x = f(p)$ not an endpoint of J. By restricting f to $\mathsf{S}_1 \backslash \{p\}$ we obtain a homeomorphism

$$\mathsf{S}_1 \backslash \{p\} \cong J \backslash \{x\}.$$

This is impossible since $\mathsf{S}_1 \backslash \{p\}$ is connected whereas $J \backslash \{x\}$ is not.

(3) For $n > 1$ Euclidean n-space R^n is not homeomorphic to the real line R. In fact, if there exists some $f \colon \mathsf{R}^n \cong \mathsf{R}$, then we would have

$$\mathsf{R}^n \backslash \{x\} \cong \mathsf{R} \backslash \{y\}$$

for $x \in \mathsf{R}^n$ and $y = f(x) \in \mathsf{R}$. This is impossible, since by 5.20 (5) the singleton $\{x\}$ does not disconnect R^n, but $\{y\}$ does disconnect R.

Having looked at subspaces and quotient spaces of connected spaces, we now turn to products of connected spaces.

5.23 Theorem. Let X and Y be connected spaces. Then $X \times Y$ is connected.

Proof. Let $a = (a_1, a_2) \in X \times Y$ and $b = (b_1, b_2) \in X \times Y$. According to 5.19, it suffices to find a connected subset of $X \times Y$ containing both a and b. Set

$$A = \{a_1\} \times Y, \qquad B = X \times \{b_2\}$$

(see Fig. 5.4). Now

$$A \cong Y, \qquad B \cong X,$$

so by hypothesis the sets A and B are both connected. Also

$$(a_1, b_2) \in A \cap B,$$

so by 5.18 the set $A \cup B$ is connected. Since $a \in A$ and $b \in B$, $A \cup B$ is the connected subset of $X \times Y$ we seek.   ▢

5.24 Corollary. Let $(X_i \mid i = 1, \ldots, n)$ be a finite family of nonempty topological spaces. Then the product space $\times_{i=1}^{n} X_i$ is connected if and only if X_i is connected for each $i = 1, \ldots, n$.

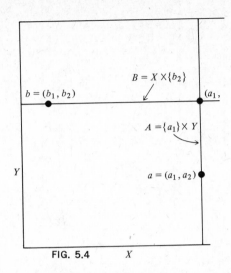

$B = X \times \{b_2\}$

$b = (b_1, b_2)$

$(a_1,$

$A = \{a_1\} \times Y$

Y

$a = (a_1, a_2)$

FIG. 5.4 X

Proof. Assume $\underset{i=1}{\overset{n}{\times}} X_i$ is connected. For each $i = 1, \ldots, n$, the ith projection

$$p_i: \overset{n}{\underset{i=1}{\times}} X_i \to X_i$$

is a continuous surjection, so by 5.8 the space X_i is connected.

The converse is proved by induction on n using 5.23 and the relation

$$\overset{n}{\underset{i=1}{\times}} X_i \cong \left(\overset{n-1}{\underset{i=1}{\times}} X_i \right) \times X_n$$

(see 3.37).   □

From 5.24 and the connectedness of the circle S_1 it follows that the n-torus

$$(\mathsf{S}_1)^n = \mathsf{S}_1 \times \mathsf{S}_1 \times \cdots \times \mathsf{S}_1 \qquad (n \text{ factors})$$

is connected.

The proof that an infinite family of connected spaces has a connected product is much harder than the proof for a finite family.

5.25 Lemma. Let x be any point in the product X of a sequence $(X_i \mid i = 1, 2, \ldots)$ of topological spaces. Then the set

$$D = \{y \in X \mid \text{for some } n \in \mathsf{N}, \ y_i = x_i \text{ for all } i > n\}$$

is dense in X.

Proof. Let U be an open neighborhood of a point z in X. We show that U intersects D. There exists an $n \geq 1$ and open sets V_1 in $X_1, \ldots, V_n$ in X_n with

$$z \in (V_1 \times \cdots \times V_n) \times (X_{n+1} \times X_{n+2} \times \cdots) \subset U.$$

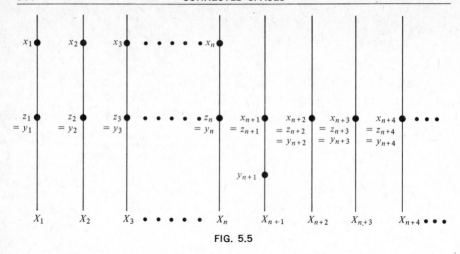

FIG. 5.5

Define a point $y \in X$ by

$$
y_i = \begin{cases} z_i & \text{if } i = 1, \dots, n, \\ x_i & \text{if } i = n+1, n+2, \dots. \end{cases}
$$

Then $y \in D \cap U$. ∎

5.26 Theorem. Let $(X_i \mid i = 1, 2, \dots)$ be a sequence of nonempty topological spaces. Then the product space $\mathsf{X}_{i=1}^{\infty} X_i$ is connected if and only if X_i is connected for all $i = 1, 2, \dots$.

Proof. Set $X = \mathsf{X}_{i=1}^{\infty} X_i$. The proof that each X_i is connected if X is connected is the same as in the finite case (5.24).

Conversely, assume that X_i is connected for all $i = 1, 2, \dots$. Choose any $x \in X$. Define

$$
C = \{ y \in X \mid \text{there is a connected } E \subset X \text{ with } x \in E, y \in E \}.
$$

Then C is connected by 5.19, so by 5.6 the subset cls C of X is also connected. We shall prove that X is connected by showing that $X = \text{cls } C$. Define D as in 5.25. Then it suffices to prove

$$
D \subset C,
$$

for then

$$
X = \text{cls } D \subset \text{cls } C \subset X.
$$

For each $n \in \mathsf{N}$, set

$$
D_n = \{ y \in X \mid y_i = x_i \text{ for all } i > n \}.
$$

Then

$$
D = \bigcup_{n=0}^{\infty} D_n,
$$

so to prove that $D \subset C$ we need to show that $D_n \subset C$ for all $n \in \mathsf{N}$. We use induction on n. First,

$$D_0 = \{x\} \subset C.$$

Now let $n \in \mathsf{N}$ and assume $D_n \subset C$. We show $D_{n+1} \subset C$.

Let $y \in D_{n+1}$. We must show $y \in C$. We have

$$y_i = x_i \qquad (i = n + 2, n + 3, \ldots).$$

Define $z \in X$ by

$$z_i = \begin{cases} y_i & \text{if } i \neq n + 1, \\ \\ x_{n+1} & \text{if } i = n + 1 \end{cases}$$

(see Fig. 5.5). Then $z \in D_n$, so by the inductive assumption $z \in C$, that is,

$$x \in A, \qquad z \in A$$

for some connected subset A of X. Now the set

$$B = \{w \in X \mid w_i = y_i \text{ for all } i \neq n + 1\}$$
$$= \{y_1\} \times \cdots \times \{y_n\} \times X_{n+1} \times \{y_{n+2}\} \times \{y_{n+3}\} \times \cdots,$$

being homeomorphic to the connected space X_{n+1}, is connected, and

$$z \in B, \qquad y \in B.$$

Then $x, y \in A \cup B$, and by 5.18 the set $A \cup B$ is connected. Hence $y \in C$. This completes the proof of the inductive step that $D_{n+1} \subset C$.   ☐

An immediate consequence of 5.26 is that the Hilbert cube

$$\left\{ x \in \mathsf{H} \colon |x_i| \leq \frac{1}{i} \ (i = 1, 2, \ldots) \right\} = \underset{i=1}{\overset{\infty}{\times}} \left[-\frac{1}{i}, \frac{1}{i} \right]$$

is connected.

EXERCISES

1. Determine which topological spaces having at most three points are connected.

2. Which of the following spaces are connected:
   (**a**) An infinite set provided with its finite-complement topology [2.3(7)]?
   (**b**) An uncountable set provided with its countable-complement topology (Exercise 2.6)?
   (**c**) The real line provided with its right-interval topology [2.18(1)]?
   (**d**) The line with two origins [2.18(3)]?
   (**e**) The half-disk space [2.20(3)]?
   (**f**) The tangent disk space (Exercise 2.27)?

3. (**a**) Show that a topological space X is connected if and only if each subset A of X with $\varnothing \neq A \neq X$ has nonempty boundary.
   (**b**) If a connected subset C of a topological space X intersects both a subset A of X and $X \backslash A$, must C intersect bdy A?

4. Given subsets A and B of a topological space X, one says that A is *separated from B in X* when

$$\text{cls } A \cap B = \varnothing = A \cap \text{cls } B.$$

(**a**) Can cls A intersect cls B when A is separated from B in X?

(**b**) Suppose the topology of X is induced by a metric d. Show that A is separated from B in X in case $d(A, B) > 0$ (Exercise 1.28). Does the converse hold?

5. (**a**) If A and B are nonempty subsets of a topological space X, prove that A is separated from B in X (Exercise 4) precisely when $\{A, B\}$ is a separation of $A \cup B$.

(**b**) Deduce from (a) that a subset C of X is connected if there do not exist two nonempty subsets A and B of X with $C = A \cup B$ and A separated from B in X.

6. Let X be a topological space with $X = A \cup B$ for nonempty subsets A and B of X.

(**a**) Can X be connected even though cls $A \cap B = \varnothing$?

(**b**) Show that X is connected if A and B are both connected and cls $A \cap B \neq \varnothing$.

7. (**a**) Let X be a totally ordered set provided with its order topology [2.52(1)]. Find a necessary and sufficient condition on the total ordering for X to be connected. (*Hint*: Which properties of the usual ordering on intervals in R were used in the proof of 5.3?)

(**b**) Apply this condition to the total ordering on R $\times$ R defined in Exercise 2.29.

8. If X is a connected space, a *cut point of X* is a point $x \in X$ such that $X \backslash \{x\}$ is disconnected. For example, $\frac{1}{2}$ is a cut point of $[0, 1]$, but 0 is a noncut point of this space.

(**a**) Let x be a cut point of a connected Hausdorff space X. If $\{U, V\}$ is a separation of $X \backslash \{x\}$, show that $U \cup \{x\}$ and $V \cup \{x\}$ are connected.

(**b**) Let X be a connected Hausdorff space containing at least two points. Prove that $X = A \cup B$ for some nonempty connected subsets A and B of X with $A \neq B$. (*Hint*: Distinguish the cases that X has or does not have a cut point.)

(**c**) In (b), can A and B always be found so that A is disjoint from B?

9. (**a**) Prove that a topological space X is connected if and only if there does not exist any continuous surjection $f: X \rightarrow \{0, 1\}$, where $\{0, 1\}$ has its discrete topology.

(**b**) Use (a) to give a new proof of 5.8.

(**c**) Suppose every continuous real-valued function on a topological space X has the "intermediate-value property" of 5.12. Show that X must be connected.

10. Let $f: \mathbb{R}^n \rightarrow \mathbb{R}$ be a continuous function, where $n \geq 2$. Suppose $f(a) < 0, f(b) > 0$ for some $a, b \in \mathbb{R}^n$. Show that $f(x) = 0$ for uncountably many points $x \in \mathbb{R}^n$.

11. If f is a homeomorphism from an interval $[a, b]$ in R onto an interval $[c, d]$ in R, show that either $f(a) = c$ and $f(b) = d$, or else $f(a) = d$ and $f(b) = c$.

12. Consider a surjection $f: [a, b] \rightarrow [c, d]$, where $a, b, c, d \in \mathbb{R}$ with $a < b, c < d$.

(**a**) Prove that f is a homeomorphism if it is either strictly increasing $[f(t) < f(s)$ whenever $t < s]$ or strictly decreasing $[f(t) > f(s)$ whenever $t < s]$.

(**b**) Is the converse of (a) true?

13. A function $f: X \rightarrow \mathbb{R}$ on a topological space X is said to be *locally constant* if each $x \in X$ has a neighborhood on which f is constant.

(**a**) Give an example of a nonconstant but locally constant continuous function $f: X \rightarrow \mathbb{R}$ if X is an appropriate subspace of R.

(**b**) Show that a locally constant continuous function $f: X \to$ R on a connected space X must be constant on X.

(**c**) Can (**b**) be generalized to spaces other than R as the codomain of f?

14. Construct a decreasing sequence of connected subsets of R^2 whose intersection is disconnected, or else prove that such a sequence cannot exist.

15. Let X be a topological space having a covering $\mathcal{Q}$ such that each $A \in \mathcal{Q}$ is connected and $A, B \in \mathcal{Q}$ with $A \neq B$ implies A is not separated from B in X (Exercise 4). Prove that X is connected.

16. A set $E \subset R^n$ is said to be *starlike* if there is a point $x \in E$ such that for each $y \in E$ the line segment joining x to y is contained in E.

(**a**) Show that each convex subset of R^n is starlike. Determine all starlike subsets of R^1. Describe several nonconvex starlike subsets of R^2 and of R^3.

(**b**) Prove that every starlike subset of R^n is connected.

17. A connected space X is said to be *unicoherent* when the intersection $A \cap B$ of any two closed connected subsets A and B of X with $X = A \cup B$ is itself connected. Verify that I is unicoherent but that S_1 is not. Do you think that S_2 is unicoherent?

18. (**a**) Let A and B be subsets of a topological space X such that $A \cup B$ and $A \cap B$ are connected. If A and B are both closed in X, prove that A and B are both connected.

(**b**) Is the hypothesis that A and B be closed really needed?

19. Is the set of points in the plane at least one of whose coordinates is irrational connected?

20. Show that the Hilbert sequence space H (1.9) is connected.

21. If $n \geq 2$, show that S_{n-1} disconnects R^n but that D^n does not.

22. Show that a circle cannot be homeomorphic to a "figure eight".

23. Show that S_1 is not homeomorphic to S_n for any $n > 1$.

24. Classify the ten digits

$$0 \quad 1 \quad 2 \quad 3 \quad 4 \quad 5 \quad 6 \quad 7 \quad 8 \quad 9$$

into distinct homeomorphism classes (that is, into equivalence classes under the relation 'is homeomorphic to').

25. Given subsets A and B of connected spaces X and Y, respectively, with $A \neq X$ and $B \neq Y$, prove that $(X \times Y) \setminus (A \times B)$ is connected.

26. Show that the Bing triangle space (Exercise 2.80) is connected. (Thus there exist countable connected Hausdorff spaces.)

27. Let X be a countable connected space.

(**a**) Show that each continuous real-valued function on X is constant.

(**b**) Prove that X cannot be metrizable if it contains more than one point. [*Hint*: If d is a metric inducing the topology of X, consider for a fixed $a \in X$ the function $x \mapsto d(x, a)$.]

28. Let (X, d) be a metric space. Given $x, y \in X$ and $\epsilon > 0$, an *ϵ-chain from x to y* is a finite family $(x_1, \ldots, x_n)$ of points of X with $x_1 = x$, $x_n = y$, and $d(x_i, x_{i+1}) < \epsilon$ for all $i = 1, \ldots, n - 1$. If $\epsilon > 0$, the space (X, d) is said to be *ϵ-chainable* when for all $x, y \in X$ there exists an ϵ-chain from x to y. Prove that (X, d) is ϵ-chainable for every $\epsilon > 0$ if X is connected. (*Hint*: Given $\epsilon > 0$ and $x \in X$, consider the set of all $y \in X$ for which there exists an ϵ-chain from x to y. *Note*: For a converse, see Exercise 30.)

Exercises 29–33 involve compactness.

29. Deduce the connectedness of $[0, 1]$ from its compactness as follows. Suppose $\{A, B\}$ is a separation of $[0, 1]$. Define $f: A \times B \to \mathbf{R}$ by $f(x, y) = |x - y|$. Then f must attain a minimum value at some point $(a, b) \in A \times B$. Consider the point $c = (a + b)/2 \in [0, 1]$.

30. Prove that a compact metric space is connected if for every $\epsilon > 0$ it is ϵ-chainable (Exercise 28).

31. Let X be a compact Hausdorff space. If $x \in X$, show that the intersection C of the collection $\mathcal{E}$ of all open, closed subsets E of X with $x \in E$ is connected. (*Hint*: Suppose $\{A, B\}$ is a separation of C. Then A and B are disjoint closed subsets of X, so there are disjoint open subsets U and V of X with $A \subset U$ and $B \subset V$. Show that $E \subset U \cup V$ for some $E \in \mathcal{E}$ and then that $E \cap U \in \mathcal{E}$. *Note*: For an application, see Exercise 56.)

32. Prove that a countable locally compact Hausdorff space cannot be connected unless it is empty or consists of a single point. (*Hint*: If X is connected and contains more than one point, then $\{x\}$ is not open in X for any $x \in X$.)

33. A *continuum* is a compact, connected Hausdorff space. Show that a continuum X containing at least two points must contain at least two noncut points (Exercise 8). [*Hint*: Assume X has a cut point x, and let $\{U, V\}$ be a separation of $X \setminus \{x\}$. Show that each of U and V contains a noncut point of X. To see that U contains a noncut point of X, assume the contrary and apply Exercise 8(a). *Note*: A theorem of R. L. Moore says that a metrizable continuum having exactly two noncut points must be homeomorphic to I.]

34. Let G be a topological group (Exercise 3.84) that is connected. If V is any neighborhood of the identity element of G, prove that G is generated by V in the sense that each $x \in G$ can be written in the form

$$v = v_1 \cdot v_2 \cdot \cdots \cdot v_n$$

for some integer $n \geq 1$ and suitable elements $v_1, \ldots, v_n$ of V. [*Hint*: Set $U = V \cap V^{-1}$, show that $\bigcup_{n=1}^{\infty} U^n$ is an open subgroup of G, where

$$U^n = \{u_1 \cdot u_2 \cdot \cdots \cdot u_n \mid u_1, \ldots, u_n \in U\},$$

and then use Exercise 3.88(c)].

2. Components and Locally Connected Spaces

In this section we examine two special kinds of connected subsets of a space: the largest connected pieces into which the space can be divided, and connected neighborhoods of points.

Even when a topological space X is disconnected, it is always possible to decompose it into pairwise disjoint connected subsets; since each singleton is connected, we can always decompose X as

$$X = \bigcup_{x \in X} \{x\}.$$

If X is discrete, this will be the only way to decompose X into connected pieces. If X is not discrete, however, there may be other—and even many other—ways of so decomposing X. For example, the space

$$X = \{x \in \mathbb{R} \mid 0 \leq x \leq 1 \text{ or } 2 < x < 3\}$$

has the following three distinct decompositions into connected subsets:

$$X = \left[0, \frac{1}{2}\right[\cup \left[\frac{1}{2}, 1\right] \cup \left]2, \frac{7}{3}\right] \cup \left]\frac{7}{3}, 3\right[,$$

$$X = \{0\} \cup]2, 3[\cup \bigcup_{n=1}^{\infty} \left]\frac{1}{n+1}, \frac{1}{n}\right],$$

$$X = [0, 1] \cup]2, 3[.$$

Geometric intuition tells us that only in the third of these decompositions have we really broken X into "the" connected pieces of X. What distinguishes this third decomposition from the other two is that neither of the connected subsets $[0, 1]$, $]2, 3[$ of X can be enlarged to a new connected subset of X.

5.27 Definition. A connected subset of a topological space X that is not properly contained in any connected subset of X is called a *component of X*.

Among all the connected subsets of a space X, a component is thus one that is *maximal* with respect to inclusion (we say "maximal" instead of "maximum" because a component need not contain every connected subset of X, it just cannot be contained in any connected subset of X different from itself).

5.28 Theorem. Each component of a topological space X is closed in X.

Proof. Let C be a component of X. Since C is connected, by 5.6 the subset cls C of X is also connected. Since $C \subset$ cls C, it follows that $C =$ cls C. Thus C is closed in X. ☐

Trivially, the empty space has $\emptyset$ as its unique component. If X is nonempty, then $\emptyset$ is not a component of X because it is properly contained in each singleton, and any singleton is connected. The following theorem says that X can really be decomposed into its components.

5.29 Theorem. Let X be a nonempty topological space. Then:

(1) Each $x \in X$ belongs to a unique component of X, namely, the union of all connected subsets of X that contain x.

(2) The collection of all components of X is a partition of X.

Proof. (1) Let $x \in X$. To see that x belongs to at most one component of X, suppose A and B are components of X with $x \in A$ and $x \in B$. By 5.18, $A \cup B$ is connected. Since $A \subset A \cup B$ and A is a component of X, $A = A \cup B$; similarly, $B = A \cup B$. Hence $A = B$.

Now define

$$C_x = \bigcup \{A \mid x \in A \subset X, A \text{ connected}\}.$$

We show that C_x is a component of X containing x. Since $\{x\}$ is connected, $\{x\} \subset C_x$, so that $x \in C_x$. By 5.18, C_x is connected. Finally, if A is a connected subset of X with $C_x \subset A$, then $x \in A$, $A \subset C_x$ by definition of C_x, and hence $C_x = A$.

(2) We have already observed that no component of X is empty. By (1), X is the union of all its components, and distinct components of X are disjoint. $\square$

According to (1), the component of X containing a given point $x \in X$ is actually the largest connected subset of X which contains x.

5.30 Corollary. A nonempty connected subset A of a topological space X is contained in a unique component of X, namely, the component of X to which each point of A belongs.

5.31 Examples

(1) A topological space X is connected precisely when X is the one and only component of X.

(2) If A is a nonempty connected subset of a topological space X that is both open and closed in X, then A is a component of X. In fact, by 5.30 the set A is contained in a component C of X, so from 5.16 it follows that $C = A$. For example, the components of $X = [0, 1] \cup {]}2, 3[$ are $[0, 1]$ and $]2, 3[$; the components of $\mathsf{R}\backslash\{0\}$ are $]{\leftarrow}, 0[$ and $]0, {\rightarrow}[$.

(3) The components of a nonempty discrete space X are just the singletons $\{x\}$ for all $x \in X$.

(4) Although the space Q is not discrete, the components of Q are the singletons $\{x\}$ for all $x \in \mathsf{Q}$. In fact, a component of Q, being a nonempty connected subset of R, is by 5.3 an interval in R, and a nonempty interval cannot be contained in Q unless it is a singleton.

Similarly, the components of the space $\mathsf{R}\backslash\mathsf{Q}$ of all irrational numbers are the singletons $\{x\}$ for all $x \in \mathsf{R}\backslash\mathsf{Q}$.

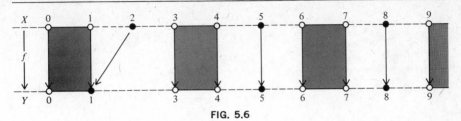

FIG. 5.6

Let $f: X \to Y$ be a continuous map. If C is a component of X, then $f(C)$ is connected, so $f(C) \subset D$ for a component D of Y. Suppose now f maps X homeomorphically onto Y. In the notation just used, $f^{-1}(D)$ is a connected subset of X with $C \subset f^{-1}(D)$, $C = f^{-1}(D)$ since C is a component of X, and hence $f(C) = D$. Thus f maps each component of X homeomorphically onto a component of Y. The same argument applies to $f^{-1}: Y \to X$. These observations establish the following result.

5.32 Theorem. Let $\mathfrak{C}$ and $\mathfrak{D}$ denote the collections of all components of topological spaces X and Y, respectively, and let

$$f: X \cong Y$$

be a homeomorphism. Then the map

$$\mathfrak{C} \to \mathfrak{D}$$

$$C \mapsto f(C)$$

is a one-to-one correspondence between $\mathfrak{C}$ and $\mathfrak{D}$ with

$$C \cong f(C) \qquad\qquad (C \in \mathfrak{C}).$$

This theorem can be used to establish that two spaces are not homeomorphic to one another, as in the following examples.

5.33 Examples

(1) The spaces
$$X = [0, 1] \cup [2, 3], \qquad Y = [0, 1] \cup [2, 3] \cup [4, 5]$$
are not homeomorphic, for X has two components and Y has three.

(2) This example is due to Kuratowski. Let

$$X = \bigcup_{n=0}^{\infty} (\,]3n, 3n + 1[\,\cup\, \{3n + 2\}),$$

$$Y = \,]0, 1] \cup \bigcup_{n=1}^{\infty} (\,]3n, 3n + 1[\,\cup\, \{3n + 2\})$$

(see Fig. 5.6). Then X is not homeomorphic to Y, for the components of X are all open intervals and singletons, and the component $]0, 1]$ of Y is not

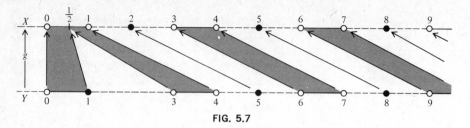

FIG. 5.7

homeomorphic to any of these. Nonetheless, there are continuous bijections

$$f: X \to Y, \qquad g: Y \to X$$

given by

$$f(x) = \begin{cases} x & \text{if } x \neq 2, \\ 1 & \text{if } x = 2, \end{cases}$$

(see Fig. 5.6) and

$$g(y) = \begin{cases} \frac{1}{2}y & \text{if } y \in \ ]0, 1], \\ \frac{1}{2}(y - 2) & \text{if } y \in \ ]3, 4], \\ y - 3 & \text{otherwise} \end{cases}$$

(see Fig. 5.7).

The largest component a space X can possibly have is the entire set X, and X is a component of X precisely when the space is connected. At the opposite extreme, the smallest components X can have, if X is nonempty, are the singletons $\{x\}$ for $x \in X$.

5.34 Definition. A topological space X is said to be *totally disconnected* if no component of X consists of more than a single point.

Since any nonempty connected subset of a topological space X is contained in some component of X, *the space X is totally disconnected if and only if the only connected subsets of X are the empty set $\varnothing$ and the singletons $\{x\}$ for $x \in X$.*

5.35 Examples

(1) Any discrete space is totally disconnected.

(2) By 5.31 (4), the nondiscrete spaces Q and $\mathsf{R} \backslash \mathsf{Q}$ are totally disconnected.

(3) The space $[0, 1] \cup [2, 3]$ is disconnected, but not totally disconnected.

(4) The Cantor discontinuum K (Example 4.13) is totally disconnected. (The reader who has not studied Chapter 4 can nonetheless read the relevant description of K in 4.13.) Just suppose K has a component containing points a and b with $a < b$. Then $[a, b] \subset \mathsf{K}$ by 5.3. We shall show this to be impossible.

Choose a positive integer n so large that

$$3^{-n} < b - a.$$

In the notation of 4.13, a and b both belong to the set K_n, which is the union of finitely many pairwise disjoint closed intervals each of length 3^{-n}. Then a and b must belong to two different ones of these intervals, so $[a, b] \not\subset K_n$. Since $\mathsf{K} \subset K_n$, we conclude that $[a, b] \not\subset \mathsf{K}$.

(5) In 1921 B. Knaster and C. Kuratowski constructed an uncountable connected subspace of the plane having the remarkable property that removing a single point from it leaves a totally disconnected space. Such a point in a connected space is called an *explosion point* (or dispersion point) of the space. We proceed to describe the Knaster-Kuratowski example.

Identify the real line R with the x-axis $\mathsf{R} \times \{0\}$ in the Euclidean plane under the homeomorphism $x \mapsto (x, 0)$. Then the Cantor discontinuum K is identified with the subset $\mathsf{K} \times \{0\}$ of the x-axis. Denote by E the set of "endpoints" of K—that is, the set of all endpoints of all the intervals comprising the sets K_n (compare Exercise 4.12)—and let

$$F = K \setminus E.$$

Let

$$p = (\tfrac{1}{2}, \tfrac{1}{2}),$$

and for each $t \in \mathsf{K}$ let L_t be the line segment in R^2 joining t to p. Form the set

$$X_E = \{ (x, y) \mid (x, y) \in L_t, t \in E, y \in \mathsf{Q} \}$$

of all points having rational ordinate and lying on the line segment joining p to some endpoint of K, and form the set

$$X_F = \{ (x, y) \mid (x, y) \in L_t, t \in F, y \notin \mathsf{Q} \}$$

of all points having irrational ordinate and lying on the line segment joining p to some nonendpoint of K. Finally, let

$$X = X_E \cup X_F$$

(see Fig. 5.8). Note that each $t \in F$ has the rational ordinate 0, so

$$F \cap X = \varnothing.$$

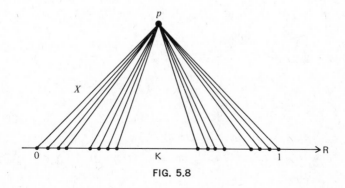

FIG. 5.8

The easy thing to show is that $X \setminus \{p\}$ is totally disconnected. Let C be a component of $X \setminus \{p\}$. Suppose, first, C intersects two line segments L_t and L_s for $t, s \in \mathsf{K}$ with $t < s$. Choose $u \in \mathsf{R}$ with

$$t < u < s, \qquad u \notin \mathsf{K};$$

such a u exists because $[t, s] \not\subset \mathsf{K}$ [compare (4)]. Then the line L in R^2 through p and u contains no points of $X \setminus \{p\}$, so the intersections with C of the two open half-planes into which L divides R^2 form a separation of C. Thus $C \subset L_t$ for some one $t \in \mathsf{K}$. Since L_t is totally disconnected [compare (2)], it follows that C consists of a single point.

The hard thing to show is that X is connected; our proof will use the Baire category theorem. Just suppose there is some separation $\{A, B\}$ of X, say with

$$p \in A.$$

Since A and B are disjoint and closed in X,

(*) $$\operatorname{cls} A \cap B = \varnothing = A \cap \operatorname{cls} B$$

(this is true whether the closures are taken in X or in R^2, but here and below all closures are taken in R^2). We shall eventually contradict (*) by showing that $X \subset \operatorname{cls} A$.

For each $t \in \mathsf{K}$ define a point

$$f(t) \in L_t$$

as follows. Set

$$f(t) = t \qquad \text{if } B \cap L_t = \varnothing.$$

If B intersects L_t, let $f(t)$ be the "least upper bound of points belonging to $B \cap L_t$", or more precisely,

$$f(t) = \text{the point } (x, y) \in L_t \text{ with } y = y_t$$

where

$$y_t = \sup \{y \mid (x, y) \in L_t \cap B \text{ for some } x\}.$$

Then

$$f(t) \neq p \qquad\qquad (t \in \mathsf{K})$$

since $p \in A$ and A is open in X.

Let $t \in \mathsf{K}$. If $f(t) \neq t$, then $f(t) \in \operatorname{cls} A \cap \operatorname{cls} B$, so $f(t) \notin X$ by (*). Thus

$$f(t) \notin X \qquad \text{or} \qquad f(t) = t.$$

Moreover, since $F \cap X = \varnothing$,

$$f(t) \in X \quad \Rightarrow \quad f(t) = t \in E.$$

Define

$$S = \{f(t) \mid t \in F\}.$$

By the preceding paragraph, $t \in F$ implies $f(t) \in L_t \setminus X$, so $f(t)$ has rational ordinate. Denoting the set of rational numbers in $]0, \tfrac{1}{2}[$ by Q, we can therefore write

$$S = S_0 \cup \bigcup_{q \in Q} S_q$$

where for each $q \in Q \cup \{0\}$,

$$S_q = \{ f(t) \mid t \in F, \text{ ordinate of } f(t) \text{ is } q \}.$$

Then

$$\text{cls } S_q \subset \mathsf{R} \times \{q\} \qquad\qquad (q \in Q \cup \{0\}).$$

Clearly

$$S_0 \subset F.$$

Let $q \in Q$. Then $S_q \subset \text{cls } A \cap \text{cls } B$, so $\text{cls } S_q \cap X = \varnothing$ by (*). Now a point $(x, y) \in L_t$ for $t \in E$ belongs to X when y is rational, and each point of $\text{cls } S_q$ has a rational ordinate. Hence

(**)
$$\text{cls } S_q \cap L_t = \varnothing \qquad\qquad (q \in Q, t \in E).$$

For each $q \in Q$ define

$$T_q = \{t \in \mathsf{K} \mid \text{cls } S_q \cap L_t \neq \varnothing\}.$$

By (**),

$$T_q \subset F \qquad\qquad (q \in Q).$$

Since $t \in F$ implies $f(t) \in S_q$ for some $q \in Q \cup \{0\}$, we have

$$F \subset S_0 \cup \bigcup_{q \in Q} T_q.$$

It follows that

(***)
$$\mathsf{K} = E \cup S_0 \cup \bigcup_{q \in Q} T_q.$$

We are going to show that S_0 is dense in K. Then the set

$$D = X \cap \bigcup_{t \in S_0} L_t$$

will be dense in X. Now $t \in S_0$ implies $f(t) = t$ and hence $X \cap L_t \subset A$ by definition of $f(t)$. Thus we will be able to conclude that $D \subset A$ and hence $X \subset \text{cls } A$, as we wished to show.

The proof that S_0 is dense in K is an application of the Baire category theorem (1.72) to K provided with its Euclidean metric. (By 1.62 and 1.68 this metric is complete, for K, being the intersection of closed subsets of R, is closed in R.) From (***) we obtain the representation

$$S_0 = \bigcap_{t \in E} (\mathsf{K} \setminus \{t\}) \cap \bigcap_{q \in Q} (\mathsf{K} \setminus T_q)$$

of S_0 as the intersection of countably many subsets of K. We must show that each of these sets is open and dense in K, or equivalently, that the complement of each is closed in K and contains no nonempty open subset of K.

If $t \in E$, then $\{t\}$ is closed in K and, by 4.13, not open in K. Now let $q \in Q$. One can show directly that T_q is closed in K (or one can argue that T_q is the image of the compact set cls S_q under the continuous map that for each $t \in \mathsf{K}$ projects all points on the line segment L_t to the point t). Since E is dense in K and disjoint from F, F cannot contain a nonempty open subset of K, so the same is true of its subset T_q.

This argument should dispel any lingering belief that specific, concrete examples are easier to understand than general, abstract theorems.

In Euclidean space R^n, any neighborhood U of any point x contains some connected neighborhood of x, for U contains some d-ball $B_\epsilon(x; d)$, where d is the Euclidean metric, and $B_\epsilon(x; d)$ is connected by 5.20 (1) because it is convex. Thus Euclidean spaces have the "local" property of having arbitrarily small connected neighborhoods of each point.

5.36 Definition. A topological space X is said to be *locally connected* when for each $x \in X$ and each neighborhood U of x there exists a connected neighborhood V of x with $V \subset U$.

Thus a locally connected space is one having at each of its points a local base consisting of connected sets.

5.37 Examples

(1) For any $n \geq 1$, Euclidean space R^n is both connected and locally connected.

(2) A discrete space X is locally connected, since for each $x \in X$ the set $\{x\}$ is a connected neighborhood of x contained in every neighborhood of x. As soon as X contains at least two points, however, it is not connected.

(3) A connected space need not be locally connected, even if it is a Hausdorff space. Hence *a Hausdorff space need not be locally connected even though each of its points has a connected neighborhood.* (Contrast this with the fact that a Hausdorff space is always locally compact if each of its points has a compact neighborhood [4.46]). For an example in the plane R^2, let X_0 be the unit interval

$$X_0 = [0, 1] \times \{0\}$$

on the x-axis, let Y_0 be the unit interval

$$Y_0 = \{0\} \times [0, 1]$$

on the y-axis, and for each positive integer n let Y_n be the vertical line segment

$$Y_n = \left\{\frac{1}{n}\right\} \times [0, 1]$$

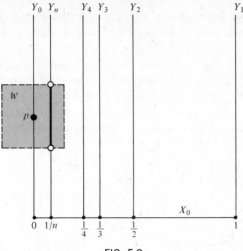

FIG. 5.9

above the point $(1/n, 0)$ of the x-axis. Consider the subspace

$$Y = X_0 \cup \bigcup_{n=0}^{\infty} Y_n$$

of $\mathbb{R}^2$ (see Figure 5.9).

The subsets X_0 and $Y_n, n = 0, 1, 2, \ldots$, of Y are all connected, being homeomorphic to $[0, 1]$. Since X_0 intersects Y_n for each n, it follows from 5.17 that Y is connected.

To see that Y is not locally connected, consider the point

$$p = (0, \tfrac{1}{2}).$$

Consider a neighborhood U of p in Y of the form

$$U = W \cap Y,$$

where

$$W = \,]{-}\epsilon, \epsilon[\times]\tfrac{1}{2} - \epsilon, \tfrac{1}{2} + \epsilon[, \qquad 0 < \epsilon < \tfrac{1}{2}.$$

Suppose U contains a connected neighborhood V of p in Y. For some sufficiently large n, V intersects Y_n. Now

$$U \cap Y_n = W \cap Y_n = \left\{\tfrac{1}{n}\right\} \times \left]\tfrac{1}{2} - \epsilon, \tfrac{1}{2} + \epsilon\right[$$

is both open and closed in U. By 5.16, $V \subset U \cap Y_n$, but this is impossible since $p \in V$.

The preceding example shows that a subspace Y of a locally connected space X need not be locally connected, even if Y is closed in X. In one case, however, we can deduce that Y is locally connected.

5.38 Proposition. Let Y be an open subspace of a locally connected space X. Then Y is locally connected.

Proof. Let U be a neighborhood in Y of a point $x \in Y$. Then U is also a neighborhood of x in X, so there exists a connected neighborhood V of x in X with $V \subset U$. However, V is a neighborhood of x in Y, too.   □

As the next example demonstrates, the continuous image of a locally connected space need not itself be locally connected.

5.39 Example. Let

$$Y = \{0\} \cup \left\{ \frac{1}{n} \middle| n = 1, 2, \ldots \right\}.$$

The discrete space N is locally connected, and the map $f \colon \mathsf{N} \to Y$ given by

$$f(n) = \begin{cases} 0 & \text{if } n = 0, \\ \dfrac{1}{n} & \text{if } n > 0 \end{cases}$$

is a continuous bijection. However, the space Y is not locally connected, for the point 0 of Y has no connected neighborhood in Y (the argument is similar to the one used in 5.37 (3)).

Local connectedness is preserved under a large class of continuous maps. To prove this we shall use the following result which relates the two ideas, local connectedness and components, being considered in this section.

5.40 Theorem. A necessary and sufficient condition for a topological space X to be locally connected is that for each open subset U of X, each component of the space U be open in X.

Proof. Assume X is locally connected. Let U be open in X, and let C be a component of U. To show C is open in X, we show it is a neighborhood of each of its points. Let $x \in C$. Since U is a neighborhood of x in X, there is a connected neighborhood V of x in X with $V \subset U$. Since V is connected and C is the component of U containing x, $V \subset C$. Hence C is a neighborhood of x in X.

Conversely, assume the condition holds. Let $x \in X$ and let W be any neighborhood of x in X. Choose an open set U with $x \in U \subset W$. Then the component of U to which x belongs is a connected open neighborhood of x in X that is contained in W.   □

5.41 Theorem. Let $f \colon X \to Y$ be a quotient map from a locally connected space X onto a topological space Y. Then Y is locally connected.

Proof. Let V be an open subset of Y and let E be a component of V. In view of 5.40, it suffices to show that E is open in Y. Since f is a quotient map, it therefore suffices to show that $f^{-1}(E)$ is open in X. Now by 5.40, each component of $f^{-1}(V)$ is open in X, so it suffices to show that $f^{-1}(E)$ is a union of components of $f^{-1}(V)$.

Let $x \in f^{-1}(E)$ and let C be the component of $f^{-1}(V)$ such that $x \in C$. We show that $C \subset f^{-1}(E)$, that is, $f(C) \subset E$. Now $f(C)$ is a connected subset of V containing $f(x)$. But E is a component of V with $f(x) \in E$. Hence $f(C) \subset E$. $\square$

5.41 Corollary. A quotient space of a locally connected space is locally connected.

For example, the torus [3.49 (4)], the Klein bottle [3.49 (6)], and the Möbius strip [3.49 (5)] are all locally connected, being quotient spaces of the square I^2.

5.42 Corollary. The image of a locally connected space under a continuous open map or a continuous closed map is locally connected.

5.43 Corollary. Local connectedness is a topological property.

The preservation of local connectedness under the formation of product spaces is left to the exercises (see Exercises 54–55).

EXERCISES

35. Show that $\mathsf{R}^2 \setminus \mathsf{S}_1$ has two components of which S_1 is the common boundary in R^2.

36. Find the components of each of the spaces in Exercise 2.

37. According to 1.20(2), a proper open subset U of R is the union of countably many pairwise disjoint open intervals and open rays. Show that these intervals and rays are the components of U.

38. Suppose a topological space X can be partitioned into a finite number n of connected sets $C_1, \ldots, C_n$ but cannot be partitioned into any smaller number of connected sets. Must $C_1, \ldots, C_n$ be the components of X?

39. Define a binary relation $\sim$ on a nonempty topological space X by the rule: $x \sim y$ if and only if there exists a connected set $C \subset X$ with $x \in C$ and $y \in C$. Verify that $\sim$ is an equivalence relation on X and that the equivalence classes of X under $\sim$ are the components of X.

40. Construct homeomorphic embeddings of each of the spaces X and Y of Example 5.33(2) into the other.

41. Show that the image of a component C of a topological space X under a continuous surjection $f: X \to Y$ need not be a component of Y.

42. Prove that the components of a product space $X_1 \times X_2$ are the sets of the form $C_1 \times C_2$, where C_1 is a component of X_1 and C_2 is a component of X_2. [*Hint:* If C is a component of $X_1 \times X_2$, then $C \subset p_1(C) \times p_2(C)$, where $p_1 : X_1 \times X_2 \to X_1$ and $p_2 : X_1 \times X_2 \to X_2$ are the projections. If C_1, C_2 are components of X_1, X_2, then $C_1 \times C_2 \subset C$ for some component C of $X_1 \times X_2$, so $C_1 = p_1(C_1 \times C_2) \subset p_1(C)$.]

43. (**a**) Generalize Exericse 42 to the product of any finite family of spaces.
(**b**) Generalize it to the product of a sequence of metrizable spaces.

44. Show that the complement of a dense open subset of $[0, 1]$ is totally disconnected.

45. In the notation of Exercise 39, prove that the quotient space $X/\sim$ is totally disconnected.

46. Prove that total disconnectedness is preserved under the formation of subspaces and product spaces. [*Note:* A quotient space of a totally disconnected space need not, however, be totally disconnected; this follows from 5.35(4) and the theorem of Alexandroff and Urysohn cited in 4.13.]

47. (**a**) Show that a Hausdorff space which is zero-dimensional (Exercise 2.78) must be totally connected.
(**b**) By considering the subspace $X \backslash \{p\}$ of the Knaster-Kuratowski space [5.35(5)], show that the converse of (a) fails. (*Note:* Compare Exercise 56.)

48. Prove or disprove: Each convex subspace of R^n is locally connected.

49. Prove that a topological space X is locally connected if and only if the collection of all its connected open subsets is a base of X.

50. Show that any manifold (3.17) is locally connected. Is the same true of any manifold-with-boundary (Exercise 3.47)?

51. Let $p = (0, 0) \in R^2$ and $L = \{0\} \times [-1, 1] \subset R^2$.
(**a**) Which, if any, of the spaces G, $G \cup \{p\}$, $G \cup L$ are locally connected if

$$G = \left\{ \left(x, \sin \frac{1}{x} \right) \middle| \; 0 < x \in R \right\}$$

(see Fig. 5.10 in Section 3).
(**b**) Repeat (a) for

$$G = \left\{ \left(x, x \sin \frac{1}{x} \right) \middle| \; 0 < x \in R \right\}.$$

52. By constructing a continuous map from the ray $[0, +\infty[$ in R onto the space of 5.37(3), show that a connected space which is the continuous image of a locally connected space need not be locally connected. (*Note:* Compare Exercise 58.)

53. Which totally disconnected spaces are locally connected?

54. (**a**) Prove that a product space $X_1 \times X_2$ is locally connected if and only if X_1 and X_2 are both locally connected. (*Hint:* Use 5.40 and Exercise 42.)
(**b**) Generalize (a) to the product of any finite family of spaces.

55. (**a**) Show that the product space $X_{n=1}^{\infty} X_n$ is not locally connected if $X_n = \{0, 1\}$ for all $n \geq 1$, but the product is locally connected if $X_1 = X_2 = \{0, 1\}$ and $X_n = \;]-1/n, 1/n[$ for all $n \geq 3$.
(**b**) What generalization of Exercise 54 does (a) suggest?

The following exercises involve compactness.

56. (**a**) Prove that the component of a compact T_2-space X containing a point $x \in X$ is the intersection of all open, closed subsets of X containing x. (*Hint*: Use Exercise 31.)

(**b**) Deduce from (a) that a compact T_2-space is zero-dimensional (Exercise 2.78) if it is totally disconnected.

57. Show that a compact locally connected space has only finitely many components.

58. Prove that a Hausdorff space which is the continuous image of a compact locally connected space is necessarily locally connected. (Hence any Hausdorff space that is the continuous image of the unit interval I is locally connected).

59. (**a**) Let (X, d) be a compact, locally connected metric space. Given any $\epsilon > 0$, establish the existence of some $\delta > 0$ such that any two points $x, y \in X$ satisfying $d(x, y) < \delta$ both belong to a connected subset C of X for which diam $C < \epsilon$.

(**b**) Is the hypothesis that X be compact really needed in (a)?

3. Path-Connected Spaces

To say a space is connected in the sense defined earlier is by no means the only way of capturing the intuitive idea that a space consists of a single piece. Another, equally natural, way is to say that one can travel continuously in the space from any given point to any other. In order to give a precise definition of this latter kind of connectedness, it is necessary first to formalize the notion of traveling continuously from one point to another.

Let x and y be points of a space X. Starting at some initial time at the point x, we want to travel continuously in the space until at some terminal time we arrive at the point y. Let us represent times by real numbers, the initial time by 0, and the terminal time by 1 (any two real numbers t_0 and t_1 with $t_0 < t_1$ would do to represent the initial and terminal times, but the choice $t_0 = 0$ and $t_1 = 1$ is a simplifying standardization). At each time t between the initial and terminal times we are to be at some point $x_t \in X$, where $x_0 = x$ and $x_1 = y$. The requirement that we move continuously means the point x_t is to be a continuous function of the time t. Thus our continuous trip from x to y may be described by the continuous function $t \mapsto x_t$ from $[0, 1]$ into X.

5.44 Definition. A *path in* a topological space X is a continuous map

$$\sigma: I \to X$$

where, as usual, I denotes the closed unit interval $[0, 1]$ in R. The points

$$x = \sigma(0), \qquad y = \sigma(1)$$

are called the *initial point* and *terminal point* of σ, respectively, and then σ is said to be a path in X *from x to y*.

The space X is said to be *path-connected* if for every $x \in X$ and $y \in X$ there exists a path in X from x to y.

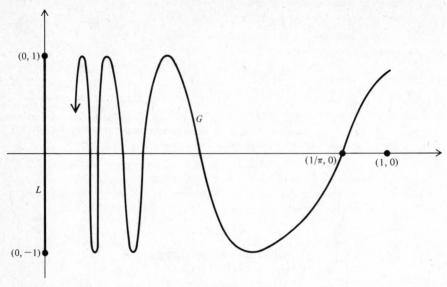

(0, 1)

G

$(1/\pi, 0)$ $(1, 0)$

L

(0, −1)

FIG. 5.10

5.45 Examples

(1) Any convex subspace X of $\mathbf{R}^n$ is path-connected. In fact, if $x, y \in X$, then $(1 - t)x + ty \in X$ for all scalars $t \in \mathsf{I}$, so the map

$$\sigma: \mathsf{I} \to X$$

$$t \mapsto (1 - t)x + ty$$

is a path in X from x to y.

In particular, the spaces $\mathbf{R}^n$, $\mathbf{B}_n$, and $\mathbf{D}_n$ are path-connected. Any interval in $\mathbf{R}$ is path-connected.

(2) Let

$$G = \left\{ \left(x, \sin \frac{1}{x} \right) \,\middle|\, 0 < x \leq 1 \right\},$$

$$L = \{ (0, y) \mid -1 \leq y \leq 1 \},$$

$$X = G \cup L$$

(see Fig. 5.10, in which different scales are used on the vertical and horizontal axes). The graph G oscillates between the lines $y = -1$ and $y = 1$ infinitely often, the oscillation becoming more rapid the closer we get to the y-axis.

The space G is connected, being the image of the interval $]0, 1]$ under the continuous map $x \mapsto (x, \sin 1/x)$; here and below we take for granted the continuity of the sine function. We shall show that X is connected but not path-connected.

That X is connected will follow from 5.6 if we can show $X \subset \mathrm{cls}\, G$, the closure being taken in $\mathbf{R}^2$ (it is easy to show $\mathrm{cls}\, G \subset X$, too). To show $X \subset$

cls G, we need only show $L \subset$ cls G. Let $(0, y) \in L$. If $0 < \epsilon < 1$, there exists $t \in [2\epsilon^{-1}, 2\epsilon^{-1} + 2\pi]$ with $\sin t = y$, and then the point

$$\left(x, \sin \frac{1}{x}\right) = \left(\frac{1}{t}, y\right) \in G$$

belongs to the Euclidean ball of radius ϵ at $(0, y)$—see Fig. 5.11. Hence $(0, y) \in$ cls G.

To see that X is not path-connected, just suppose there exists a path σ in X from $(0, 0)$ to $(1/\pi, 0)$. Denote by p and q the restrictions to X of the projections from R^2 onto R. Then the maps

$$p \circ \sigma : \mathsf{I} \to \mathsf{R}, \qquad q \circ \sigma : \mathsf{I} \to \mathsf{R}$$

are continuous. Let us assume that $\sigma(s)$ does not remain in L for an entire interval $[0, t]$ with $t > 0$, that is,

$$0 < t < 1 \quad \Rightarrow \quad p(\sigma(s)) > 0 \quad \text{for some } s \in \,]0, t].$$

Since $q \circ \sigma$ is continuous at 0 and $q(\sigma(0)) = 0$, there exists $0 < b < 1$ such that

(*) $\qquad\qquad 0 < t < b \quad \Rightarrow \quad q(\sigma(t)) \in \,] - \tfrac{1}{2}, \tfrac{1}{2}[.$

By our assumption, such a b exists with

$$p(\sigma(b)) > 0.$$

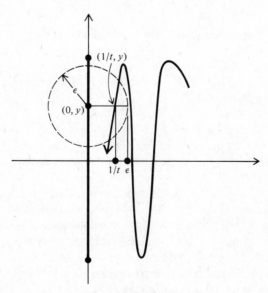

FIG. 5.11

Since $p \circ \sigma$ is continuous on $[0, b]$, there exists $0 < t < b$ with

$$p(\sigma(t)) = \frac{2}{n\pi}$$

for some positive integer n. Then

$$|q(\sigma(t))| = \left| \sin \frac{1}{2/n\pi} \right| = 1,$$

which contradicts (*).

If our assumption concerning σ does not hold, define

$$a = \sup \{t \in [0, 1] \mid p(\sigma(s)) = 0 \text{ for all } s \in \,]0, t]\}.$$

By continuity of $p \circ \sigma$ at a, $p(\sigma(a)) = 0$; since $p(\sigma(1)) = 1/\pi > 0$, $a < 1$. We now argue as in the preceding paragraph, but using the continuity of $p \circ \sigma$ at a instead of at 0.

Path-connectedness is a topological property. This is a consequence of the following, more general fact.

5.46 Theorem. Let $f: X \to Y$ be a continuous surjection from a path-connected space X onto a topological space Y. Then Y is path-connected.

Proof. Let $y_0, y_1 \in Y$. Choose $x_0, x_1 \in X$ with $y_0 = f(x_0)$, $y_1 = f(x_1)$. There is a path σ in X from x_0 to x_1. Then $f \circ \sigma$ is a path in Y from y_0 to y_1. ☐

From 5.45 (1) and 5.46 it follows, for example, that the circle S_1, the Möbius strip [3.49 (5)], and the Klein bottle [3.49 (6)] are path-connected.

5.47 Theorem. Every path-connected space is connected.

Proof. Let X be path-connected. If $x, y \in X$, there is a path σ in X from x to y, and then $\sigma(\mathsf{I})$ is a connected subset of X containing both x and y. By 5.19, X is connected. ☐

Note that the preceding proof does not say that the path-connected space X itself is a continuous image of I (a noncompact path-connected space, such as R, could not be a continuous image of the compact space I).

Example 5.45 (2) shows it is not true, conversely, that every connected space is path-connected. Nonetheless, there is a large class of topological spaces, including manifolds, for which connectedness is equivalent to path-connectedness. To prove this we need two technical lemmas.

5.48 Lemma. Let σ be a path in a topological space X from a point x to a point y, and let τ be a path in X from y to a point z. Then the map $\omega: \mathsf{I} \to X$

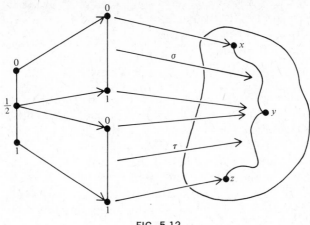

FIG. 5.12

given by

$$\omega(t) = \begin{cases} \sigma(2t) & \text{if } 0 \le t \le \tfrac{1}{2}, \\ \tau(2t - 1) & \text{if } \tfrac{1}{2} \le t \le 1 \end{cases}$$

is a path in X from x to z. We call ω the *product of σ and τ*.

Thus we may travel continuously in X from x to z in unit time by traversing in succession, with twice the original speed, the given routes from x to y and from y to z, passing through the point y at time $\tfrac{1}{2}$ (see Fig. 5.12).

Proof of 5.48. Observe first that ω is a well-defined map on I because $\sigma(2 \cdot 0) = y = \tau(2 \cdot 0 - 1)$. Also, $\omega(0) = \sigma(0) = x$ and $\omega(1) = \tau(1) = z$. The restrictions

$$\omega \mid [0, \tfrac{1}{2}], \qquad \omega \mid [\tfrac{1}{2}, 1]$$

of ω are continuous, for the first is the composite of the homeomorphism $t \mapsto 2t$ of $[0, \tfrac{1}{2}] \to [0, 1]$ with σ, and the second is the composite of the homeomorphism $t \mapsto 2t - 1$ of $[\tfrac{1}{2}, 1] \to [0, 1]$ with τ. Since $[0, \tfrac{1}{2}]$ and $[\tfrac{1}{2}, 1]$ are closed subsets of I, it follows from the gluing lemma that ω is continuous. $\square$

The second technical lemma says that by reversing direction on a path from a point x to a point y we obtain a path from y to x.

5.49 Lemma. Let σ be a path in a topological space X from a point x to a point y. Then the map $\tau \colon \mathsf{I} \to X$ given by

$$\tau(t) = \sigma(1 - t)$$

is a path in X from y to x. We call τ the *inverse of σ*.

5.50 Theorem. Let X be a connected space each point of which has at least one path-connected neighborhood. Then X is path-connected.

Proof. Fix $x \in X$. Define

$$K = \{y \in X \mid \text{there exists a path in } X \text{ from } x \text{ to } y\}.$$

It suffices to show $K = X$, for then if $y, z \in X$ there will exist paths in X from x to y and from x to z, and hence by 5.48 and 5.49 a path in X from y to z. Since the constant map from I to X with value x is a path in X from x to x, $x \in K$. Hence $K \neq \varnothing$. Since X is connected, to show $K = X$ it will suffice to show that K is both open and closed in X.

We show K is open in X. Let $y \in K$. By hypothesis there exists some path-connected neighborhood U of y in X. If $z \in U$, there is a path in U—and therefore in X—from y to z, and there is a path in X from x to y since $y \in K$; by 5.48 there is a path in X from x to z, that is, $z \in K$. Hence $U \subset K$.

We show K is closed in X. Let $y \in X \setminus K$. There exists a path-connected neighborhood V of y in X. Just suppose V intersects K, and let $z \in V \cap K$. Since $z \in K$, there exists a path from x to z; since $z \in V$, there exists a path from z to y. By 5.48, there exists a path from x to y, that is, $y \in K$. This is a contradiction. ☐

5.51 Corollary. Let X be a topological manifold. Then a necessary and sufficient condition for X to be path-connected is that it be connected.

Proof. If $x \in X$, then there is a neighborhood U of x in X with $U \cong \mathbf{R}^n$ for some n, and U is path-connected since $\mathbf{R}^n$ is. Sufficiency now follows from 5.50. Necessity follows from 5.47. ☐

An application of 5.51 is that for $n \geq 1$ the n-sphere $\mathbf{S}_n$ is path-connected, for by 5.20 (3) and 3.16 (12) the n-sphere is a connected manifold. Of course, we could also prove that $\mathbf{S}_n$ is path-connected the hard way, by explicitly constructing a path between any two points.

A given path $\sigma: I \to X$ from a point x to a point $y \neq x$ may describe a motion that crosses itself, doubles back on itself, or even stops awhile at some intermediate point. Such excess motion is avoided when σ is an embedding, for then σ is injective (since I is compact, a continuous injection from I to X is automatically an embedding whenever X is separated). A path σ in X from x to y that is actually an embedding of I into X is called an *arc in X from x to y*. When for any two distinct points x, y of X there is an arc in X from x to y, the space X is said to be *arc-connected*. It can be proved that a path-connected Hausdorff space is arc-connected; the proof uses R. L. Moore's characterization of I mentioned in Exercise 33—see Hocking and Young [17, Thm. 2-27] or Willard [34, 31.6].

According to Theorem 5.46, a continuous image of the closed unit interval I is path-connected. Which spaces *are* continuous images of I? With the excep-

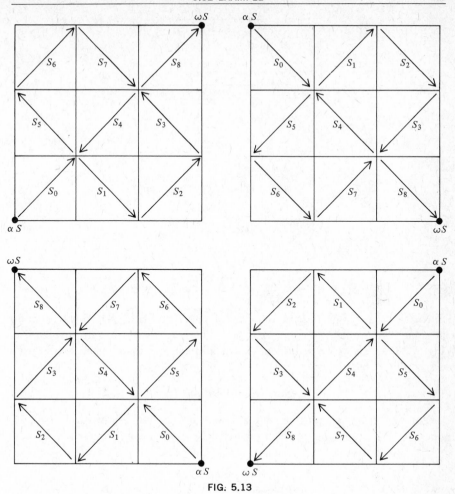

FIG. 5.13

tion of a one-point space, such a space seems to be "one-dimensional": it takes just a single continuously varying parameter $0 \leq t \leq 1$ to describe its points. It seems intuitively obvious that the space $I \times I$ requires not one, but two continuously varying parameters $0 \leq t \leq 1, 0 \leq s \leq 1$ to describe its points, so that $I \times I$ is "two-dimensional". This intuition is wrong! In 1890, G. Peano published the following surprising example of a "plane-filling curve", a continuous map from I *onto* $I \times I$.

5.52 Example. First we describe a method for subdividing a square into smaller squares. Let S be any square in the plane, having its sides parallel to the axes, in which two diagonally opposite vertices αS, ωS have been designated. Divide S into 9 congruent squares. There is a unique way of numbering these squares $S_0, \ldots, S_8$ and of designating two opposite vertices $\alpha S_i, \omega S_i$ of each S_i in such a way that S_0 and S_1 share a common vertical

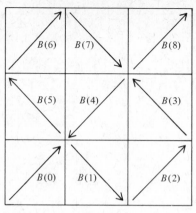

FIG. 5.14

side and that

(*) $\alpha S_0 = \alpha S, \quad \omega S_i = \alpha S_{i+1} \quad (i = 0, \ldots, 7), \quad \omega S_8 = \omega S.$

Since there are four possible choices of the pair $(\alpha S, \omega S)$, there are four possible configurations for $S_0, \ldots, S_8$. These configurations are depicted in Fig. 5.13 where we have indicated αS_i and ωS_i by drawing an arrow from αS_i to ωS_i. Observe that

(**)   diam $(S_i) = \frac{1}{3}$ diam (S) $(i = 0, \ldots, 8).$

Starting with the square $B = I \times I$ with designated vertices $\alpha B = (0, 0)$, $\omega B = (1, 1)$, apply the preceding construction to obtain 9 squares

$$B(0) = B_0, \ldots, B(8) = B_8$$

with each $B(i)$ having two designated opposite vertices $\alpha B(i)$, $\omega B(i)$ (see Fig. 5.14). Next, apply the construction to each of the squares $B(i)$ to obtain squares

$$B(i, 0) = B(i)_0, \ldots, B(i, 8) = B(i)_8 \quad (i = 0, \ldots, 8),$$

yielding $9^2 = 81$ smaller squares $B(i, j)$ in all, each with two designated opposite vertices $\alpha B(i, j)$, $\omega B(i, j)$ (see Fig. 5.15). Continue this process repeatedly, so that with the 9^n squares $B(i_1, \ldots, i_n)$ already determined for all choices of $i_1, \ldots, i_n \in \{0, \ldots, 8\}$, apply the construction to each of these squares to obtain squares

$$B(i_1, \ldots, i_n, 0) = B(i_1, \ldots, i_n)_0, \ldots,$$

$$B(i_1, \ldots, i_n, 8) = B(i_1, \ldots, i_n)_8.$$

Each of the $9^n \cdot 9 = 9^{n+1}$ squares so obtained has, of course, two designated opposite vertices.

For each $n \geq 1$ and each n-tuple $(i_1, \ldots, i_n)$ in $\{0, \ldots, 8\}$ we have a square $B(i_1, \ldots, i_n)$. By (*),

$$\alpha B(i_1, \ldots, i_n, 0) = \alpha B(i_1, \ldots, i_n),$$

$$\omega B(i_1, \ldots, i_n, 8) = \omega B(i_1, \ldots, i_n),$$

$$\omega B(i_1, \ldots, i_n, i) = \alpha B(i_1, \ldots, i_n, i+1) \quad (i = 0, \ldots, 7).$$

From (**),

$$\text{diam } B(i_1, \ldots, i_n) = \frac{\sqrt{2}}{3^n}.$$

Now

$$B(i_1, \ldots, i_n, i) \subset B(i_1, \ldots, i_n) \quad (i = 0, \ldots, 7).$$

Hence from the nested set theorem (1.71) it follows that for each infinite

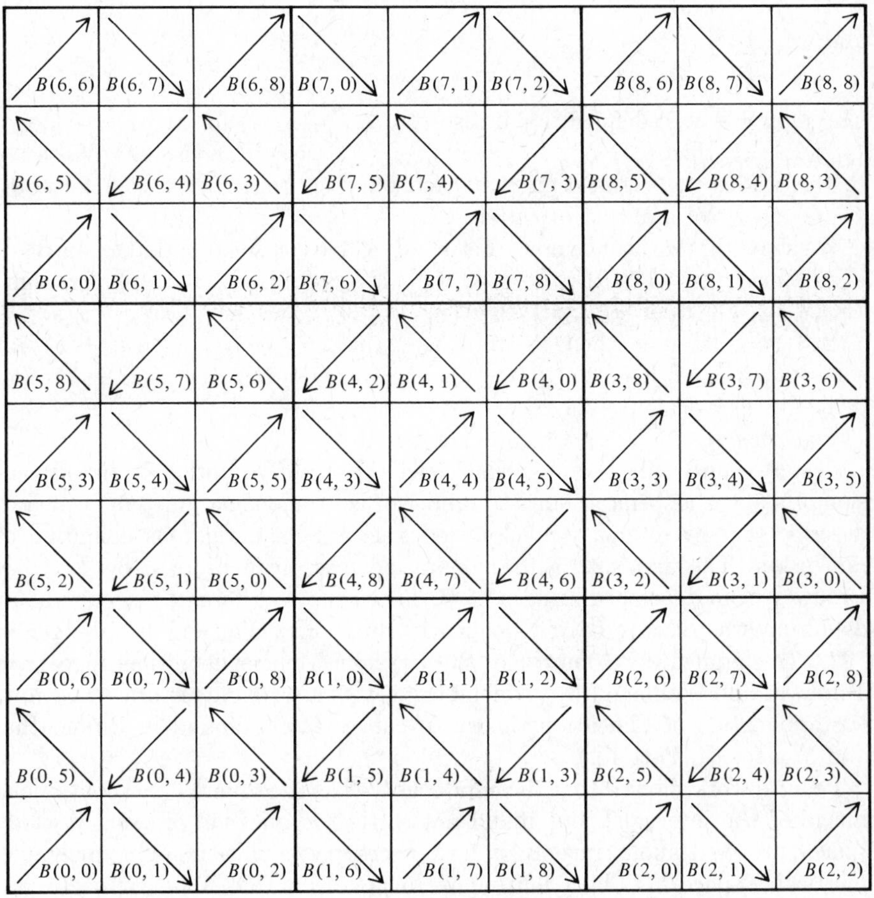

FIG. 5.15

sequence $(i_1, i_2, \ldots)$ in $\{0, \ldots, 8\}$ there is a *unique* point

$$x(i_1, i_2, \ldots) \in \bigcap_{n=1}^{\infty} B(i_1, \ldots, i_n).$$

Divide the interval I into 9 intervals $A(0), \ldots, A(8)$ each of length 9^{-1}, numbered from left to right. Next, for each $i \in \{0, \ldots, 8\}$, divide $A(i)$ into 9 intervals $A(i, 0), \ldots, A(i, 8)$ each of length 9^{-2}, numbered from left to right. Continuing this process, we obtain, for each $n \geq 1$, 9^n intervals $A(i_1, \ldots, i_n)$ each of length 9^{-n}. For each infinite sequence $(i_1, i_2, \ldots)$ in $\{0, \ldots, 8\}$ there is a *unique* point

$$t(i_1, i_2, \ldots) \in \bigcap_{n=1}^{\infty} A(i_1, \ldots, i_n).$$

Define a map $\sigma: \mathsf{I} \to \mathsf{I} \times \mathsf{I}$ as follows. Let $t \in \mathsf{I}$. There is a sequence $(i_1, i_2, \ldots)$ in $\{0, \ldots, 8\}$ with

$$t = t(i_1, i_2, \ldots),$$

and we set

$$\sigma(t) = x(i_1, i_2, \ldots).$$

The map σ is well-defined, for if also $t = t(j_1, j_2, \ldots)$ for another sequence $(j_1, j_2, \ldots)$, then $x(i_1, i_2, \ldots) = x(j_1, j_2, \ldots)$ (see Exercise 69). We leave to the reader the verification that each $x \in \mathsf{I} \times \mathsf{I}$ is $x(i_1, i_2, \ldots)$ for some $(i_1, i_2, \ldots)$, so that σ is surjective.

We show that σ is continuous. Let $\epsilon > 0$. Choose n so large that $2\sqrt{2}/3^n < \epsilon$. Suppose $t, s \in \mathsf{I}$ with $|t - s| < 9^{-n}$. If $t, s \in A(i_1, \ldots, i_n)$ for some $(i_1, \ldots, i_n)$, then $\sigma(t), \sigma(s) \in B(i_1, \ldots, i_n)$, whence $\|\sigma(t) - \sigma(s)\| \leq \sqrt{2}/3^n < \epsilon$. If not, then $t \in A(i_1, \ldots, i_n)$, $s \in A(i_1, \ldots, i_n + 1)$ for some $(i_1, \ldots, i_n)$, $\sigma(t) \in B(i_1, \ldots, i_n)$, $\sigma(s) \in B(i_1, \ldots, i_n + 1)$, and hence $\|\sigma(t) - \sigma(s)\| \leq 2\sqrt{2}/3^n < \epsilon$.

Peano's unexpected example shattered the intuitive notion of the dimension of a space as being the least number of continuous parameters needed to describe the space, and it precipitated a search for a rigorous definition of dimension. The first satisfactory definition was given in 1913 by L. E. J. Brouwer, who developed an idea of H. Lebesgue. A different, but equivalent definition was discovered independently by P. Urysohn and K. Menger in 1922. (A detailed development of the Urysohn-Menger definition appears in Hurewicz and Wallman [18], which begins with a readable historical account; the application of dimension theory to curves can be found in Blumenthal and Menger [4, Part 4].)

Peano's work showed that the cube I^3 as well as the square I^2 is a continuous image of the interval I, but it did not settle the question of exactly which spaces are continuous images of I. A necessary condition for a nonempty metrizable space X to be a continuous image of I is that X be compact, connected, and locally connected (see Theorems 4.17, 5.8, and 5.41). In 1914–

1920, H. Hahn and S. Mazurkiewicz proved that this condition is also sufficient. Now a Hausdorff space that is the continuous image of a compact
metrizable space is itself metrizable (see Exercise 4.59 and the paragraph
following 4.28). Hence we may state the *Hahn-Mazurkiewicz theorem* as:

> *A nonempty Hausdorff space is a continuous image of the closed unit
> interval* $[0, 1]$ *if and only if it is compact, connected, locally con-
> nected, and metrizable.*

For a proof, see Willard [34, section 31]; the idea of the proof is used in our
Exercise 71.

EXERCISES

60. Must a subset of $\mathbf{R}^n$ be path-connected if it is starlike (Exercise 16)?

61. Is the space

$$G = \left\{ \left(x, x \sin \frac{1}{x} \right) \, \middle| \, 0 < x \leq 1 \right\}$$

path-connected? the space $G \cup \{(0, 0)\}$?

62. Construct a path-connected subset of the plane containing two points a and b
having the property that no path in the subset from a to b has finite length.

63. Discuss the path-connectedness of product spaces.

64. Let $X \subset \mathbf{R}^n$. If $x, y \in X$, a *polygonal path in X from x to y* is the union of finitely
many line segments $L(x_0, x_1), \ldots, L(x_{n-1}, x_n)$ contained in X, where $L(x_i, x_{i+1})$
denotes the line segment joining x_i to x_{i+1}, such that $x_0 = x$ and $x_n = y$.
  (**a**) Show that a polygonal path in X from x to y is the range of some path in
  X from x to y (in the sense of 5.44).
  (**b**) Call X *polygonally connected* if for any points $x, y \in X$ there exists a polygonal
  path in X from x to y. From (a) and 5.47 it follows that X is connected if it is
  polygonally connected. Prove that the converse holds in case X is open in $\mathbf{R}^n$
  but otherwise need not hold.

65. Let X be a nonempty open connected subset of the plane $\mathbf{R}^2$. Prove that if x, y, z
are any three distinct points of X, then there is a path σ in X from x to y with $z \notin \sigma(1)$.
Use this to show that X is not homeomorphic to any subset of the line $\mathbf{R}$.

66. Let X be a nonempty space. A *path-component* of X is a path-connected subset
of X that is not properly contained in any other path-connected subset of X.
  (**a**) Formulate and prove the analog of Theorem 5.29 for path-components.
  Prove that two points x, y of X belong to the same path-component of X if and
  only if there exists a path in X from x to y.
  (**b**) Is the analog of Theorem 5.28 for path-components true?

67. A space X is said to be *locally path-connected* if each neighborhood of each point
$x \in X$ contains a path-connected neighborhood of x.
  (**a**) Prove that X is locally path-connected if for each neighborhood U of each
  point $x \in X$ there is a neighborhood V of x such that $V \subset U$ and such that for
  each $y \in V$ there is a path in V from x to y.

(**b**) What can be said about the path-components (Exercise 66) of a locally path-connected space?

(**c**) Construct a path-connected space that is not locally path-connected.

68. Give an example of a path-connected space that is not arc-connected. (By the discussion following 5.51, such a space will not be a Hausdorff space.)

69. Let $t \in I$. In the notation of 5.52, how many sequences $(i_1, i_2, \ldots)$ can there be with $t = t(i_1, i_2, \ldots)$? For which t is there a unique such sequence? (*Hint*: Use base 9 expansions.)

70. (**a**) Show that Peano's map $\sigma: I \to I \times I$ (Example 5.52) is not injective by exhibiting $t, s \in I$ with $t \neq s$ but $\sigma(t) = \sigma(s)$.

(**b**) Can any continuous map from I onto $I \times I$ be injective?

(**c**) Can any map from I onto $I \times I$ be injective?

71. Let $g: K \to I \times I$ be a continuous surjection [see Exercise 4.14(b)]. Construct a continuous extension $f: I \to I \times I$ of g to I. [*Hint*: Being open in R, the set $I \backslash K$ is by 1.20(2) the union of a sequence $J_1, J_2, \ldots$ of pairwise disjoint open intervals. Make f linear on each J_i. *Note*: To prove the Hahn-Mazurkiewicz theorem we would first obtain a continuous map from K onto the given locally connected, compact, connected metrizable space and then would extend this to a continuous map on I.]

72. Construct a continuous map from I onto I^3 by using the method of either Example 5.52 or Exercise 71.

73. A space X is said to be *one-dimensional* if X is not zero-dimensional (Exercise 2.78) and if each neighborhood of each point $x \in X$ contains a neighborhood V of x with bdy V zero-dimensional. Verify that the following are one-dimensional: a polygonal path in R^2 (Exercise 64); the space X of Example 5.45(2); any manifold that is one-dimensional in the sense of 3.17. (*Note*: This definition of 'one-dimensional', due to Urysohn and Menger, serves to distinguish I^2 from I. It is possible to prove, for example, that a compact connected subset of the plane R^2 is one-dimensional if and only if it is nowhere dense in R^2—see Blumenthal and Menger [4, Section 14]).

4. Homotopy

The disk

$$D_2 = \{x \in R^2 : ||x|| \leq 1\}$$

and the annulus

$$A = \{x \in R^2 : \tfrac{1}{2} \leq ||x|| \leq 1\}$$

do not appear to be homeomorphic (see Fig. 5.16), yet none of the topological properties we have already studied distinguish between these two spaces: both are path-connected, locally connected, compact, separable, and metrizable (and hence also connected, locally compact, second-countable, first-countable, and Hausdorff).

It is intuitively plausible that the disk D_2 can be continuously shrunk to a point but the annulus A cannot, that D_2 cannot be continuously deformed into a circle but A can be, and that D_2 does not have a hole but A does. The elaboration of these ideas forms part of homotopy theory, which deals with

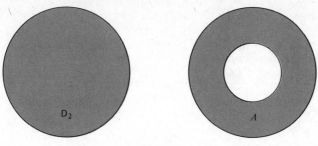

FIG. 5.16

the questions of whether one space can be continuously deformed into another space and whether one map can be continuously transformed into another map.

What should we mean by saying that a space X can be continuously shrunk to a point $y \in X$? We cannot mean simply that X can be mapped continuously onto $\{y\}$, for that is always possible. We mean, instead, that a continuous process can be performed over a time interval during which each point $x \in X$ moves to y. As usual, we parametrize time by real numbers $0 \leq t \leq 1$, with 0 being the initial and 1 the terminal time. If we denote by $h(x, t)$ the position of point $x \in X$ at time $t \in [0, 1]$, then the continuous process is to satisfy $h(x, 0) = x$ and $h(x, 1) = y$. Hence shrinking X continuously to a point may be defined as follows.

5.53 Definition. A *contraction of* a topological space X *to* a point $y \in X$ is a continuous map

$$h: X \times I \to X$$

such that

$$h(x, 0) = x, \qquad h(x, 1) = y \qquad\qquad (x \in X).$$

If some contraction of X to some point of X exists, then X is said to be *contractible*.

5.54 Examples

(1) Any convex subset X of R^n for $n \geq 1$ is contractible, for if we choose any $y \in X$, then the map

$$h: X \times I \to X$$

$$(x, t) \mapsto (1 - t)x + ty$$

is a contraction of X to y; observe that h moves each $x \in X$ to y along the line segment joining x to y. In particular, the spaces R^n, D_n, and B_n are all contractible.

(2) The subspace $X = \mathsf{S}_n \backslash \{p\}$ of the n-sphere S_n, where p is the north pole, is not convex (unless $n = 0$). However, X is contractible. To see this

we use a homeomorphism

$$f : X \cong \mathsf{R}^n$$

[see 3.16 (11)] and a contraction

$$H : \mathsf{R}^n \times \mathsf{I} \to \mathsf{R}^n$$

of R^n to the origin 0 to form the map

$$h : X \times \mathsf{I} \to X$$

given by

$$h(x, t) = f^{-1}(H(f(x), t)).$$

Then h is a contraction of X to the point $f^{-1}(0) \in X$.

It is a fact that the entire n-sphere S_n is not contractible. The case $n = 1$ will be proved later. The case $n = 0$ follows from the next proposition.

5.55 Proposition. Each contractible space is path-connected.

Proof. Let $h : X \times \mathsf{I} \to X$ be a contraction of the space X to the point $y \in X$. In view of 5.48 and 5.49, to show X is path-connected it suffices to show that for each $x \in X$ there is a path in X from x to y. But for $x \in X$, the continuous map

$$\sigma : \mathsf{I} \to X$$

$$t \mapsto h(x, t)$$

is a path in X from $x = h(x, 0)$ to $y = h(x, 1)$.   ☐

The choice of the point to which a contractible space is shrunk is immaterial.

5.56 Proposition. Let z be any point in a contractible space X. Then there is a contraction of X to z.

Proof. There is a contraction $h : X \times \mathsf{I} \to X$ of X to some point $y \in X$. To construct a contraction of X to z, first we move each $x \in X$ to y along the path $t \mapsto h(x, t)$ at twice the original speed, so that x arrives at y at time $t = \frac{1}{2}$; then we move y to z backward along the path $t \mapsto h(z, t)$, again at twice the original speed. Formally, we define $H : X \times \mathsf{I} \to X$ by

$$H(x, t) = \begin{cases} h(x, 2t) & \text{if } 0 \le t \le \frac{1}{2}, \\ h(z, 2 - 2t) & \text{if } \frac{1}{2} \le t \le 1. \end{cases}$$

The map H is well-defined, for on

$$X \times \{\tfrac{1}{2}\} = (X \times [0, \tfrac{1}{2}]) \cap (X \times [\tfrac{1}{2}, 1])$$

we have

$$h(x, 2 \cdot \tfrac{1}{2}) = y = h(z, 2 - 2 \cdot \tfrac{1}{2}).$$

By the gluing lemma 3.10, H is continuous, for its restrictions to the closed subspaces $X \times [0, \frac{1}{2}]$ and $X \times [\frac{1}{2}, 1]$ of $X \times I$ are continuous, each being a composition of continuous maps. Finally,

$$H(x, 0) = h(x, 0) = x, \qquad H(x, 1) = h(z, 0) = z$$

for each $x \in X$. $\square$

Explicit use of the gluing lemma, such as was made in the preceding proof, will usually be omitted below.

A space that is not contractible can sometimes still be continuously deformed onto a proper subspace without the points of the subspace being disturbed.

5.57 Definition. Let E be a subspace of a topological space X. A *deformation retraction of X onto E* is a continuous map

$$h \colon X \times I \to X$$

such that

$$h(x, 0) = x, \qquad h(x, 1) \in E \qquad\qquad (x \in X),$$
$$h(y, 1) = y \qquad\qquad (y \in E).$$

When some deformation retraction of X onto E exists, E is called a *deformation retract of X*.

5.58 Examples

(1) The circle

$$S_1 = \{x \in \mathsf{R}^2 \colon ||x|| = 1\}$$

is a deformation retract of the annulus

$$A = \{x \in \mathsf{R}^2 \colon \tfrac{1}{2} \leq ||x|| \leq 1\},$$

for we may slide each $x \in A$ toward the point $||x||^{-1}x \in S_1$ along the line through the origin and x (see Fig. 5.17). Formally, a deformation retraction

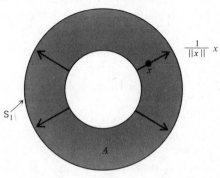

FIG. 5.17

$h: A \times I \to A$ of A onto S_1 is defined by the formula

$$h(x, t) = (1 - t)x + t\left(\frac{1}{||x||} x\right).$$

The same formula defines a deformation retraction of the "punctured disk" $\mathsf{D}_2 \setminus \{(0, 0)\}$ onto S_1.

(2) The map $h: \mathsf{R}^n \times I \to \mathsf{R}^n$ defined by

$$h(x, t) = \begin{cases} (1 - t)x + t\left(\dfrac{1}{||x||} x\right) & \text{if } ||x|| \geq 1 \\ x & \text{if } ||x|| \leq 1 \end{cases}$$

is a deformation retraction of Euclidean n-space R^n onto the n-disk D_n.

(3) A contraction of a space X to a point $y \in X$ is just a deformation retraction of X onto its subspace $\{y\}$.

A deformation retraction $h: X \times I \to X$ of a space X onto a subspace E gives rise to the map $x \mapsto h(x, 1)$ of X onto E which leaves each $x \in E$ fixed.

5.59 Definition. Let E be a subspace of a topological space X. A *retraction of X onto E* is a continuous map

$$r: X \to E$$

such that

$$r(x) = x \qquad\qquad (x \in E)$$

(whence r maps X onto E); in other words, a retraction of X onto E is a continuous extension to X of the identity map of E. When some retraction of X onto E exists, E is called a *retract of X*.

A deformation retract E of X is thus a retract of X (but a deformation retraction of X onto E is not a retraction of X onto E). However, a retract of X need not be a deformation retract of X: Any one-point subspace of any space X is a retract of X, but a one-point subspace cannot be a deformation retract of X unless X is contractible.

The next proposition explains why Examples 5.58 did not include, for example, a deformation retraction of the plane R^2 onto the open ball B_2.

5.60 Proposition. A retract E of a Hausdorff space X is closed in X.

Proof. Let $r: X \to E$ be a retraction. Take $x \in X \setminus E$. Then $r(x) \neq x$, so $r(x)$ and x have disjoint neighborhoods V and W in X. Since r is continuous at x, there is a neighborhood U_1 of x with $r(U_1) \subset V$. Then the neighborhood $U = U_1 \cap W$ of x is disjoint from E; in fact, $y \in U \cap E$ implies both $y = r(y) \in V$ and $y \in W$. $\quad\square$

Earlier we saw that the disk D_2 is contractible, and later we shall prove that the circle S_1 is not contractible. From the next result it will then follow that S_1 is not a retract of D_2.

5.61 Proposition. A retract of a contractible space is itself contractible.

Proof. Let $c: X \times I \to X$ be a contraction of the space X to a point $z \in X$, and let $r: X \to E$ be a retraction of X onto a subspace E. Define $h: E \times I \to E$ by

$$h(x, t) = r(c(x, t)).$$

Then $x \in E$ implies

$$h(x, 0) = r(x) = x, \qquad h(x, 1) = r(z),$$

so h is a contraction of E to $r(z) \in E$. $\quad\square$

A deformation retraction $h: X \times I \to X$ of X onto $E \subset X$ may be viewed as a continuous family of paths $t \mapsto h(x, t)$ in X ending in E, one path for each $x \in X$ (compare the proof of 5.55). It may also be viewed as a continuous family of continuous maps

$$h_t: X \to X$$

$$x \mapsto h(x, t),$$

one map for each $t \in I$, where h_0 is the identity map of X and h_1 sends X onto E while leaving each $x \in E$ fixed. From the latter viewpoint, h may then be regarded as a continuous process that deforms the map h_0 into the map h_1. We are now ready to generalize.

5.62 Definition. Let $f, g: X \to Y$ be continuous maps. A continuous map

$$h: X \times I \to Y$$

is called a *homotopy from f to g* when

$$h(x, 0) = f(x), \qquad h(x, 1) = g(x) \qquad\qquad (x \in X),$$

and then the notation

$$h: f \simeq g$$

is used. If such an h exists, we say that f is *homotopic to g* and write $f \simeq g$.

In terms of the maps

$$h_t: X \to Y$$

$$x \mapsto h(x, t)$$

defined for all $t \in I$, the condition a continuous map $h: X \times I \to Y$ must satisfy to be a homotopy from f to g is simply that

$$h_0 = f, \qquad h_1 = g.$$

5.63 Examples

(1) Let $j: E \to X$ be the inclusion map of a subspace E of a space X into X, and let $i: X \to X$ be the identity map. Then a deformation retraction of X onto E is just a homotopy $h: i \simeq j \circ r$ for some retraction $r: X \to E$.

(2) A space X is contractible precisely when the identity map of X is homotopic to a constant map of X into X.

(3) Any two continuous maps $f, g: X \to Y$ from the same space X into a contractible space Y are homotopic. In fact let $c: Y \times I \to Y$ be a contraction of Y to $z \in Y$. Define $h: X \times I \to X$ by

$$h(x, t) = \begin{cases} c(f(x), 2t) & \text{if } 0 \le t \le \frac{1}{2}, \\ c(g(x), 2 - 2t) & \text{if } \frac{1}{2} \le t \le 1. \end{cases}$$

Then $h: f \simeq g$.

In particular, any two continuous maps from the same space into R^n, or into a convex subspace of R^n, are homotopic.

(4) The identity map of S_1 is homotopic to the reflection $x \mapsto -x$ of S_1 about the origin. To construct a homotopy h, notice that the two maps are rotations of S_1 through angles of 0 and π, and for each $t \in I$ take h_t to be the rotation through an angle of $t\pi$. More precisely, represent each $x \in \mathsf{S}_1$ in the form

$$x = (\cos \pi s, \sin \pi s)$$

with $s \in \mathsf{R}$ and set

$$h(x, t) = (\cos \pi(s + t), \sin \pi(s + t))$$

(of course, it must be checked that this definition of $h(x, t)$ is independent of the particular s used to represent x).

5.64 Theorem. The homotopy relation $\simeq$ is an equivalence relation on the set of all continuous maps from a topological space X to a topological space Y.

Proof. Reflexivity. Let $f: X \to Y$ be a continuous map. Then the map $h: X \times I \to Y$ defined by $h(x, t) = f(x)$ is a homotopy from f to f.

Symmetry. Let $h: f \simeq g$ be a homotopy between continuous maps $f, g: X \to Y$. Then the map $H: X \times I \to Y$ defined by

$$H(x, t) = h(x, 1 - t)$$

is a homotopy from g to f, so $g \simeq f$.

Transitivity. Let

$$F: f_0 \simeq f_1, \qquad G: f_1 \simeq f_2$$

be homotopies, where $f_0, f_1, f_2: X \to Y$ are continuous. Define $H: X \times I \to Y$

by

$$H(x, t) = \begin{cases} F(x, 2t) & \text{if } 0 \le t \le \frac{1}{2}, \\ G(x, 2t - 1) & \text{if } \frac{1}{2} \le t \le 1. \end{cases}$$

Then $H: f_0 \simeq f_2$ since $H_0 = F_0$ and $H_1 = G_1$. Hence $f_0 \simeq f_2$. ☐

Aside from contractions and deformation retractions, the homotopies in which we are most interested are deformations of paths that leave endpoints fixed. Suppose σ and τ are two paths in a space X, both from a point x_0 to a point x_1. To deform σ into τ while leaving x_0 and x_1 fixed is to prescribe a continuous family $(h_t \mid t \in \mathsf{I})$ of paths in X from x_0 to x_1 with $h_0 = \sigma$ and $h_1 = \tau$.

5.65 Definition. Let $\sigma, \tau: \mathsf{I} \to X$ be paths in a topological space X with

$$\sigma(0) = \tau(0) = x_0, \qquad \sigma(1) = \tau(1) = x_1.$$

We call a homotopy $h: \sigma \simeq \tau$ for which

$$h_t(0) = x_0, \qquad h_t(1) = x_1 \qquad\qquad (t \in \mathsf{I})$$

a *homotopy from σ to τ relative to* $\{0, 1\}$ and write

$$h: \sigma \sim \tau.$$

When such an h exists we say that σ is *homotopic to τ relative to* $\{0, 1\}$ and write $\sigma \sim \tau$.

A homotopy from σ to τ relative to $\{0, 1\}$ is thus just a continuous map

$$h: \mathsf{I} \times \mathsf{I} \to X$$

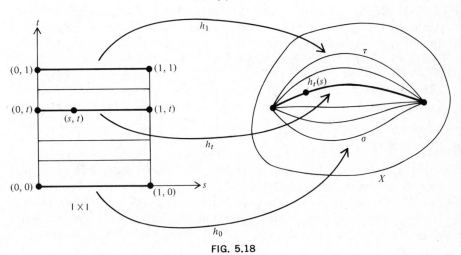

FIG. 5.18

on the square $I \times I$ for which

$$h(s, 0) = \sigma(s), \qquad h(s, 1) = \tau(s) \qquad\qquad (s \in I),$$

$$h(0, t) = x_0, \qquad h(1, t) = x_1 \qquad\qquad (t \in I)$$

(see Fig. 5.18). The first variable, which we denote by s, parametrizes each individual path h_t from x_0 to x_1, and the second variable, which we denote by t, parametrizes the family of paths h_t.

If in the proof of 5.64 we start with homotopies relative to $\{0, 1\}$ between paths having the same endpoints, then the homotopies constructed there will also be relative to $\{0, 1\}$. Hence *the relative homotopy relation $\sim$ is an equivalence relation on the set of all paths in a space X from a point x_0 to a point x_1.*

5.66 Examples

(1) Let σ and τ be paths from a point x_0 to a point x_1 in a convex subspace X of R^n. Then $\sigma \sim \tau$, a homotopy h from σ to τ relative to $\{0, 1\}$ being the map

$$h(s, t) = (1 - t)\sigma(s) + t\tau(s)$$

which from time $t = 0$ to time $t = 1$ moves $\sigma(s)$ along a straight line to $\tau(s)$.

(2) Let σ be any path in a space X. A strictly increasing function $\alpha: I \to I$ of I onto I yields a new path $\tau = \sigma \circ \alpha$ in X from $\sigma(0)$ to $\sigma(1)$ by a "change in parameter". The formula

$$h(s, t) = \sigma((1 - t)s + t\alpha(s))$$

defines a homotopy $h: \sigma \sim \tau$ relative to $\{0, 1\}$ with $h_t(I) = \sigma(I)$ for each $t \in I$.

Examination of Fig. 5.18 suggests that for a homotopy $h: \sigma \sim \tau$ between two paths joining the same points, the paths h_t must move through the points "surrounded by" $\sigma(I) \cup \tau(I)$ as t varies from 0 to 1. Hence the requirement that any two paths in a space X with the same endpoints be homotopic relative to $\{0, 1\}$ says that the space X have no "holes".

5.67 Definition. A topological space X is *simply connected* if X is path-connected and if $\sigma \sim \tau$ whenever σ and τ are paths in X with $\sigma(0) = \tau(0)$, $\sigma(1) = \tau(1)$.

Example 5.66 (1) together with 5.45 (1) says that any convex subset of R^n is simply connected. In particular, the plane R^2 and the ball B_2 are simply connected. It is an extremely attractive conjecture—and is true—that neither a circle nor an annulus is simply connected, but the proofs will take some effort.

Simply connected open sets in the plane play an important role in the theory of line integrals (see Apostol [1, Chapter 10 and pp. 383–385]). Suppose

$$F(x, y) = (P(x, y), Q(x, y))$$

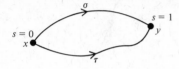 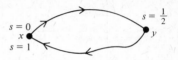

FIG. 5.19

is a continuously differentiable vector field on a connected open set S in the plane satisfying the condition $\partial P/\partial y = \partial Q/\partial x$ on S. If S is simply connected, then the value of the line integral $\int F \cdot d\alpha$ along a piecewise smooth path α in S from a point (x_0, y_0) to a point (x_1, y_1) will be independent of the particular path from (x_0, y_0) to (x_1, y_1) used, and F will be the gradient of some scalar field on S. (Physically, if F represents a force field on S, this says that F will be conservative.)

Two paths σ, τ from a point x in a space X to a point y in X may be combined into a path ω from x back to x that traverses σ followed by τ (see Fig. 5.19)—take ω to be the product of σ with the inverse of τ (5.48, 5.49); that is,

$$\omega(s) = \begin{cases} \sigma(2s) & \text{if } 0 \leq s \leq \tfrac{1}{2}, \\ \tau(2 - 2s) & \text{if } \tfrac{1}{2} \leq s \leq 1. \end{cases}$$

Conversely, any path ω from x to x can be obtained in this manner starting with paths σ, τ from x to $y = \omega(\tfrac{1}{2})$, namely,

$$\sigma(s) = \omega(\tfrac{1}{2}s) \qquad\qquad (0 \leq s \leq 1),$$

$$\tau(s) = \omega(1 - \tfrac{1}{2}s) \qquad\qquad (0 \leq s \leq 1).$$

Hence simple connectedness should be expressible purely in terms of paths that begin and end at the same point.

5.68 Definition. Let x be a point in a topological space X. A *loop at x in X* is a path in X from x to x—that is, a continuous map $\omega \colon \mathsf{I} \to X$ with $\omega(0) = \omega(1) = x$. The *null loop ν_x at x* is the constant path in X with $\nu_x(s) = x$ for all $s \in \mathsf{I}$.

A loop ω at x is *nullhomotopic* if $\omega \sim \nu_x$.

Loosely speaking, a loop at x in X is nullhomotopic when it can be shrunk within the space X to the single point x through a continuous family of loops at x (see Fig. 5.20).

5.69 Theorem. A necessary and sufficient condition for a path-connected space X to be simply connected is that each loop at each point of X be nullhomotopic.

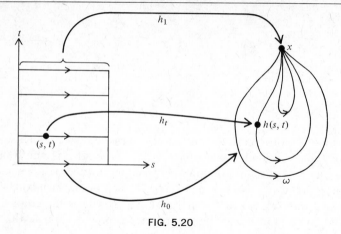

FIG. 5.20

Proof. Necessity. Assume X is simply connected. Let ω be a loop at a point x in X. Then both ω and the null loop ν_x are paths in X from x to $y = x$. Hence $\omega \sim \nu_x$.

Sufficiency. Assume the condition holds. Let σ and τ be paths in X from x to y. We must show $\sigma \sim \tau$. As in the discussion preceding 5.68, define a loop ω at x by

$$\omega(s) = \begin{cases} \sigma(2s) & \text{if } 0 \le s \le \tfrac{1}{2}, \\ \tau(2 - 2s) & \text{if } \tfrac{1}{2} \le s \le 1. \end{cases}$$

By assumption there exists some homotopy $h : \omega \sim \nu_x$ from ω to the null loop ν_x at x. Then on the boundary of $I \times I$ the map h takes values

$$h(s, 0) = \sigma(2s) \quad (0 \le s \le \tfrac{1}{2}), \qquad h(s, 0) = \tau(2 - 2s) \quad (\tfrac{1}{2} \le s \le 1),$$

$$h(s, 1) = x \qquad\qquad (s \in I),$$

$$h(0, t) = h(1, t) = x \qquad\qquad (t \in I),$$

so h takes the constant value x on the top and vertical sides of $I \times I$ (see Fig. 5.21).

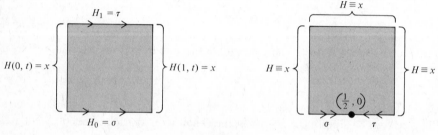

FIG. 5.21

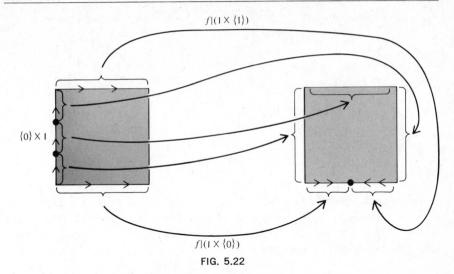

FIG. 5.22

The homotopy $H: \sigma \sim \tau$ we seek is to take on the boundary of $I \times I$ the values

$$H(s, 0) = \sigma(s) \qquad\qquad (s \in I),$$

$$H(s, 1) = \tau(s) \qquad\qquad (s \in I),$$

$$H(0, t) = x, \qquad H(1, t) = y \qquad\qquad (t \in I)$$

(see Fig. 5.21). Hence it will suffice to construct a continuous map $f: I \times I \to I \times I$ for which

$$f(s, 0) = (\tfrac{1}{2}s, 0), \qquad f(s, 1) = (1 - \tfrac{1}{2}s, 0) \qquad (s \in I),$$

$$f(\{0\} \times I) = (\{0\} \times I) \cup (I \times \{1\}) \cup (\{1\} \times I)$$

(see Fig. 5.22) and then to set $H = h \circ f$.

One continuous map $f: I \times I \to I \times I$ having the required properties is depicted in Fig. 5.23. For each $t \in I$, this f maps the horizontal segment $I \times \{t\}$ linearly onto the line segment that joins $(\tfrac{1}{2}, 0)$ to a point on the boundary of $I \times I$ and that makes an angle of πt with the line segment joining $(0, 0)$ to $(\tfrac{1}{2}, 0)$, the point $(1, t)$ being mapped to $(\tfrac{1}{2}, 0)$. ☐

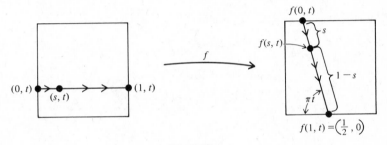

FIG. 5.23

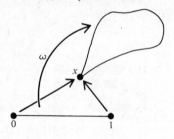

ω(I) a simple closed curve

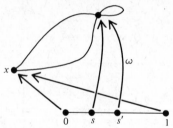

ω(I) not a simple closed curve

FIG. 5.24

A loop ω at a point x in a space X has the same range as a continuous map $f: \mathsf{S}_1 \to X$ sending $(1, 0)$ to x, and vice versa. In fact, if $p: \mathsf{I} \to \mathsf{S}_1$ is the quotient map $s \mapsto (\cos 2\pi s, \sin 2\pi s)$, the unique map f satisfying $f \circ p = \omega$ (see 3.47 and 3.49 (1)) has range

$$f(\mathsf{S}_1) = \omega(\mathsf{I}).$$

Suppose now X is Hausdorff and the loop ω does not pass through any point except x more than once, that is, $\omega(s) \neq \omega(s')$ whenever $s, s' \in \,]0, 1[$ with $s \neq s'$. In this event f is injective and maps S_1 homeomorphically onto $f(\mathsf{S}_1)$. In other words $\omega(\mathsf{I})$ is then a *simple closed curve*—a homeomorphic image of the circle (see Fig. 5.24).

Let us now specialize to the plane. We want to assert that a simple closed curve C in the plane divides the plane into two regions, one inside and the other outside C. Surely this assertion is geometrically evident when C is a circle, a square, or a triangle. But is it evident for more general C? Which

FIG. 5.25

points are "inside" the simple closed curve of Fig. 5.25 and which are "outside" it?

The precise assertion we want is the *Jordan curve theorem*:

> Each simple closed curve C in the plane $\mathbb{R}^2$ has a complement $\mathbb{R}^2 \setminus C$ consisting of exactly two components B and U, one of them bounded and the other unbounded, with $C = $ bdy $B \cap$ bdy U their common boundary.

The explicit statement of this as a theorem requiring proof is due to C. Jordan, who in 1893 gave an incomplete proof. The first correct proof was given only in 1905, by O. Veblen. Known proofs require methods outside the scope of this book; an accessible one having a very geometric flavor may be found in Hall and Spencer [15, Chapter 5, Sections 3–4].

Let us now combine the preceding ideas to give a nice geometric criterion for the simple connectedness of certain spaces. Consider a subspace X of the plane that is open and connected (and hence by 5.51 path-connected). For X to be simply connected, according to Theorem 5.69 it is necessary that each loop ω in X whose range is a simple closed curve be nullhomotopic; then the region of the plane inside $\omega(\mathsf{I})$ ought to be contained in X. And this is correct (although we offer no proof): X is simply connected if and only if for each simple closed curve $C \subset X$, the bounded component of $\mathbb{R}^2 \setminus C$ is contained in X. Thus a simply connected region in the plane is just a region having no holes.

Relying on 5.69, in treating simple connectedness we shall actually work with loops and not arbitrary paths. For example, we next explain why our definition of a simply connected space stipulated that the space be path-connected.

5.70 Theorem. Let X be path-connected. Assume there is some point $x \in X$ such that each loop at x is nullhomotopic. Then X is simply connected.

Proof. Let ω be a loop at an arbitrary point y in X. We wish to show $\omega \sim \nu_y$. To utilize the hypothesis concerning x we use ω to construct a loop at x. Since X is path-connected, there is a path σ in X from x to y. By traversing first σ, then ω, and finally σ in the reverse direction (see Fig. 5.26) we obtain

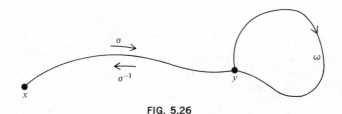

FIG. 5.26

a loop $\sigma\omega\sigma^{-1}$ at x defined by

$$\sigma\omega\sigma^{-1}(s) = \begin{cases} \sigma(3s) & \text{if } 0 \leq s \leq \frac{1}{3}, \\ \omega(3s-1) & \text{if } \frac{1}{3} \leq s \leq \frac{2}{3}, \\ \sigma(3-3s) & \text{if } \frac{2}{3} \leq s \leq 1. \end{cases}$$

By hypothesis there exists a homotopy

$$h : \sigma\omega\sigma^{-1} \sim \nu_x.$$

In the same way we just used σ to transform the loop ω at y into the loop $\sigma\omega\sigma^{-1}$ at x, so we now use σ^{-1} to transform the loops $\sigma\omega\sigma^{-1}$ and ν_x into loops $\sigma^{-1}(\sigma\omega\sigma^{-1})\sigma$ and $\sigma^{-1}\nu_x\sigma$ at y. Formally,

$$\sigma^{-1}(\sigma\omega\sigma^{-1})\sigma(s) = \begin{cases} \sigma(1-3s) & \text{if } 0 \leq s \leq \frac{1}{3}, \\ \sigma\omega\sigma^{-1}(3s-1) & \text{if } \frac{1}{3} \leq s \leq \frac{2}{3}, \\ \sigma(3s-2) & \text{if } \frac{2}{3} \leq s \leq 1, \end{cases}$$

$$\sigma^{-1}\nu_x\sigma(s) = \begin{cases} \sigma(1-3s) & \text{if } 0 \leq s \leq \frac{1}{3}, \\ x & \text{if } \frac{1}{3} \leq s \leq \frac{2}{3}, \\ \sigma(3s-2) & \text{if } \frac{2}{3} \leq s \leq 1. \end{cases}$$

By squeezing $h : \mathsf{I} \times \mathsf{I} \to X$ onto the rectangle $\left[\frac{1}{3}, \frac{2}{3}\right] \times \mathsf{I}$ we obtain a homotopy

(1) $h' : \sigma^{-1}(\sigma\omega\sigma^{-1})\sigma \sim \sigma^{-1}\nu_x\sigma$

given by

$$h'(s, t) = \begin{cases} \sigma(1-3s) & \text{if } 0 \leq s \leq \frac{1}{3}, \\ h(3s-1, t) & \text{if } \frac{1}{3} \leq s \leq \frac{2}{3}, \\ \sigma(3s-2) & \text{if } \frac{2}{3} \leq s \leq 1 \end{cases}$$

(see Fig. 5.27).

To complete the proof we shall show

(2) $\sigma^{-1}\nu_x\sigma \sim \nu_y,$

(3) $\sigma^{-1}(\sigma\omega\sigma^{-1})\sigma \sim \omega.$

Since $\sim$ is an equivalence relation on the set of all loops at y, it will then follow from (1), (2), and (3) that $\omega \sim \nu_y$.

We show (2). A homotopy

$$F : \sigma^{-1}\nu_x\sigma \sim \nu_y$$

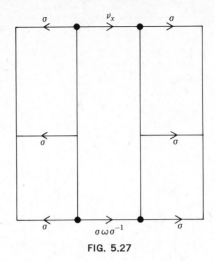

FIG. 5.27

may be constructed as suggested by Fig. 5.28. The square $| \times |$ is divided
into five triangular regions as shown. On regions I, III, and V, F is assigned
the constant value y. On the horizontal line segment of region II at a height
of t—that is, on the line segment joining $(t/2, t)$ to $((t + 2)/6, t)$—F takes
in order from right to left the same values as σ takes from left to right on
$[0, 1]$. Now the linear map carrying $t/2$ to 1 and $(t + 2)/6$ to 0 has the form

$$u = \frac{3}{t - 1}\left[s - \frac{t + 2}{6}\right].$$

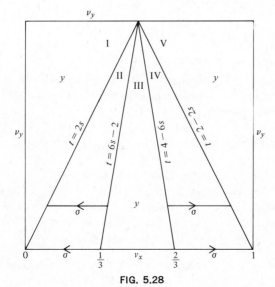

FIG. 5.28

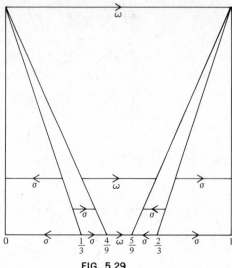

FIG. 5.29

Hence

$$F(s, t) = \left(\sigma \left(\frac{3}{t-1} \left[s - \frac{t+2}{6} \right] \right), t \right) \qquad \text{if } 6s - 2 \leq t \leq 2s.$$

Similarly, on region IV

$$F(s, t) = \left(\sigma \left(\frac{3}{1-t} \left[s - \frac{4-t}{6} \right] \right), t \right) \qquad \text{if } 4 - 6s \leq t \leq 2 - 2s.$$

To show (3) we construct a homotopy

$$G: \sigma^{-1}(\sigma\omega\sigma^{-1})\sigma \sim \omega$$

as suggested by Fig. 5.29. □

The preceding proof contains a hidden bonus. It allows us to expand our repertory of simply connected spaces beyond convex subsets of $\mathbf{R}^n$ to arbitrary contractible spaces.

5.71 Theorem. Every contractible space is simply connected.

Proof. Assume X is contractible. According to 5.55, X is already path-connected. Let ω be a loop at an arbitrary point $y \in X$. We shall show $\omega \sim \nu_y$.

It is tempting to take for the desired homotopy $g: \omega \sim \nu_y$ the map $g: \mathsf{I} \times \mathsf{I} \to X$ given by

$$g(s, t) = c(\omega(s), t),$$

where $c: X \times \mathsf{I} \to X$ is a contraction of X to y. *This would be incorrect.* To be sure, $g(s, 0) = \omega(s)$ and $g(s, 1) = y$ for all $s \in \mathsf{I}$, so that g is a homotopy from ω to ν_y. Unfortunately, we do not know that g takes the constant value y on the vertical sides $\{0\} \times I$, $\{1\} \times I$ of $I \times I$—in other words, that g is a homotopy relative to $\{0, 1\}$.

Our strategy for a correct proof is to construct a certain loop σ at y for which $\sigma\omega\sigma^{-1} \sim \nu_y$; here the loop $\sigma\omega\sigma^{-1}$ is defined as in the proof of 5.70 (with $x = y$). The proof of 5.70 will then allow us to conclude $\omega \sim \nu_y$.

To construct σ we shall first construct a homotopy

$$h: \omega \simeq \nu_y$$

(*not* a homotopy relative to $\{0, 1\}$) for which

$$(*) \qquad\qquad\qquad h(0, t) = h(1, t) \qquad\qquad (t \in I).$$

Then the restriction of h to the left edge $\{0\} \times I$ of $I \times I$ may be regarded as a loop at y, and σ is defined to be the inverse of this loop. Thus

$$\sigma(t) = h(0, 1 - t) \qquad\qquad (t \in I)$$

(see Fig. 5.30).

Once all this has been done it is an easy task to construct a homotopy F: $\sigma\omega\sigma^{-1} \sim \nu_y$ in the way suggested by Fig. 5.31. Analytically, we take on the rectangle $[0, \frac{1}{3}] \times I$

$$F(s, t) = \begin{cases} y & \text{if } t \geq 3s, \\ \sigma(3s - t) & \text{if } t \leq 3s, \end{cases}$$

on the rectangle $[\frac{1}{3}, \frac{2}{3}] \times I$

$$F(s, t) = h(3s - 1, t),$$

and on the rectangle $[\frac{2}{3}, 1] \times I$

$$F(s, t) = \begin{cases} \sigma(3 - t - 3s) & \text{if } t \leq 3 - 3s, \\ y & \text{if } t \geq 3 - 3s. \end{cases}$$

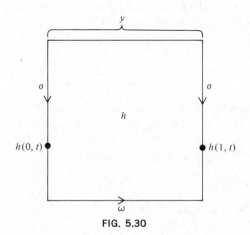

FIG. 5.30

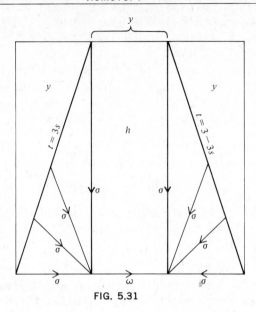

FIG. 5.31

To construct $h: \omega \simeq \nu_y$ satisfying (*) we regard ω as a map from the circle S_1 into X sending $(1, 0) \in S_1$ to y. More precisely, if $p: I \rightarrow S_1$ is the quotient map

$$p(s) = (\cos 2\pi s, \sin 2\pi s) \qquad\qquad (s \in I),$$

let

$$\omega': S_1 \rightarrow X$$

be the unique continuous map such that

$$\omega' \circ p = \omega$$

(of course, ω' is not actually a loop at y since its domain is not I). Recall that X is contractible. By 5.63 (3) the map ω' is homotopic to the constant map taking value y on S_1. Hence there is a homotopy

$$h': S_1 \times I \rightarrow X$$

with

$$h'(z, 0) = \omega'(z), \qquad h'(z, 1) = y \qquad\qquad (z \in S_1).$$

By cutting the cylinder $S_1 \times I$ along the line segment $\{(1, 0)\} \times I$ and unwrapping it, we obtain the desired homotopy $h: \omega \simeq \nu_y$. Formally, we define $h: I \times I \rightarrow X$ by

$$h(s, t) = h'(p(s), t).$$

To check (*) we compute

$$h(0, t) = h'(p(0), t) = h'(p(1), t) = h(1, t). \qquad \square$$

EXERCISES

74. Construct a contraction of the subspace

$$X = (I \times \{0\}) \cup (\{0\} \times I) \cup \bigcup_{n=1}^{\infty} \left(\left\{ \frac{1}{n} \right\} \times I \right)$$

of the plane to the point $(0, 1)$.

75. (**a**) Prove that each subset of R^n that is starlike (Exercise 16) is contractible, thereby extending Example 5.54(1).
(**b**) Describe or draw several contractible subsets of the plane R^2 that are not starlike.

76. Is contractibility a topological property?

77. Must the product of two contractible spaces be contractible?

78. Given a space X, the *cone over* X is the space KX obtained from $X \times I$ by collapsing the set $X \times \{1\}$ to a point (compare Exercise 3.105). Show that X can be embedded in KX and that KX is contractible.

79. (**a**) Construct a deformation retraction of the square $[-1, 1] \times I$ onto the set $(\{-1, 1\} \times I) \cup ([-1, 1] \times \{0\})$ consisting of its sides and bottom. [*Hint*: Move each point of the square along the line passing through the point and drawn from the point $(0, 2)$; you will want to distinguish the case that the line intersects the sides from the case that it intersects the bottom of the square].
(**b**) Construct a deformation retraction of the solid cylinder $D_2 \times I$ onto the set $(S_1 \times I) \cup (D_2 \times \{0\})$.
(**c**) Generalize (a) and (b) to arbitrary dimension.

80. Find a subset of the Möbius strip M [3.49(5)] homeomorphic to S_1 that is a deformation retract of M.

81. (**a**) If $A \subset X$ is a retract of X and $B \subset A$ is a retract of A, show that B is a retract of X.
(**b**) Does the analog of (a) for deformation retracts hold?

82. Prove that a subspace E of a space X is a retract of X if and only if each continuous map $f: E \rightarrow Y$ into each topological space Y has a continuous extension to X.

83. Call a space Y "nice" if any continuous map $f: A \rightarrow Y$ on any closed subspace A of any topological space X can be extended continuously to X. Prove that a retract E of a nice space Y is also nice.

84. Show that a space Y is contractible if each two continuous maps $f, g: X \rightarrow Y$ from each space X into Y are homotopic to one another.

85. Let $f, f': X \rightarrow Y$ and $g, g': Y \rightarrow Z$ be continuous maps with $f \simeq f'$ and $g \simeq g'$. Deduce that $g \circ f \simeq g' \circ f'$. (You may wish to consider first the special cases $f = f'$ and $g = g'$ and then to exploit 5.64.)

86. (**a**) Let $f, g: X \rightarrow S_n$ be continuous maps such that for all $x \in X, f(x)$ is not antipodal to $g(x)$, that is, $f(x) \neq -g(x)$. Show that $f \simeq g$. [*Hint*: Project the line segment joining $f(x)$ and $g(x)$ onto S_n.]
(**b**) If $f: X \rightarrow S_n$ is a continuous map with $f(X) \neq S_n$, show that f is homotopic to a constant map.

87. Prove that a continuous map $f: S_n \to Y$ is homotopic to a constant map if and only if f has a continuous extension to D_{n+1}. (For inspiration, draw pictures for $n = 1$.)

88. Suppose a continuous map $f: S_n \to Y$ is homotopic to a constant map $c: S_n \to Y$. Given any $z \in S_n$, construct a homotopy h from f to a constant map with value some $y \in Y$ such that $h(z, t) = y$ for all $t \in I$.

89. Draw diagrams illustrating the proof that $\sim$ is an equivalence relation on. the set of all paths in a space X from a point x_0 to a point x_1.

90. (**a**) May the complement of a simple closed curve in the Möbius strip have just one component? two components?
(**b**) Assuming the Jordan curve theorem, what can you say about the complement of a simple closed curve in the sphere S_2?

91. Does 5.70 remain true if the hypothesis that X be path-connected is dropped (and the requirement of path-connectedness is omitted from the definition of simple connectedness)?

92. Prove that a path-connected space X is simply connected precisely when each continuous $f: S_1 \to X$ has a continuous extension to D_2.

93. Let σ and τ be arbitrary paths in a path-connected space X, not necessarily joining the same points. Prove that σ is homotopic to τ (not necessarily relative to $\{0, 1\}$, of course). [*Hint:* First construct a continuous map h' from $(I \times \{0, 1\}) \cup (\{0\} \times I)$ into X with $h'(s, 0) = \sigma(s)$ and $h'(s, 1) = \tau(s)$ for all $s \in I$. Then extend h' to $I \times I$.]

In exercises 94–98, $\sigma \cdot \tau$ denotes the product of the paths σ and τ as defined in 5.48, and σ^{-1} denotes the inverse of the path σ as defined in 5.48.

94. Let σ, σ' be paths from x to y and τ, τ' be paths from y to z in a space X with $\sigma \sim \sigma'$ and $\tau \sim \tau'$. Show that $\sigma \cdot \tau \sim \sigma' \cdot \tau'$.

95. Let σ, τ, η be paths in a space X from x to y, from y to z, and from z to w, respectively. Show that $(\sigma \cdot \tau) \cdot \eta \sim \sigma \cdot (\tau \cdot \eta)$.

96. Let σ be a path from x to y in a space X. Show:
(**a**) $\sigma \cdot \nu_y \sim \sigma$ and $\nu_x \cdot \sigma \sim \sigma$.
(**b**) $\sigma \cdot \sigma^{-1} \sim \nu_x$ and $\sigma^{-1} \cdot \sigma \sim \nu_y$

Exercises 97–98 assume an elementary knowledge of groups and their homomorphisms.

97. Given a point x of a space X, let $\Omega(X, x)$ be the set of all loops at x in X and let $\pi(X, x)$ be the quotient set $\Omega(X, x)/\sim$ where $\sim$ is the equivalence relation of homotopy of loops relative to $\{0, 1\}$. For $\omega \in \Omega(X, x)$, denote by $[\omega]$ the equivalence class of ω under $\sim$. If $a, b \in \pi(X, x)$, then $a = [\sigma]$, $b = [\tau]$ for some loops σ, τ, and we define $a \cdot b = [\sigma \cdot \tau]$; according to Exercise 94 this definition of $a \cdot b$ is independent of the choice of the particular representatives σ of a, τ of b. In this way, a binary operation $\cdot$ on the set $\pi(X, x)$ is obtained.
(**a**) Explain why this operation makes $\pi(X, x)$ into a group. [The group $\pi(X, x)$ is called the *fundamental group of X at x* and was introduced in 1895 by H. Poincaré.]
(**b**) Express the simple connectedness of X as a statement about the groups $\pi(X, x)$ for $x \in X$.

98. Let σ be a path from x_0 to x_1 in a space X. Then for each loop ω at x_0, the path

$(\sigma^{-1} \cdot \omega) \cdot \sigma$ is a loop at x_1; if ω' is another loop at x_0 with $\omega \sim \omega'$, then by Exercises 94–96 we have $(\sigma^{-1} \cdot \omega) \cdot \sigma \sim (\sigma^{-1} \cdot \omega') \cdot \sigma$. Hence there is a well-defined map

$$\sigma_*: \pi(X, x_0) \to \pi(X, x_1)$$

such that

$$\sigma_*([\omega]) = [(\sigma^{-1} \cdot \omega) \cdot \sigma].$$

(a) Show that σ_* is a group homomorphism.
(b) By using similarly the path $\tau = \sigma^{-1}$ from x_1 to x_0 we obtain a group homomorphism

$$\tau_*: \pi(X, x_1) \to \pi(X, x_0).$$

Prove that the maps σ_* and τ_* are inverses of one another and hence that σ_* is a group isomorphism of $\pi(X, x_0)$ with $\pi(X, x_1)$.
(c) Interpret your answer to Exercise 91 in the present context.

99. (This exercise shows that, under some restrictions on the spaces involved, a homotopy really is a continuous family of continuous maps.) Let (X, d) be a compact metric space, let (Y, d') be any bounded metric space, and let $h: X \times I \to Y$ be a continuous map. Then for each $t \in I$, the map $h_t: X \to Y$ is continuous, of course. Denote the set of all continuous maps from X to Y by $C(X, Y)$. Provide $C(X, Y)$ with its topology of uniform convergence. Prove that the map

$$I \to C(X, Y)$$

$$t \to h_t$$

is continuous.

100. In the notation of Exercise 99, let $f, g: X \to Y$ be continuous maps. Show that $f \simeq g$ if and only if there is a path in the space $C(X, Y)$ from f to g.

5. Simple Connectedness and the Circle

We conclude our excursion into homotopy theory with the promised proof that the circle S_1 is not simply connected and with the deduction of several important consequences including the fundamental theorem of algebra.

Consider the loop $s \mapsto (\cos 2\pi s, \sin 2\pi s)$ at $(1, 0)$ in S_1 that wraps around S_1 once in the counterclockwise direction. We will have proved that S_1 is not simply connected if only we can establish the fact that this loop is not nullhomotopic. However geometrically evident this fact may seem, its proof is surprisingly difficult. To approach the proof we study homotopies of loops in S_1 by studying homotopies of paths in R.

For the remainder of this section we denote by p the quotient map

$$p: R \to S_1$$

$$s \mapsto (\cos 2\pi s, \sin 2\pi s).$$

We have $p(n) = (1, 0)$ for every integer n. Then any path σ in R from 0 to an integer n induces a loop $p \circ \sigma: I \to S_1$ at $(1, 0)$ in S_1. For example, the path $\sigma(s) = s$ from 0 to 1 in R induces the loop considered in the preceding

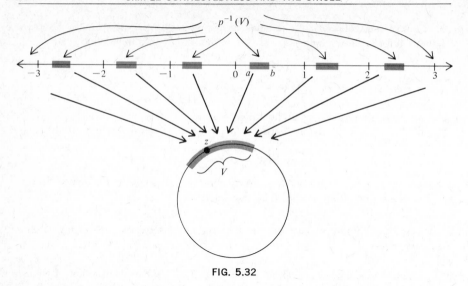

FIG. 5.32

paragraph, and the path $\sigma(s) = 0$ from 0 to 0 induces the null loop at $(1, 0)$. More generally, for each positive integer n the path $\sigma(s) = ns$ from 0 to n in R induces a loop at $(1, 0)$ in S_1 that wraps around the circle n times in the counterclockwise direction.

Conversely, let a loop τ at $(1, 0)$ in S_1 be given. We wish to show that τ is induced in the preceding manner by some unique path σ in R with $\sigma(0) = 0$. (Caution: The path σ need not be a loop at 0 in R, as the preceding examples demonstrate.) In other words, we want to *lift* τ in a unique way to a continuous map $\sigma \colon \mathsf{I} \to \mathsf{R}$ with $\sigma(0) = 0$ making the diagram below commutative.

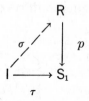

To proceed we need the elementary property that p is a *covering map*: Each $z \in \mathsf{S}_1$ has an open neighborhood V in S_1 that is *evenly covered* in the sense that $p^{-1}(V)$ is the union of a collection of disjoint open subsets of R each of which p maps homeomorphically onto V. To see this, for a given $z \in \mathsf{S}_1$ take V to be any proper open arc $p(\,]a, b[\,)$ in S_1 containing z. Since

$$p(s) = p(s') \qquad \Leftrightarrow \qquad s - s' \in \mathsf{Z},$$

we have

$$p^{-1}(V) = \bigcup_{n \in \mathsf{Z}} \{n + s \mid a < s < b\}$$

with each of the translates $\{n + s \mid a < s < b\}$ of $]a, b[$ being open in R and mapped homeomorphically onto V by p (see Fig. 5.32).

5.72 Lemma (path-lifting property). Let $\tau \colon I \to S_1$ be any path in S_1 and let $u \in R$ be any point with

$$p(u) = \tau(0).$$

Then there is a *unique* path $\sigma \colon I \to R$ in R such that

$$p \circ \sigma = \tau, \qquad \sigma(0) = u.$$

Proof. Existence. We shall suitably choose points

(*) $$0 = s_0 < s_1 < \cdots < s_i < s_{i+1} < \cdots < s_n = 1$$

in I and then successively define σ on each of the subintervals $\{0\} = [0, s_0]$, $[0, s_1], \ldots, [0, s_n] = I$ of I.

The points $s_0, \ldots, s_n$ of I satisfying (*) are so chosen that for each $i = 0, \ldots, n-1$,

$$\tau([s_i, s_{i+1}]) \subset V_i$$

for some evenly covered open subset V_i of S_1. This choice can be made using the compactness of I as follows. Each $z \in S_1$ has an evenly covered open neighborhood V_z. By 4.41 the open cover $\{\tau^{-1}(V_z) \mid z \in S_1\}$ of I has a Lebesgue number $\epsilon > 0$. Arbitrarily choose $s_0, \ldots, s_n$ in I satisfying (*) with $|s_{i+1} - s_i| < \epsilon$ for each i, and for each i take $V_i = V_z$ for some $z \in S_1$ with $[s_i, s_{i+1}] \subset \tau^{-1}(V_z)$.

Using induction we now show that for each $i = 0, \ldots, n$ there is a continuous map

$$\sigma_i \colon [0, s_i] \to R$$

such that

(**) $$p \circ \sigma_i = \tau \mid [0, s_i], \qquad \sigma_i(0) = u.$$

Then $\sigma = \sigma_n$ will be the desired lifting of τ. To begin the induction simply set $\sigma_0(0) = u$.

Now let $0 \le i < n$ and assume a map $\sigma_i \colon [0, s_i] \to R$ satisfying (**) exists. The next map $\sigma_{i+1} \colon [0, s_{i+1}] \to R$ is obtained by gluing together σ_i and a continuous map $\eta \colon [s_i, s_{i+1}] \to R$ such that

$$\eta(s_i) = \sigma_i(s_i), \qquad p \circ \eta = \tau \mid [s_i, s_{i+1}].$$

Such an η exists by the choice of s_i and s_{i+1}. In fact, since $p\sigma_i(s_i) = \tau(s_i)$, we have $\sigma_i(s_i) \in p^{-1}(V_i)$. Since V_i is evenly covered, there is a unique open set U in R containing $\sigma_i(s_i)$ and mapped homeomorphically by p onto V_i. Denote by p^{-1} the inverse of the homeomorphism $U \cong V_i$ defined by p. Finally, set

$$\eta(s) = p^{-1}(\tau(s)) \qquad (s \in [s_i, s_{i+1}]).$$

Uniqueness. Suppose $\sigma, \sigma' \colon I \to R$ are two paths with

$$p \circ \sigma = \tau = p \circ \sigma', \qquad \sigma(0) = u = \sigma'(0).$$

We shall show that the subset

$$E = \{s \in I \mid \sigma(s) = \sigma'(s)\}$$

of I equals I. Now $E \neq \varnothing$ because $0 \in E$, and by continuity of σ and σ' the set E is closed in I. Since I is connected, it remains only to show that E is open in I.

Let $s \in E$. Choose an evenly covered open neighborhood V of $p\sigma(s)$ in S_1. There is an open subset U of R containing $\sigma(s)$ which p maps homeomorphically onto V. Then the open subset

$$W = \sigma^{-1}(U) \cap (\sigma')^{-1}(U)$$

of I contains s. Moreover, $W \subset E$, for $t \in W$ implies $\sigma(t), \sigma'(t) \in U$, by hypothesis $p(\sigma(t)) = p(\sigma'(t))$, and hence $\sigma(t) = \sigma'(t)$ because p is one-to-one on U. $\quad \square$

We are now ready for the crucial result, which says that a homotopy in the circle can be lifted to a homotopy in the line.

5.73 Lemma (homotopy-lifting property). Let

$$h: \omega_0 \sim \omega_1$$

be a homotopy of loops ω_0, $\omega_1: I \to S_1$ at $(1, 0)$. Let σ_0, $\sigma_1: I \to R$ be the unique paths with

$$p \circ \sigma_0 = \omega_0, \qquad \sigma_0(0) = 0 = \sigma_1(0), \qquad p \circ \sigma_1 = \omega_1.$$

Then

$$\sigma_0(1) = \sigma_1(1)$$

and there is a homotopy $H: \sigma_0 \sim \sigma_1$ with $p \circ H = h$.

Proof. We shall use the method of the preceding proof to obtain a continuous map $H: I \times I \to R$ such that

$$H_0 = \sigma_0, \qquad p \circ H = h.$$

We can then conclude the proof by repeatedly using the uniqueness assertion of 5.72 as follows. The path $t \mapsto H(0, t)$ and the null loop ν_0 at 0 in R satisfy

$$p \circ H(0, t) = h(0, t) = (1, 0) = p \circ \nu_0(t) \qquad (t \in I),$$

$$H(0, 0) = \sigma_0(0) = 0 = \nu_0(0),$$

so by 5.72,

$$H(0, t) = 0 \qquad (t \in I).$$

In particular, $H(0, 1) = 0$. Next, the paths H_1 and σ_1 in R satisfy

$$p \circ H_1(s) = pH(s, 1) = h(s, 1) = \omega_1(s) = p \circ \sigma_1(s) \qquad (s \in I),$$

$$H_1(0) = H(0, 1) = 0 = \sigma_1(0),$$

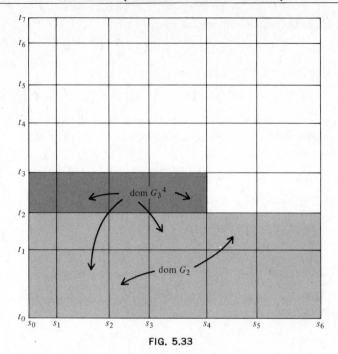

FIG. 5.33

so $H_1 = \sigma_1$, that is,

$$H(s, 1) = \sigma_1(s) \qquad\qquad (s \in I).$$

Next, the path $t \mapsto H(1, t)$ and the null loop $t \mapsto \sigma_0(1)$ at $\sigma_0(1)$ in R satisfy

$$p \circ H(1, t) = h(1, t) = (1, 0) = \omega_0(1) = p(\sigma_0(1)) \qquad (t \in I),$$

$$H(1, 0) = H_0(1) = \sigma_0(1),$$

so

$$H(1, t) = \sigma_0(1) \qquad\qquad (t \in I).$$

Hence

$$\sigma_0(1) = H(1, 1) = \sigma_1(1)$$

and $H \colon \sigma_0 \sim \sigma_1$.

To construct H, as in the proof of 5.72 we divide the square $I \times I$ into rectangles determined by points

$$0 = s_0 < \cdots < s_i < s_{i+1} < \cdots < s_n = 1,$$

$$0 = t_0 < \cdots < t_j < t_{j+1} < \cdots < t_m = 1$$

in such a way that for each $i = 0, \ldots, n - 1$ and $j = 0, \ldots, m - 1$,

$$h([s_i, s_{i+1}] \times [t_j, t_{j+1}]) \subset V_{ij}$$

for some evenly covered subset V_{ij} of S_1. Starting with the map $G_0 \colon I \times \{0\} \to \mathsf{R}$ defined by $G_0(s, 0) = \sigma_0(s)$, we use induction to show that for each

$j = 0, \ldots, m$ there is a continuous map

$$G_j : I \times [t_0, t_j] \to R$$

on the horizontal strip $I \times [t_0, t_j]$ with

$$G_j(s, 0) = \sigma_0 \quad (s \in I), \qquad p \circ G_j = h \mid (I \times [t_0, t_j]).$$

Suppose $0 \le j < m$ and that G_j has already been shown to exist. To show that G_{j+1} exists, we use induction to show that for each $i = 0, \ldots, n$ there is a continuous map

$$G^i_{j+1} : (I \times [t_0, t_j]) \cup ([s_0, s_i] \times [t_j, t_{j+1}]) \to R$$

that extends G_j and that satisfies $p \circ G^i_{j+1}(s, t) = h(s, t)$ on its domain (the domains of G_j and G^i_{j+1} are indicated in Fig. 5.33). This induction is similar to the one in the proof of 5.72.   []

Our main theorem is an immediate consequence of 5.73.

5.74 Theorem. Let n and m be integers with $n \ne m$, and let $\sigma_n, \sigma_m : I \to R$ be the paths given by

$$\sigma_n(s) = ns, \qquad \sigma_m(s) = ms \qquad\qquad (s \in I).$$

Then the loops $p \circ \sigma_n$ and $p \circ \sigma_m$ at $(1, 0)$ in S_1 are not homotopic relative to $\{0, 1\}$.

In particular, the null loop $p \circ \sigma_0$ at $(1, 0)$ in S_1 is not homotopic relative to $\{0, 1\}$ to the loop $p \circ \sigma_n$ whenever $n \ne 0$. At last we have proved the following theorem.

5.75 Theorem. The circle S_1 is not simply connected.

5.76 Corollary (the no-retract theorem). The circle S_1 is not a retract of the disk D_2.

Proof. According to 5.54 (1), the disk D_2 is contractible. If S_1 were a retract of D_2, then by 5.61 the circle S_1 would be contractible, and hence by 5.71 the circle S_1 would be simply connected.   []

Recall that a point x of a space X is called a *fixed-point* of a map $f : X \to X$ when $f(x) = x$. Earlier we deduced from the connectedness of I that each continuous map $f : I \to I$ has a fixed-point (see 5.14). From the no-retract theorem we now deduce that each continuous map $f : I^2 \to I^2$ from the square I^2 to itself has a fixed-point. Given a continuous map $f : I^2 \to I^2$, the map $g : D_2 \to D_2$ defined by $g = h^{-1} \circ f \circ h$, where $h : D_2 \cong I^2$, is also continuous, and then $h(y)$ will be a fixed-point of f whenever $y \in D_2$ is a fixed-point of g. Hence it suffices to consider maps from D_2 to D_2.

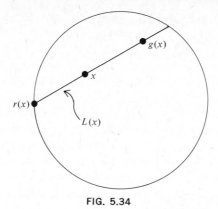

FIG. 5.34

5.77 Theorem (Brouwer fixed-point theorem for dimension 2).
Each continuous map $g \colon D_2 \to D_2$ has a fixed-point.

Proof. Just suppose $g(x) \neq x$ for all $x \in D_2$. Then for each $x \in D_2$ there is a unique line $L(x)$ passing through the distinct points x and $g(x)$, and $L(x)$ intersects S_1 in exactly two points; denote by $r(x)$ that one of these two points which lies on the same side of $g(x)$ as does x (see Fig. 5.34). We shall show that the map $r \colon D_2 \to S_1$ thus obtained is a retraction, thereby contradicting the no-retract theorem. By construction, $r(x) = x$ whenever $x \in S_1$.

It remains only to establish the continuity of r. This is an easy consequence of the continuity of g. For each x the point $r(x)$ is the unique $y = tx + (1 - t)g(x)$ for which $t \geq 1$ and $||y|| = 1$. By explicitly solving the quadratic equation $||y|| = 1$ in t we see that its root t satisfying $t \geq 1$ depends continuously on x and $g(x)$, and hence on x alone. Therefore $r(x)$ depends continuously on x. $\Box$

Brouwer's theorem has a surprising physical interpretation. Imagine a circular dish filled with some liquid. Suppose the liquid is stirred for a while and suppose when all motion stops the molecules originally on the surface, and only those molecules, remain on the surface. Then some molecule on the surface will be where it originally was.

One mathematical application of Brouwer's theorem concerns a system

$$(*) \qquad \begin{cases} h_1(x_1, x_2) = 0 \\[2mm] h_2(x_1, x_2) = 0 \end{cases}$$

of two simultaneous equations, where h_1 and h_2 are continuous real-valued functions of two real variables. Introduce the auxiliary map f defined by

$$f(x_1, x_2) = (h_1(x_1, x_2) + x_1, \, h_2(x_1, x_2) + x_2).$$

A point $x = (x_1, x_2) \in \mathbf{R}^2$ will be a solution of (*) if and only if it is a fixed-point of the map f. If $f(E) \subset E$ for some subset E of $\mathbf{R}^2$ homeomorphic to D_2, then f will have a fixed-point $x \in E$ that will be a solution of the original system (*) in E.

Our final application of the main theorem 5.74 is a topological proof of the fundamental theorem of algebra, which says that each polynomial having real or complex numbers as its coefficients necessarily has at least one complex root. This theorem was first proved in 1799 by Karl Friedrich Gauss in his doctoral dissertation. A number of different proofs are possible (Gauss himself gave four), including one using Liouville's theorem about "entire" complex functions (see Duncan [8, p. 183] or any other book on complex analysis) and one using only minimal amounts of analysis and topology (see Lang [23, Chapter VIII, Section 3]).

It might seem strange that our topological considerations concerning the line, plane, and circle should serve to prove an algebraic result. The explanation is that the set $\mathbf{C}$ of complex numbers is precisely the set of points of the plane $\mathbf{R}^2$: a complex number $z = x + iy$ with real part x and imaginary part y is just the ordered pair (x, y) of real numbers. Then the modulus $|z|$ of the complex number $z = x + iy$ is just the norm $\| (x, y) \|$ of the point $(x, y) \in \mathbf{R}^2$.

Along with some elementary facts about the algebra of complex numbers, with which we assume the reader is familiar, we shall have need of De Moivre's theorem:

$$(\cos 2\pi t + i \sin 2\pi t)^n = \cos 2\pi n t + i \sin 2\pi n t$$

for each positive integer n and each real number t. In the notation of this section, De Moivre's theorem says

$$z = p(t) \quad \Rightarrow \quad z^n = p(nt) \qquad (t \in \mathbf{R}; n = 1, 2, \ldots).$$

5.78 Fundamental Theorem of Algebra. Let

$$f(z) = a_0 z^n + a_1 z^{n-1} + \cdots + a_{n-1} z + a_n$$

be a polynomial of degree $n \geq 1$ with complex coefficients $a_0, a_1, \ldots, a_n$. Then $f(z) = 0$ for some complex number z.

Proof. The leading coefficient $a_0 \neq 0$ because f has degree n, and $f(z) = 0$ if and only if $a_0^{-1} f(z) = 0$. Hence we may assume without loss of generality that $a_0 = 1$.

We begin by comparing the behavior of the functions $z \mapsto f(z)$ and $z \mapsto z^n$ for large values of z. For

$$|z| > c = \max \{1, |a_1| + |a_2| + \cdots + |a_n|\}$$

we have

$$
\begin{aligned}
|f(z) - z^n| &= |a_1 z^{n-1} + a_2 z^{n-2} + \cdots + a_n| \\
&\leq |a_1| \cdot |z|^{n-1} + |a_2| \cdot |z|^{n-2} + \cdots + |a_n| \\
&= |z|^{n-1} \left\{ |a_1| + \frac{|a_2|}{|z|} + \cdots + \frac{|a_n|}{|z|^{n-1}} \right\} \\
&\leq |z|^{n-1} \{ |a_1| + |a_2| + \cdots + |a_n| \} \\
&< |z|^n = |z^n|,
\end{aligned}
$$

so that the line segment joining $f(z)$ and z^n cannot pass through the origin. (Already this shows that $f(z) \neq 0$ for $|z| > c$.)

Next we compare the behavior of $z \mapsto f(z)$ and $z \mapsto z^n$ on the circle $\{rp(s) \mid s \in I\}$ in the complex plane having center at the origin and radius $r > 0$. The two maps

$$
s \mapsto f(rp(s)), \qquad s \mapsto [rp(s)]^n = r^n p(ns)
$$

from I to $C = R^2$ are loops in C at the points $f(r)$, r^n. The map $h \colon I \times I \to C$ given by

$$
h(s, t) = t r^n p(ns) + (1 - t) f(rp(s))
$$

is a homotopy between these two maps satisfying

$$
h(0, t) = h(1, t) = t r^n + (1 - t) f(r) \qquad\qquad (t \in I).
$$

According to the preceding paragraph, moreover,

$$
(*) \qquad\qquad r > c \quad\Rightarrow\quad h(s, t) \neq 0 \quad \text{for all } s \in I, t \in I.
$$

The loops in C just considered may be transformed into loops ω_r, τ at $1 = (1, 0)$ in S_1 through the "normalizations"

$$
\omega_r(s) = \frac{|f(r)|}{f(r)} \cdot \frac{f(rp(s))}{|f(rp(s))|}, \qquad \tau(s) = p(ns).
$$

Then

$$
r > c \quad\Rightarrow\quad \omega_r \sim \tau,
$$

for in view of $(*)$ we obtain a homotopy $H \colon \omega_r \sim \tau$ by setting

$$
H(s, t) = \frac{|h(0, t)|}{h(0, t)} \cdot \frac{h(s, t)}{|h(s, t)|}.
$$

For the first time now, suppose $f(z) \neq 0$ for all $z \in C$. Then the loop ω_r is nullhomotopic for each $r > 0$, a homotopy from ω_r to the null loop at $(1, 0)$ in S_1 being

$$
G(s, t) = \frac{|f(tr)|}{f(tr)} \cdot \frac{f(trp(s))}{|f(trp(s))|}.
$$

Since $\omega_r \sim \tau$ for $r > c$, we conclude that τ is nullhomotopic. This contradicts 5.74. ∎

EXERCISES

101. If C is a proper connected subset of S_1, show that p maps each of the components of $p^{-1}(C)$ homeomorphically onto C.

102. Generalize the uniqueness assertion of 5.72 as follows: Let $f, g: X \to R$ be continuous maps on a connected Hausdorff space X such that $p \circ f = p \circ g$ and $f(x_0) = g(x_0)$ for some $x_0 \in X$. Then $f = g$.

103. Let $f, g: X \to S_1$ be continuous maps on a topological space X such that $|f(x) - g(x)| < 2$ for all $x \in X$. Suppose there is a continuous $F: X \to R$ with $f = p \circ F$. Show that then $g = p \circ G$ for some continuous $G: X \to R$.

104. Let $f: X \to S_1$ be a continuous map on a compact space X. Prove that a necessary and sufficient condition for f to be homotopic to a constant map is that $f = p \circ F$ for some continuous $F: X \to R$. [*Hint*: For necessity, let $H: X \times I \to S_1$ be a homotopy from a constant map to f. By uniform continuity, there is a $\delta > 0$ such that $|H_t(x) - H_s(x)| < 2$ whenever $x \in X$ and $|t - s| < \delta$. Now use Exercise 103 repeatedly, starting with a lifting of the constant map H_0.]

105. Establish the uniqueness of the homotopy $H: \sigma_0 \sim \sigma_1$ with $p \circ H = h$ in 5.73.

106. For a loop ω in S_1 at $(1, 0)$, the *degree of* ω is the integer $\deg \omega = \sigma(1)$ where σ is the unique path in R with $p \circ \sigma = \omega$ and $\sigma(0) = 0$. (Intuitively, $\deg \omega$ is the net number of times ω winds around the circle in the counterclockwise direction.) According to 5.73, if ω, ω' are loops in S_1 at $(1, 0)$ with $\omega \sim \omega'$, then $\deg \omega = \deg \omega'$. Establish the converse.

107. (**a**) If ω, ω' are loops at $(1, 0)$ in S_1, show that

$$\deg(\omega \cdot \omega') = \deg \omega + \deg \omega'.$$

(Here $\omega \cdot \omega'$ is the product in the sense of 5.48.)

(**b**) Use (a) and Exercise 106 to construct a group isomorphism from the fundamental group $\pi(S_1, (1, 0))$ (see Exercise 97) to the additive group Z of integers.

108. Given a positive integer n, prove the equivalence of the following conditions: (i) S_{n-1} is not a retract of D_n; (ii) each continuous map $f: D_n \to D_n$ has a fixed-point; (iii) the identity map of S_{n-1} is not homotopic to a constant map. [*Note*: The way Brouwer's fixed-point theorem is usually proved for arbitrary dimension n is to establish the truth of (iii) by the methods of "homology theory"—see Keesee [20, pp. 111–114] or another book on algebraic topology.]

109. Generalize the fundamental theorem of algebra by showing that a continuous function $f: C \to C$ must vanish at some $z \in C$ provided that $\lim_{z \to \infty} z^{-n} f(z) \neq 0$ for some positive integer n.

110. (**a**) Let $p_1: X_1 \times X_2 \to X_1$, $p_2: X_1 \times X_2 \to X_2$ be the projections on the product of two topological spaces. If ω_0, ω_1 are loops at a point of $X_1 \times X_2$, prove that $\omega_0 \sim \omega_1$ if and only if $p_1 \circ \omega_0 \sim p_1 \circ \omega_1$ and $p_2 \circ \omega_0 \sim p_2 \circ \omega_1$.

(**b**) Deduce that the torus is not simply connected.

111. (**a**) Suppose a path-connected space X has simply connected open subsets U_1 and U_2 such that $X = U_1 \cup U_2$ and $\varnothing \neq U_1 \cap U_2$ is path-connected. Prove that X must be simply connected. [*Hint*: If ω is a loop in X at a point $x \in U_1 \cap U_2$, then the open cover $\{\omega^{-1}(U_1), \omega^{-1}(U_2)\}$ of I has a Lebesgue number.]

(**b**) Deduce that the n-sphere S_n is simply connected for every $n > 1$.

Guide to the Exercises

This guide indicates exercises containing definitions, results, and examples that are needed for solving subsequent exercises or for reading subsequent portions of the text. Not listed here are exercises needed for immediately following exercises.

Exercise	Needed for Exercises	Needed for Text
1.4	1.43	
1.9	1.25	
1.12		after 1.68, 3.67(2)
1.13	1.29, 1.63, 1.87, 1.89, 2.12	
1.14	1.29, 1.63, 1.87, 1.90	
1.15	1.30, 1.59, 1.71	
1.23		after 3.22
1.28	1.37, 4.22, 4.54, 5.4	
1.29	2.12	
1.33	1.69	
1.34	1.50, 1.60, 1.73	after 3.24
1.37	4.58	
1.38	1.90, 4.60	3.22(1)
1.42	1.54, 1.55	
1.48		3.35
1.51	1.66	
1.52	1.66	
1.55	3.47	
1.56		3.11(2)
1.66		after 4.42
1.68	1.90	
1.73		4.40
1.87	1.90	
2.3	2.21	
2.6	2.21, 2.74, 4.4, 4.61, 5.2	3.54
2.10	2.19, 2.20, 2.31, 2.32, 3.49, 3.76, 3.119, 4.3	
2.12	4.3	

Exercise	Needed for Exercises	Needed for Text
2.20	2.76, 3.57	3.22(3)
2.24	2.45	
2.27	2.34, 2.56, 3.5, 4.4, 4.61, 5.2	
2.29	2.69, 5.7	
2.32	2.61, 3.49, 3.76, 3.107, 3.119	
2.40		4.38
2.49	3.17	
2.60	2.80, 3.49, 3.77, 3.107, 3.119, 4.16	
2.71	2.78	3.42, 3.67(1)
2.78	5.47, 5.56, 5.73	
2.79	3.6	
2.80	5.26	
3.28	4.21	
3.47	3.52, 3.98, 5.50	
3.48	3.98	
3.57	4.24, 4.31	
3.84	3.121, 3.123, 5.34	
3.88	5.34	
3.90	3.103	
3.98	3.112, 3.114, 3.115	
3.117	3.121, 4.28, 4.59, 4.85	
3.127	3.148	
4.11	4.42	
4.23	4.31	
4.27	4.59, 4.85	
4.37	4.76	
5.4	5.15	
5.8	5.33	
5.16	5.60, 5.75	
5.31	5.56	
5.39	5.45	
5.42	5.54	
5.64	5.73	
5.97	5.107	

Bibliography

1. T. M. Apostol, *Calculus*, vol. 2, 2nd ed., Blaisdell, Waltham, Mass., 1969.

2. G. Birkhoff and S. MacLane, *A survey of modern algebra*, 3rd ed., Macmillan, New York, 1965.

3. D. W. Blackett, *Elementary topology*: *A combinatorial and algebraic approach*, Academic Press, New York, 1967.

4. L. M. Blumenthal and K. Menger, *Studies in geometry*, Freeman, San Francisco, 1970.

5. R. P. Boas, Jr., *A primer of real functions*, Carus Mathematical Monographs, vol. 13, Math. Assoc. Amer., 1960.

6. N. Bourbaki, *General topology*, 2 vol., Addison-Wesley, Reading, Mass., 1966.

7. J. Dugundji, *Topology*, Allyn and Bacon, Boston, 1966.

8. J. Duncan, *The elements of complex analysis*, Wiley, New York and London, 1968.

9. M. Eisenberg, *Axiomatic theory of sets and classes*, Holt, Rinehart and Winston, New York, 1971.

10. W. W. Fairchild and C. Ionescu Tulcea, *Sets*, Saunders, Philadelphia, 1970.

11. W. H. Fleming, *Functions of several variables*, Addison-Wesley, Reading, Mass., 1965.

12. D. J. Foulis, *Fundamental concepts of mathematics*, Prindle, Weber & Schmidt, Boston, 1969.

13. R. R. Goldberg, *Methods of real analysis*, Blaisdell, Waltham, Mass., 1964.

14. J. Greever, *Theory and examples of point-set topology*, Brooks/Cole, Belmont, Calif., 1967.

15. D. W. Hall and G. L. Spencer II, *Elementary topology*, Wiley, New York, 1955.

16. F. Hausdorff, *Set theory*, 3rd ed., Chelsea, New York, 1957.

17. J. G. Hocking and G. S. Young, *Topology*, Addison-Wesley, Reading, Mass., 1961.

18. W. Hurewicz and H. Wallman, *Dimension theory*, Princeton Math. Series, vol. 4, Princeton Univ. Press, Princeton, N.J., 1941.

19. H. G. Jacob and D. W. Bailey, *Linear algebra*, Houghton Mifflin, Boston, 1971.

20. J. W. Keesee, *An introduction to algebraic topology*, Brooks/Cole, Belmont, Calif., 1970.

21. J. L. Kelley, *General topology*, Van Nostrand, New York, 1955.

22. D. L. Kreider, R. G. Kuller, and D. R. Ostberg, *Elementary differential equations*, Addison-Wesley, Reading, Mass., 1968.

23. S. Lang, *Analysis I*, Addison-Wesley, Reading, Mass., 1968.

24. S. Lefschetz, *Introduction to topology*, Princeton Math. Series, vol. 11, Princeton Univ. Press, Princeton, N.J., 1949.

25. J. H. Manheim, *The genesis of point set topology*, Macmillan, New York, 1964.

26. W. S. Massey, *Algebraic topology: An introduction*, Harcourt Brace Jovanovich, New York, 1967.

27. W. S. Massey, "Topology, Algebraic," in *Encyclopaedia Britannica*, 1970.

28. G. McCarty, *Topology: An introduction with application to topological groups*, McGraw-Hill, New York, 1967.

29. R. A. Rosenbaum, *Introduction to projective geometry and modern algebra*, Addison-Wesley, Reading, Mass., 1963.

30. K. T. Smith, *Primer of modern analysis*, Bogden & Quigley, Tarrytown-on-Hudson, N.Y., 1971.

31. L. A. Steen and A. J. Seebach, Jr., *Counterexamples in topology*, Holt, Rinehart and Winston, New York, 1970.

32. C. T. C. Wall, *A geometric introduction to topology*, Addison-Wesley, Reading, Mass., 1972.

33. R. L. Wilder, "Topology, General," in *Encylopaedia Britannica*, 1970.

34. S. Willard, *General topology*, Addison-Wesley, Reading, Mass., 1970.

The set-theoretic preliminaries we have sketched in Chapter 0 are treated in greater detail, at an elementary level, in Fairchild and Ionescu Tulcea [10] and Foulis [12]. A full axiomatic development of set theory is given by Eisenberg [9]; a condensed axiomatic development appears in the Appendix to Kelley [21].

For an overview of topology, see Wilder [33] and Massey [27].

Accounts of the historical development of topology appear in the several Historical Notes of Bourbaki [6] and in Manheim [25]. Extensive historical notes, with references to the original sources, are included at the end of Willard [34]. Numerous citations of the literature appear in the footnotes of Greever [14].

Greever [14] and Steen and Seebach [31] are replete with examples, especially pathological ones.

Standard advanced texts in general topology are Bourbaki [6], Dugundji [7], Hocking and Young [17], Kelley [21], and Willard [34]. Our own preference, for sheer readability as well as comprehensiveness, is Willard [34]. Kelley [21] contains the best exercises, some of which develop entire theories.

Our discussion of homotopy just touched upon one of the concerns of algebraic topology. Dugundji [7] and Hocking and Young [17] both contain some algebraic topology. Fuller, but still introductory, treatments may be found in Blackett [3], Keesee [20], Lefschetz [24], Massey [26], and Wall [32].

The following additional readings can be used for individual projects; others should be suggested by references throughout the text.

Affine topologies. The topology on the plane consisting of the "linearly open" sets, as defined in Exercise 2.9, is an example of an affine topology. The peculiar properties such topologies can have are discussed in V. L. Klee, *Some finite-dimensional affine topological spaces*, Portugal. Math. *14* (1955), 27–30. A related paper is C. A. Kottman, *A characterization of the Euclidean topology among the affine topologies*, Israel J. Math. *10* (1971), 212–217.

Brouwer's fixed-point theorem. Brouwer's theorem and its relatives in low dimensions are treated in A. W. Tucker, *Some topological properties of disk and sphere*, Proc. First Canadian Math. Congress (Montreal, 1945), 285–309. Brouwer's theorem in arbitrary dimension is derived from a combinatorial lemma in H. W. Kuhn, *Some combinatorial lemmas in topology*, IBM J. Res. Dev. *4* (1960), 518–524.

Compactness. The thesis that compactness is a generalization of finiteness is illustrated in E. Hewitt, *The rôle of compactness in analysis*, Amer. Math. Monthly *67* (1960), 499–516. At least the first two examples of Hewitt should be accessible to readers of this text. A much more complete treatment of Hewitt's second example, the Stone-Weierstrass approximation theorem, is M. H. Stone, *A generalized Weierstrass approximation theorem*, in *Studies in Modern Analysis*, Studies in Math., vol. 1, Math. Assoc. Amer., 1962, pp. 30–87.

Connected sets in the plane. Weird connected subsets of the plane which arise as graphs of solutions to Cauchy's functional equation are discussed in F. B. Jones, *Connected and disconnected plane sets and the functional equation $f(x) + f(y) = f(x + y)$*, Bull. Amer. Math. Soc. *48* (1942), 115–120.

Fixed-point property. A space X is said to have the "fixed-point property" when each continuous map $f: X \to X$ has at least one fixed-point. An expository account, with many examples, is R. H. Bing, *The elusive fixed point property*, Amer. Math. Monthly *76* (1969), 119–132. Bing discusses the famous unsolved problem of whether a continuum in the plane that does not disconnect the plane must necessarily have the fixed-point property.

Invertible spaces. The n-sphere has the property that the complement of each of its nonempty open subsets U can be mapped into U by a suitable homeomorphism of the sphere. Topological features and examples of other spaces with this property are discussed in P. H. Doyle and J. G. Hocking, *Invertible spaces*, Amer. Math. Monthly *68* (1961), 959–965.

Tychonoff's theorem. The product of an arbitrary family of compact spaces is compact. Most advanced texts give proofs of this theorem, all of which depend on some variant of the axiom of choice. A self-contained proof which begins with the axiom of choice is given on pages 25–28 of Hocking and Young [17]. It is interesting that Tychonoff's theorem is logically equivalent to the axion of choice—see J. L. Kelley, *The Tychonoff product theorem implies the axiom of choice*, Fund. Math. *37* (1950), 75–76.

Wild spheres. At the end of Section 2 of Chapter 3 we exhibited a pair of subspaces of Euclidean 3-space that are homeomorphic to one another and yet have the property that no homeomorphism of 3-space maps one of the subspaces onto the other. Surprisingly, there exist such pairs with one of the subspaces being the ordinary 2-sphere, and then the other is called a "wild 2-sphere". Examples appear in R. H. Bing, *Spheres in E^3*, Amer. Math. Monthly *71* (1964), 353–364. Bing includes here an interesting proof of Tietze's extension theorem for metrizable spaces.

List of Symbols

SYMBOL	MEANING	PAGE
$\Rightarrow$	implies	1
$\Leftrightarrow$	if and only if	1
$\in$	element of	1
$\notin$	not an element of	1
$=$	equals	2
$\neq$	not equal to	2
$\{\dots\}$	set consisting of ...	2
$\{x \mid P\}, \{x{:}P\}$	set of x such that P	2
$\{x \in X \mid P\}, \{x \in X{:}P\}$	set of x belonging to X such that P	2
$\varnothing$	empty set	2
N	set of natural numbers	3
Z	set of integers	3
Q	set of rational numbers	3
R	set of real numbers	3
C	set of complex numbers	3
I	closed unit interval	3
$\subset$	contained in (subset of)	3
$\supset$	contains	3
$\mathcal{P}(X)$	power set	4
$A \cup B$	union of two sets	4
$A \cap B$	intersection of two sets	4
$X \setminus A$	complement	5
(x, y)	ordered pair	5
$X \times Y$	product of two sets	6
xRy	x is related to y by R	6
$f{:}\, X \to Y$	map from X to Y	7
$x \mapsto y$	x is mapped to y	7

SYMBOL	MEANING	PAGE
$f: X \to Y$ $\quad x \mapsto f(x)$	map from X to Y sending x to $f(x)$	7
$|x|$	absolute value	8
$f|_E, f \mid E$	restriction of map	8–9
$g \circ f$	composite of maps	9
$f(A)$	image of set	11
$f^{-1}(D)$	inverse image of set	11
$f^{-1}(y)$	inverse image of singleton	11
f^{-1}	inverse of bijection	13
$(x_i \mid i \in I)$	family	16
$(x_1, x_2, \ldots, x_n)$	n-tuple	16
$(x_0, x_1, \ldots)$	sequence	16
X^n	set of n-tuples in X	16
R^n	n-dimensional Euclidean space	17
$x + y$	sum of two vectors	17
αx	product of scalar and vector	17
$-x$	negative of vector	17
X^I	set of families in X indexed by I	18
$\times_{i=1}^n X_i,$ $\quad X_1 \times X_2 \times \cdots \times X_n$	product of n-tuple of sets	18
$\times_{i=1}^\infty X_i, X_1 \times X_2 \times \cdots$	product of sequence of sets	19
$\times_{i \in I} X_i$	product of family of sets	19
$\bigcup_{i \in I} X_i$	union of family of sets	20
$\bigcap_{i \in I} X_i$	intersection of family of sets	20
$\bigcup_{i=1}^n X_i,$ $\quad X_1 \cup X_2 \cup \cdots \cup X_n$	union of n-tuple of sets	20
$\bigcup_{i=1}^\infty X_i, X_1 \cup X_2 \cup \cdots$	union of sequence of sets	20
$\bigcup \mathcal{C}$	union of collection of sets	21
$\bigcap \mathcal{C}$	intersection of collection of sets	21
$\leq$	total ordering	32
$\max E, \max_{1 \leq i \leq n} x_i$	maximum	32
$\min E, \min_{1 \leq i \leq n} x_i$	minimum	32
$x < y$	$x \leq y$ and $x \neq y$	33
$]a, \to[, \;]\leftarrow, a[$	open rays	33
$[a, \to[, \;]\leftarrow, a]$	closed rays	33
$]a, b[$	open interval	33
$[a, b]$	closed interval	33
$]a, b], [a, b[$	half-open intervals	33
$\sup A$	supremum (least upper bound)	34
$\inf A$	infimum (greatest lower bound)	34
$\sim$	equivalence relation	42
$x/\sim$	equivalence class	43
$X/\sim$	quotient set	43

SYMBOL	MEANING	PAGE
$\|x\|$	Euclidean norm	49
d_∞	max metric	54, 55
H	Hilbert sequence space	58
$\|x\|_2$	Hilbert norm	58
d_2	metric on H	59
δ	discrete metric	60
$B_\epsilon(x; d)$	d-ball of radius ϵ at x	64
$D_\epsilon(x; d)$	d-disk of radius ϵ at x	64
$S_\epsilon(x; d)$	d-sphere of radius ϵ at x	64
$d(x, A)$	distance from point to set	75
$d(A, B)$	distance between two sets	78
diam A	diameter	82
$-\infty, +\infty$	points in extended real line	83
R*	extended real line	83
bdy A	boundary	147
int A	interior	150
cls A	closure	152
$f: X \cong Y$	homeomorphism from X to Y	190
$X \cong Y$	X is homeomorphic to Y	190
B_n	n-ball	193
I^n	n-cube	194
D_n	n-disk	194
S_n	n-sphere	195
I^∞	Hilbert cube	204
∂M	boundary of manifold-with-boundary	209
RP_n	real projective n-space	247
X/A	quotient space obtained by collapsing A to a point	250
$X \cup_f Y$	quotient space obtained by attaching X to Y by f	252
$(x_i \mid i \in I) \to x,$ $x_i \to x$	convergent net	261
$\lim (x_i \mid i \in I),$ $\lim_{i \in I} x_i, \lim x_i$	limit of net	265
$\lim_{t \to x} f(t)$	limit of map	272
K	Cantor discontinuum	288
X_∞	one-point compactification	321
$h: f \simeq g$	homotopy between maps	377
$f \simeq g$	homotopic maps	377
$h: \sigma \sim \tau$	homotopy between paths	379
$\sigma \sim \tau$	homotopic paths	379
ν_x	null loop at x	381
$p: R \to S_1$	covering map	393

Index

Index

For d-ball, *d*-open sets, (d, d')-continuous maps, *etc., see the entries* Ball, Open sets, Continuous maps, *etc.*

A

Absolute value, 8, 183
Alexandroff compactification, 321–322
  See also One-point compactification
Almost all indices, 213
Antipodal map, 198
Arc, 366
  See also Path
Arc-connected space, 366
  See also Path-connected space
Archimedean property, 35
Arcwise connected space (*see* Arc-connected space)
Attaching map, 252
Axiom of choice, 28, 217, 228

B

Baire category theorem, 119, 156, 327
Ball, 64, 66, 68, 151
  homeomorphic to R^n, 196
  See also N-ball

Banach contraction mapping principle, 126
Base, 162–164
  countable, 167
  and local bases, 163
  of product space, 214, 231
  of subspace, 171
  See also Second-countable space
Basic open set, 214, 218
Bijection, 12
Bing triangle space, 174–175, 347
Bolzano-Weierstrass theorem, 307
Borsuk-Ulam theorem, 338
Boundary, 147, 153, 157
  of manifold-with-boundary, 209
  of product, 229
Bounded functions, 62, 116
  on compact space, 280, 294
Bounded metric, 82, 85, 113–114
Bounded metric space, 82
  vs. totally bounded metric space, 313
Bounded set, 81